Advances of
Atoms and Molecules
in Strong Laser Fields

Advances of
Atoms and Molecules
in Strong Laser Fields

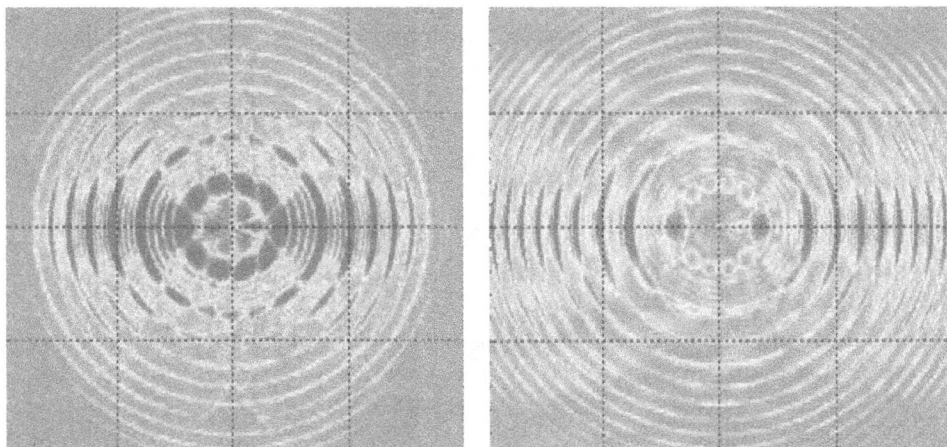

Yunquan Liu

Peking University, China

World Scientific

NEW JERSEY · LONDON · SINGAPORE · BEIJING · SHANGHAI · HONG KONG · TAIPEI · CHENNAI · TOKYO

Published by

World Scientific Publishing Co. Pte. Ltd.

5 Toh Tuck Link, Singapore 596224

USA office: 27 Warren Street, Suite 401-402, Hackensack, NJ 07601

UK office: 57 Shelton Street, Covent Garden, London WC2H 9HE

British Library Cataloguing-in-Publication Data
A catalogue record for this book is available from the British Library.

ISBN 978-981-4696-38-8

In-house Editor: Song Yu

Typeset by Stallion Press
Email: enquiries@stallionpress.com

Preface

In parallel to the invention of first ultrashort laser in the mid-sixties, the field of research "atoms and molecules in strong laser fields" has evolved considerably during the last two decades owing to the rapid technological development of lasers as well as the advanced experimental techniques. Recent developments have shown that high intensity lasers open the way to realize pulses with the shortest durations to date, giving birth to the field of attosecond science (1 asec = 10^{-18}s). Irradiating atomic or molecular targets by an intense laser field leads to many highly nonlinear phenomena. "Atoms and molecules in strong laser fields" is a fascinating field in atomic, molecular and optical (AMO) physics, and it has progressed considerably during the last decade. In view of the rapid achievement in both experimental and theoretical studies of atoms and molecules in strong laser fields, it would now be very desirable to present a more comprehensive and advanced reviews text which contains all the new and important progress.

Depending on the laser wavelength and laser intensity, atomic and molecular physics in strong laser fields is essentially different. At near-infrared and visual wavelength where the commercial Ti:sapphire femtosecond lasers cover, multiphoton ionization and tunneling ionization usually dominates. On the other hand, the Free Electron Laser (FEL) facilities, emerging as a rapid developing technology in advanced light sources in the last decade, provides the coherent, brilliant and ultrashort light pulse which, is tunable from Terahertz to X-ray regime, and now opens up new vistas in the study of the atomic and molecular physics.

This book is about recent progress of atoms and molecules in strong laser fields. The goal is to give an up-to-date introduction to the broad range of strong-field AMO (atomic molecular and optical) physics, i.e., tunneling ionization, multiphoton ionization, photoelectron holography, electron correlation, controlling molecular dissociation, high harmonic generation, attosecond science, and X-FEL laser physics, etc. As an expanding field, a lot of intriguing progress has been achieved. It is 'no way' to collect all of them. It is expected that there will be a lot of activities in this research area in the near future.

I would like to take this opportunity to thank all the authors for their important contributions. It is hoped that this collection will be useful not only to active researchers in the strong-field community but also to other scientists in biology and chemistry.

Yunquan Liu
Peking University

Contents

Preface   v

1. Long Range Ionic Potential Effect on Strong-Field Tunneling   1

   Xufei Sun, Min Li, Chengyin Wu, Yunquan Liu

   1 Introduction   1
   2 Experimental Technique   4
   3 Low Energy Structure   5
   4 Local Ionization Suppression   8
   5 Coulomb Asymmetry   14
   6 Conclusion   20
   Acknowledgment   20
   References   20

2. Photoelectron Interference and Photoelectron Holography   25

   Min Li, Qihuang Gong, and Yunquan Liu

   1 Introduction   25
   2 Semiclassical Model with Implementing Interference Effect   27
      2.1 Intercycle interference and intracycle interference   27
      2.2 Forward rescattering photoelectron interference   29
      2.3 Backward rescattering photoelectron
           interference   31
   3 Classical-Quantum Correspondence of above Threshold
      Ionization   38
   4 Conclusion   48
   References   48

3. Dissociation of Hydrogen Molecular Ions in Strong
    Laser Fields   51

    Feng He

    1. Introduction   51
    2. Numerical Methods   52
    3. Dissociation in Strong Laser Fields   55
        A   Electron localization   56
        B   Dissociation in long wavelengths and short
              wavelengths   59
        C   Dissociation in short and intense laser fields 64
        D   Control vibrational states   67
        E   Rescattering dissociation   70
        F   Two dimensional wave packet propagation   74
    Conclusion   76
    Acknowledgment   77
    References   77

4. Nonsequential Double Ionization of Atoms in
    Strong Laser Field   81

    Difa Ye, Libin Fu and Jie Liu

    1   Introduction   82
    2   Semiclassical Model   83
    3   The Surprising Fingerlike Structure in NSDI 84
    4   NSDI Below the Recollision Threshold   90
    5   Atomic Species Dependence of NSDI   100
    Acknowledgment   109
    References   109

5. Few-Photon Double Ionization of He and H_2 111

    Wei-Chao Jiang, Wei-Hao Xiong, Ji-Wei Geng, Qihuang Gong,
    and Liang-You Peng

    1   Introduction   111
    2   Numerical Implementation to Solve Two-Electron TDSE   112
        2.1   General technologies in solving TDSE 112
              2.1.1   Time-propagation method   113
              2.1.2   Space discretization method 115

2.2 Methods to solve TDSE of helium 118
 2.2.1 Time-dependent close coupling (TDCC)
    scheme 118
 2.2.2 Extraction of physical observables 120
2.3 Methods to solve TDSE of H_2 molecule 122
 2.3.1 Introduction of the prolate spheroidal
    coordinates 122
 2.3.2 TDSE of H_2 molecule in the prolate spheroidal
    coordinates 123
 2.3.3 Extraction of physical observable 128
3 Applications of TDSE Methods on Double Ionization
 Processes 129
 3.1 One-photon and two-photon double ionization
  cross sections 129
   3.1.1 Definitions for photon-double ionization cross
     sections 129
   3.1.2 Numerical results for electron spectrums of helium
     atom and H_2 molecule 131
   3.1.3 Recoil ion spectrum of photon double ionization of
     helium 136
 3.2 Electron correlation in the joint angular distributions 141
 3.3 Double ionization of helium by double attosecond
  pulses 146
References 151

6. Probing Orbital Symmetry of Molecules Via
Alignment-Dependent Ionization Probability and High-Order
Harmonic Generation by Intense Lasers   157

Song-Feng Zhao, Xiao-Xin Zhou and C. D. Lin

1 Introduction 158
2 Theoretical Methods 160
 2.1 Construction of one-electron potentials of linear
  molecules 160
 2.2 Calculation of molecular wavefunction with correct
  asymptotic tail by solving the time-independent
  Schrödinger equation 162
 2.3 The MO-ADK and MO-PPT models 164

2.4   Calculation of the orientation-dependent ionization probability
      of molecules with the MO-SFA model   166
2.5   The molecular Lewenstein model for HHG from molecules .   166
3   Results and Discussions   167
3.1   The one-electron potentials for Cl_2   167
3.2   Extracting structure parameters for several highly occupied
      orbitals of linear molecules   167
3.3   Comparison of alignment dependent ionization probabilities
      between the MO-ADK model and other more elaborate
      calculations   169
3.4   Comparison with experiments   171
3.5   Alignment dependence of ionization rates from HOMO,
      HOMO-1, and HOMO-2 orbitals   171
3.6   Probing the shape of the ionizing molecular orbitals with the
      orientation-dependent ionization rates   172
3.7   Examination of the validity of the MO-ADK and MO-PPT
      models   173
3.8   Probing the molecular orbital with the alignment-dependent
      HHG signals   175
4   Conclusions   178
Acknowledgments   179
References   179

7. High-order Harmonic Generation Driven by Sub-Cycle
    Shaped Laser Field   185

    Yinghui Zheng, Zhinan Zeng, Pengfei Wei, Jing Miao,
    Ruxin Li, and Zhizhan Xu

1   Introduction   185
2.1   Two pulses scheme   187
2.2   Multi-color field scheme   191
Acknowledgment   201
References   201

8. Imaging Ultra-fast Molecular Dynamics
    in Free Electron Laser Field   205

    Y. Z. Zhang and Y. H. Jiang

1. Introduction   205
2. The Temporal Characteristic and Resolution
    of Free Electron Laser   209

3. The EUV Pump/Probe Measurement 212
   3.1. The split-mirror scheme 212
   3.2. Tracing the nuclear wave-packet in D_2^+
        by EUV time-resolved experiment 213
   3.3. Nuclear wavepacket dynamics in excited
        states of N_2 215
   3.4. Ultrafast photoisomerization of acetylene cation 217
   3.5. Time-resolved interatomic Coulombic decay
        (ICD) in Ne_2 218
   3.6. Electron rearrangement dynamics
        in dissociating I_2^{n+} 221
4. Optical-pump — X-ray-probe (OPXP) Experiments 222
   4.1. Probing a prototypical photoinduced
        ring opening 223
   4.2. Imaging charge transfer in iodomethane 225
5. Summary and Outlook 228
Acknowledgments 230
References 230

Chapter 1

Long Range Ionic Potential Effect
on Strong-Field Tunneling

Xufei Sun, Min Li, Chengyin Wu, Yunquan Liu

Department of Physics and State Key Laboratory for Mesoscopic Physics,
Peking University, Beijing 100871, China
yunquan.liu@pku.edu.cn

The tunneling ionization as one of the fundamental quantum processes exposing of atoms in strong laser fields, has been investigated intensively in recent decades. The tunneling ionization has intrigued many important implications. As the first stage, the post ionization effects have been revealed in high-resolution experiments. All of those experiments indicate the importance of the long range potential on the tunneling ionization. In this chapter, we will overview the long-range potential effect on strong-field tunneling. Especially, the long range ionic potential has significant effect on the low-energy photoelectron energy structures ("ionization surprise"), the ionization asymmetry and the local ionization suppression.

1. Introduction

The pioneering theoretical description of the photoionization of atoms and ions exposed to high-intensity laser radiation is underlain by the Keldysh theory proposed in 1964 [1]. The ionization potential I_p of an atom is defined as the energy required to lift its most loosely bound electron into the continuum (a bound-continuum transition). When irradiated by light of sufficiently high frequency ω, this is possible via a single-photon transition [2]. If increasing the laser intensity ionization is also possible in lower frequency fields. Here, ionization takes place via the absorption of n photons of energy $\hbar\omega$, such that $(n-1)\hbar\omega < Ip < n\hbar\omega$. The corresponding multiphoton ionization rate can be obtained by treating the field as a small perturbation with respect to the Coulomb potential of the atom [3, 4].

The associated theoretical framework is called lowest order perturbation theory and results in an intensity scaling that takes the form of a power law. However, with increasing intensity, higher order terms become significant requiring an extension of perturbative theory.

Above-threshold ionization (ATI), as one of the cornerstones in strong-field community, has attracted an abundance of attention. This intriguing phenomenon is characterized by a series of peaks separated by the laser frequency in the electron energy spectrum and the relative yield of each peak drops exponentially with increasing the electron energy. Beyond that, a photoelectron plateau structure with an approximately constant amplitude is discovered from $\sim 2U_\mathrm{p}$ extending up to an abrupt cutoff at around $10U_\mathrm{p}$ [U_p is the ponderomotive energy, $U_p = F^2/4\omega^2$, where F and ω are the laser field strength and laser frequency, respectively. Atomic units (a.u.) are used throughout unless specified] [5].

Alternatively, increasing the laser intensity, the potential barrier of an atom is suppressed drastically and electrons can easily escape from the atom through tunneling, as seen in Fig. 1. Usually, multiphoton ionization and tunneling ionization are distinguished by the Keldysh parameter γ [6] ($\gamma = \sqrt{2I_p/U_p}$, I_p: the ionization potential). Tunneling ionization will dominate if $\gamma < 1$, while multiphoton ionization prevails when $\gamma > 1$.

A thorough understanding of atomic ionization in strong fields is essential for further explorations and diverse applications. Electron tunneling is a key process when atoms and molecules are exposed to strong laser fields. Strong-field tunneling ionization has triggered a rich set of physical phenomena that are closely associated with the flourishing attosecond physics. Strong laser field ionization now provides a sophisticated method to image and probe the atomic and molecular quantum processes. Recent advances in strong field physics have opened a window to precisely measure the delay time and the initial coordinates of quantum tunneling [7, 8].

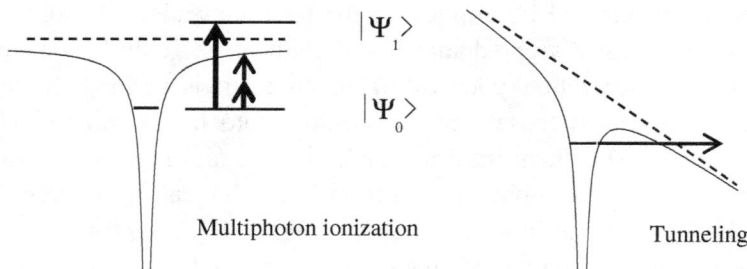

Fig. 1. The illustration of multiphoton ionization and tunneling ionization when atoms with a ionization potential of I_p in strong laser fields.

For the case of linearly polarized light the motion of atomic or molecular fragments emerging from ionization can be naturally separated into a longitudinal part, which is parallel to the electric field polarization, and a part transverse to the field. Under these assumptions analytical expressions for the final momentum distributions along longitudinal direction $w(p_{||})$ and perpendicular or transverse' directions $w(p_\perp)$ to the field polarization were obtained [9]:

$$w(p_{||}) \propto \exp\left(-\frac{p_{||}^2 \omega^2 (2I_p)^{3/2}}{3E^3}\right)$$

$$w(p_\perp) \propto \exp\left(-\frac{p_\perp^2 (2I_p)^{1/2}}{E}\right)$$

The Simple-man model is very fundamental for the understanding on strong-field ionization. In this model, strong-field ionization takes place in two-step, e.g., the electron is initially released through tunnelling and then classically propagates in the laser field. During the propagation process, the influence of the atomic potential is neglected. This model was later improved by taking into account of recollision effect (three-step Simple-man model) [10, 11]. In this three-step model, the tunnelled electron is accelerated in the laser field, then changes its travelling direction and finally is scattered upon the nucleus, giving rise to high-order above-threshold ionization (HATI) for elastic scattering with $10Up$ cut off [12], nonsequential double ionization (NSDI) for inelastic scattering [13] and high-harmonic generation (HHG) for recombination [14].

According to the Simple-man model, the electrons released at different tunnelling phase can be clarified into direct electrons and rescattered electrons in each half laser cycle. The electron tunnelled before the laser maximum will be directly pulled away by the laser field, so those electrons are termed as "direct electrons". Alternatively, the electrons released after the laser maximum will be scattered by the nucleus one or several times and they are termed as "rescattered electrons". As the basic processes, electron direct tunneling and rescattering lie in the very heart of attosecond physics [15, 16]. Both direct tunneling and rescattering effects have been success-fully used to resolve molecular orbitals [17, 18].

Many strong-field phenomena appearing in atomic and molecular ionization by intense laser fields are explained within the framework of the strong-field approximation (SFA). In this approximation, the motion of an electron after ionization is assumed to be influenced by the electric field of the laser pulse only. The Coulomb field of the ionized atom is neglected as it quickly decreases with increasing distance of the electron from the ion. If considering the long-range Coulomb potential, the electron classical motion will be modified, as shown in Fig. 2. Once the electron is set free at the tunnel exit, its motion is dominated

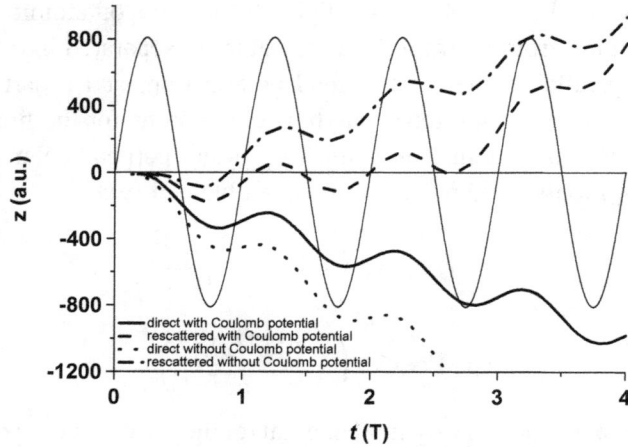

Fig. 2. The electron trajectories of direct electrons and rescattered electrons with and without considering the long-range Coulomb potential.

by the interaction with the laser field even though modified by the long-range ionic Coulomb potential, which both may lead to various photoionization effects. In this chapter, we will review the recent progress on the long range Coulomb potential on the tunnelling ionization, which benefits from the recent high-resolution experimental efforts.

2. Experimental Technique

All the experimental results presented in this chapter were measured with the combination of a strong laser system and a state-of-art technique in atomic physics, i.e., cold-target recoil ion momentum spectroscopy (COLTRIMS) [19]. In the experiments, we employed the 25 fs, 795 nm pulses from a Ti:Sa laser system with 3 kHz repetition rate, amplified pulse energy up to 0.8 mJ. In this amplified laser system, the initial pulses of ∼ 4 nJ energy and 12 fs duration were delivered by a broadband Kerr-lense mode-locked Ti:Sa oscillator with 75 MHz repetition rate. The air and material dispersion was controlled by a set of dispersive mirrors. The pulses were amplified with the so-called 'chirped-pulse amplification' technique [20], i.e., they were stretched to 6–7 ps in a 10 mm thick glass block before being sent to a multipass amplifier system. Stretching was needed in order to reduce the peak intensity of the pulse and, thus, to stay below the damage threshold of the amplifier crystal. The beam passed 9 times through the amplification medium (Ti:Sa crystal pumped by a 3 kHz Nd:YAG laser of 5–6 mJ pulse energy and ∼ 160 ns duration). After the fourth path the repetition rate was reduced from 75 MHz to 3 kHz by means of a Pockels cell. Due to the gain profile of the crystal, the spectrum of the amplified pulses was narrowed to ∼ 70 nm FWHM. The dispersion introduced by a glass block as well as by the other optical elements was compensated by a prism compressor. In addition, in order to compensate the third-order dispersion of the whole laser system, so-called TOD ('third-order dispersion') mirrors were introduced in the oscillator region. The wavelength acceptance of the compressor further reduced the pulse spectral width to ∼ 40 nm FWHM, resulting in the production of Fourier-limited 25 fs (FWHM) pulse with a central wavelength of ∼ 790 nm and a pulse energy up to 800 μJ.

Along with the impressive advances in femtosecond laser technology, within last 20 years tremendous progress has been achieved in the development of novel multi-particle detection techniques for atomic and molecular physics. This culminated in the realization of the so-called 'reaction microscopes', which were nicknamed as 'bubble chambers of atomic and molecular physics'. These machines, based on a combination of COLTRIMS and novel photoelectron imaging spectrometers, enable coincident measurements of several charged reaction fragments (ions and electrons) with excellent momentum and energy resolution, and with detection solid angle close to 4π. Often so-called 'kinematically complete' experiments, where full three-dimensional momentum vectors of all final-state reaction products are determined, become feasible. The main results presented in this chapter, were obtained with a reaction microscope with photoelectron momentum resolution ∼ 0.05 a.u. (atomic units) along the time-of-flight direction and ∼0.08 a.u. along the transverse direction. The electric (∼5 V/cm) and magnetic (∼8 G) fields were applied along the time-of-flight axis. Ions and electrons were measured with two position-sensitive microchannel plate (MCP) detectors respectively. From the time-of-flight and position on the detectors, the full momentum vectors of particles were calculated. In the off-line analysis, the photoelectrons were selected in coincidence with their singly charged parent atomic ions.

3. Low Energy Structure

Recently, high-resolution experiments on above threshold ionization [21] provided evidences for the resonant origin of detailed narrow structures within the plateau, in agreement with a series of numerical simulations [22–24]. A comprehensive numerical analysis of the problem [25] revealed a close relation between the resonant-like behaviour within the plateau and the existence of electron trajectories which lead to multiple recollisions with the parent ion ([26]). The significance of multiphoton resonances for "super-ponderomotive" photoelectrons in the tunnelling regime was predicted for the case of He [27], but until now was not confirmed experimentally.

The new phenomenon, dubbed as "ionization surprise" [28, 29] that in the tunnel regime an unexpected structure in the electron-energy spectrum was observed at small energies, which could not be explained within the SFA but could be seen in full quantum mechanical [30, 31] and also classical calculations [31, 32]. First high-resolution measurements on single ionization clearly demonstrated deviations from the smooth Gaussian shape, manifested as a small deep at zero in the electron (and, thus, ion) momentum distribution parallel to the laser polarization for Ne [33]. For the case of He this was predicted within the semiclassical model [34], where tunnelling with subsequent classical propagation of the emitted electron in the combined laser field and Coulomb field of the ion was considered. While for the case when Coulomb interaction was taken into account, a clear minimum is observed, the calculation for which the Coulomb interaction has been switched off yielded a smooth Gaussian-like curve. In the same work the authors predicted some deviations from the Gaussian shape also for the transverse momentum distribution.

Later, the origin of the low-energy structure (LES) [30, 31] was analyzed in detail and was shown to be a largely classical effect of the combined laser and Coulomb fields on the electronic motion after ionization. More recently, it was shown that the formation of the LES is due to a bunching or focusing of electrons during "soft recollisions" [16, 35]. At turning points of the quiver motion of the electron in the vicinity of the ionic core ($z \sim 0$ in combination with momentum $p_{\parallel} \sim 0$), In contrast to the prediction of the standard tunneling picture within the framework of strong-field approximation, it has been discovered that low-energy photoelectrons in the above threshold ionization process possess nontrivial behavior. For example, the momentum distribution of photoelectrons shows some intriguing low-energy structures, which deviated obviously from the Simpleman's perspective. The low energy structure observed in the momentum distributions along the laser polarization has been intensively investigated recently and its mechanism is still in debate [36–41]. Semiclassical model has been applied to explore this low-energy structure and revealed the essential role of the long range Coulomb potential in its production.

In Figs. 3(a)–3(c), the low-energy part of the measured photoelectron spectra (collected in a solid angle of 5–8 degrees in the field direction) of noble gas atoms at different laser wavelengths and intensities are illustrated. The parameters chosen here guarantee that the process is in the tunneling regime ($\gamma < 1$). Several above threshold ionization peaks can be clearly seen in the spectrum for wavelength of 800 nm. With increasing wavelength, the ATI peaks become less pronounced and can be hardly distinguished in the spectrum for 1800 nm. On the other hand, the LES becomes noticeable in the long-wavelength spectra. It was denoted this as the high-energy low-energy structure (HLES), in order to distinguish it from the

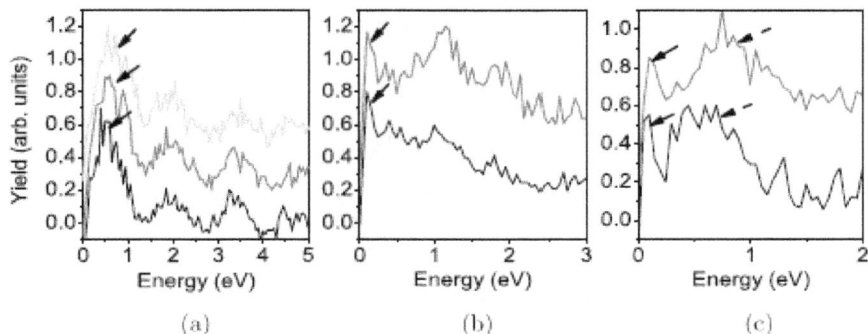

Fig. 3. Low-energy photoelectron energy spectra of Ne (a), Kr (b), and Xe (c) in infrared laser fields. For visual convenience, the VLES and HLES are marked by solid and dashed arrows in all figures of energy spectrum, respectively.

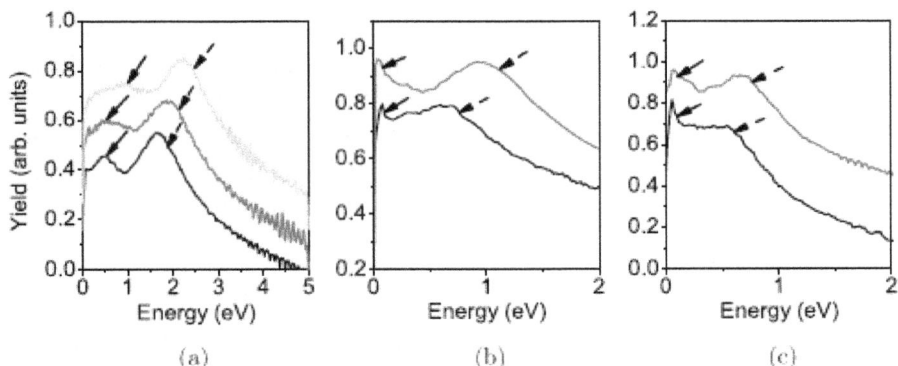

Fig. 4. The semi-classical simulation of low energy structure. The VLES and HLES are marked by solid and dashed arrows in all figures of energy spectrum, respectively.

very-low-energy structure. (For visual convenience, the HLES and the VLES are marked by dashed and solid arrows, respectively, in all figures of energy spectra.) It is intriguing that all spectra show a VLES below 1 eV, which is located at about 0.5 eV for 800 nm and 0.1–0.2 eV for 1320 nm and 1800 nm, respectively.

In order to reveal the low energy structures, we have performed the semiclassical simulation, The simulation results are shown in Fig. 4, and exhibit the following main features: all energy spectra show a VLES above 1 eV and a HLES, which develops and shifts with intensity; the position of the VLES locates at about 0.5 eV and shifts slightly with intensity for 800 nm, while for 1320 and 1800 nm, it locates at about 0.1 eV independently of intensity; all LMDs show a DHS and, on each side, the hump consists of one pronounced peak at low intensity or two or more peaks at high intensity; the peaks beside the minimum in the LMD correspond to the VLES in

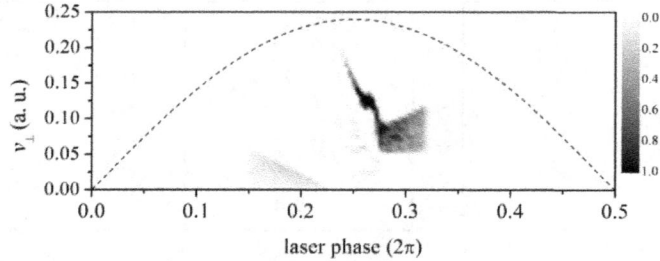

Fig. 5. The distribution of electrons in the LES in phase space (the initial transverse momentum versus the ionization phase) within the electron energy interval (0, 20) eV, with shadow coded probability. The dashed line indicates the laser field.

the ATI spectrum. Clearly, the main features shown in the semiclassical simulation are qualitatively consistent with the experimental results.

We have further traced the initial momentum of those HLES and VLES. As seen in Fig. 5, (i) a large number of electrons with a large initial transverse momentum $v_\perp$ is concentrated after the interaction in the phase space with a low transverse momentum. This is usually termed as Coulomb focusing (CF) [42]. The CF is largest near the laser peak and decreases with the increasing of the ionization phase. (ii) the decrease of $v_\perp$ is not monotonic but shows a steplike slope change. We state that this slope change is responsible for the appearance of peaks in the LES.

4. Local Ionization Suppression

Atomic stabilization in superintense high-frequency fields has been studied theoretically for decades (see, e.g., [43] for a review), which has led to a wealth of profound and intrigue concepts in strong-field physics [44–46]. It is usually discussed in intense high-frequency laser fields in the Kramers-Henneberger (KH) frame, i.e., the moving coordinate frame of a free electron responding to a monochromatic laser field. In the KH frame, the ground state wave function of the atom splits into two non-overlapping peaks and the atom becomes stabilized against ionization when the laser frequency is higher than the bound state frequency of the atom [47]. Recent theoretical investigations further reveal that this concept is not exclusively associated with high frequencies, as widely assumed [48]. The stabilization can also occur when the field frequency is sufficiently large compared to some typical atomic excitation energy. Lacking experimental evidence, this theory is not yet solidified. On the other hand, in an intense low-frequency (e.g., in the infrared regime) laser field the tunneling limit of multiphoton ionization is more appropriately described by the Keldysh theory [49]. Along this, an effect of ionization suppression associated with tunneling was predicted at some specific field strengths [50]. The controversial issue in the deduction however has been argued [51], in which the dependence of the ionization rate on the laser field is found to be monotonic. In contrast to the high-frequency multiphoton ionization, the low-frequency atomic stabilization in the tunneling regime is a more subtle and unsettled question that needs further experimental and theoretical investigation.

Figure 6 shows two-dimensional photoelectron angular distributions in momentum space (P_z, $P_\perp$) of Xe and Kr in the intensities of $0.2 \sim 1 \times 10^{14}$ W/cm^2 at 1320 nm. The P_z and $P_\perp(P_\perp = \sqrt{P_x^2 + P_y^2})$ represent the momentum distribution parallel and perpendicular to the laser polarization axis respectively. One may firstly observe the "fan-out" structures on the PADs. Indeed, there is a large body of theoretical literature on the origin of resonant structures [52] in multiphoton ionization regime with both the classical and quantum models [53, 54], which is beyond the scope of this paper. The striking finding in Fig. 6 is that the relative yield of near-threshold electrons are more suppressed when the laser intensity increases.

In order to achieve deep insight into the ionization dynamics behind the local ionization suppression, we have performed semi-classical electron ensemble simulations including the tunneling effect [55, 56]. Briefly, in the model the electron initial position along the laser polarization direction is derived from the Landau's effective potential theory [57]. The tunneled electrons have initially zero longitudinal velocity and a Gaussian-like transverse velocity distribution. Each electron trajectory

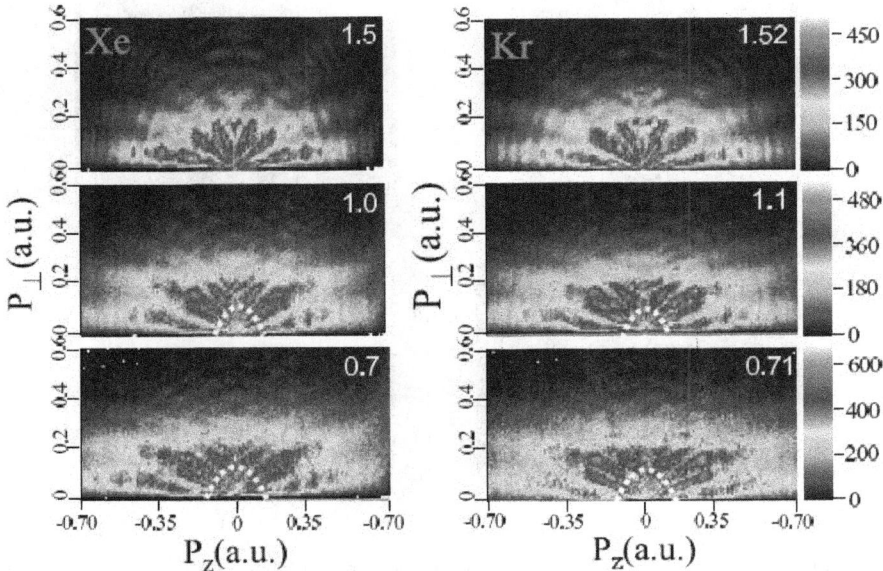

Fig. 6. Experimental two-dimensional PADs in momentum space (P$_z$, P$_\perp$) of Kr and Xe in the intensities of $0.2 \sim 1 \times 10^{14}$ W/cm^2 at 1320 nm. The Keldysh parameters are labeled at the top right corner.

is weighted by the ADK ionization rate $\varpi(t_0, v_\perp^i) = \varpi_0(t_0)\varpi_1(v_\perp^i)$ [58]. $\varpi_1(v_\perp^i) \propto v_\perp^i \exp[\sqrt{2I_p}(v_\perp^i)^2/|\varepsilon(t_0)|]$ is the distribution of initial transverse velocity, and $\varpi_0(t_0) = |(2I_p)^2/\varepsilon(t_0)|^{2/\sqrt{2I_p}-1} \exp[-2(2I_p)^{3/2}/|3\varepsilon(t_0)|]$ depends on the field phase ωt_0 at the instant of ionization as well as the atomic potential I_p.

After the tunneling the electron evolution in the combined oscillating laser field and Coulomb field is solved via the Newtonian equation, $\ddot{\vec{r}} = -\vec{r}/r^3 - \varepsilon \cos \omega t \vec{e}_z$, where the polarization direction is along the z axis, r is the distance between the electron and nucleus, ε and ω are the amplitude and frequency of laser field respectively. The dressed energy of the electron in an oscillating laser field takes the form of $E = \frac{1}{2}(\dot{\vec{r}} + \frac{\varepsilon}{\omega}\sin \omega t \vec{e}_z)^2 - \frac{1}{r}$. In the simulation the laser field has a constant amplitude with ten cycles and is ramped off within three laser cycles.

The simulation results of two-dimensional momentum distributions of $(P_z, P_\perp)$ for Xe atoms at the intensity of 6×10^{13} W/cm^2 and the wavelength of 1320 nm are presented in Fig. 7(a). Generally, the simulation result agrees with the experimental results qualitatively except for the interference patterns. In particular, it reproduces the phenomenon of local ionization suppression in PADs at the origin.

In order to trace the suppressed events, we illustrate the energy distribution of all electrons that tunnel from the first half of the laser cycle $\omega t \in (-\pi/2, \pi/2)$,

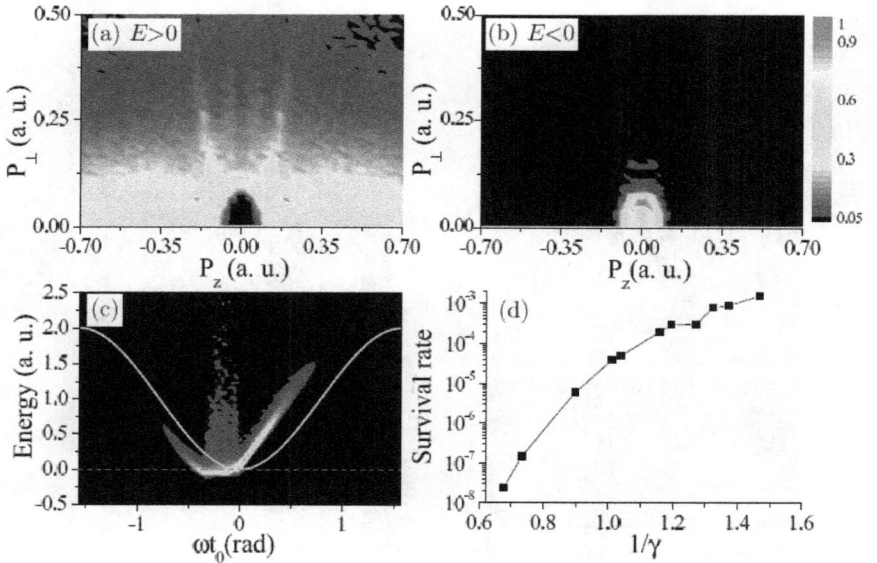

Fig. 7. Simulation results of two-dimensional momentum distribution (P$_z$, P$_\perp$) of the ionized electrons (a) and the unionized electrons (b) for Xe at the intensity of 6×10^{13} W/cm^2 at 1300 nm. (c) The energy distribution of all electrons vs. the initial phase. The solid curve shows the prediction of the simple-man model. (d) The survival rate with respect to the inverse Keldysh parameter.

regardless of whether the final energy is positive or negative, in Fig. 7(c). It exhibits that, after the laser pulse there are a large number of tunneled electrons with the final negative energies within $(-0.01, 0)$ a.u., which means that those electrons are finally bounded by the atomic potential. The binding energies of those Rydberg states ($E < 0.01$ a.u.) are much smaller than the photon energy of the low-frequency light (for 1320 nm, the photon energy is ~ 0.04 a.u.). It has been found experimentally that a large number of excited neutral atoms can survive in strong laser fields [59]. In Fig. 7(b), we show that the two-dimensional momentum distribution of those electrons with the negative energy (the survival events), which can really make up the suppressed yield of those low-energy ionized electrons. The fact that a substantial part of the tunneled electrons end up in the bound states should affect the momentum spectrum of ionized electrons, particularly the low-energy part. Moreover, the survival rate increases with the inverse Keldysh parameter as presented in Fig. 7(d). This is consistent with the experimental observation that the relative yield of near-threshold momentum electrons are more suppressed in deep tunneling regime.

As shown in Fig. 7(c), in the presence of the Coulomb field, the energies of the electrons released at the rising front of the laser cycle are depressed and the energies of the electrons released at the descending front are enhanced, as compared with the prediction by the simple-man model [60] (the solid curve in Fig. 7(c)). Some electrons tunneled from the phase region slightly before the field maximum can achieve much higher energy than $2\,Up$, which are resulted from the electron chaotic motions [55].

In order to shed more light on the local ionization suppression effect, we now consider the subcycle ionization dynamics. The ionization (survival) rate from the first half of the laser cycle is shown in Fig. 8(a). Without the Coulomb field, the tunneling probability in a linearly polarized field is obtained from the ADK formula. Electrons released from the rising front of the laser cycle will drift away from their parent ions and never return to the vicinity of the nucleus. However, if the Coulomb potential is present, the final ionization rate deviates dramatically from the ADK rate. The suppressed yield is result from several distinct phase regions at the rising front of a laser cycle, leading to the survival rate given by the solid green curve in Fig. 8(a). On a closer inspection, the two-dimensional momentum distribution of $(P_z, P_\perp)$ is shown when the electrons are launched within the rising front of the first half laser cycle $\omega t \in (-\pi/2, 0)$ in Fig. 8(b).

To address why those electrons released from those time windows remain unionized, we consider the role of the initial transverse velocity $v_\perp^i$ of the tunneled electrons. We illustrate the distributions of the initial transverse velocities of the tunneled electrons that contribute to the ionization rate and survival rate with respect

(a)

(b)

(c)

(d)

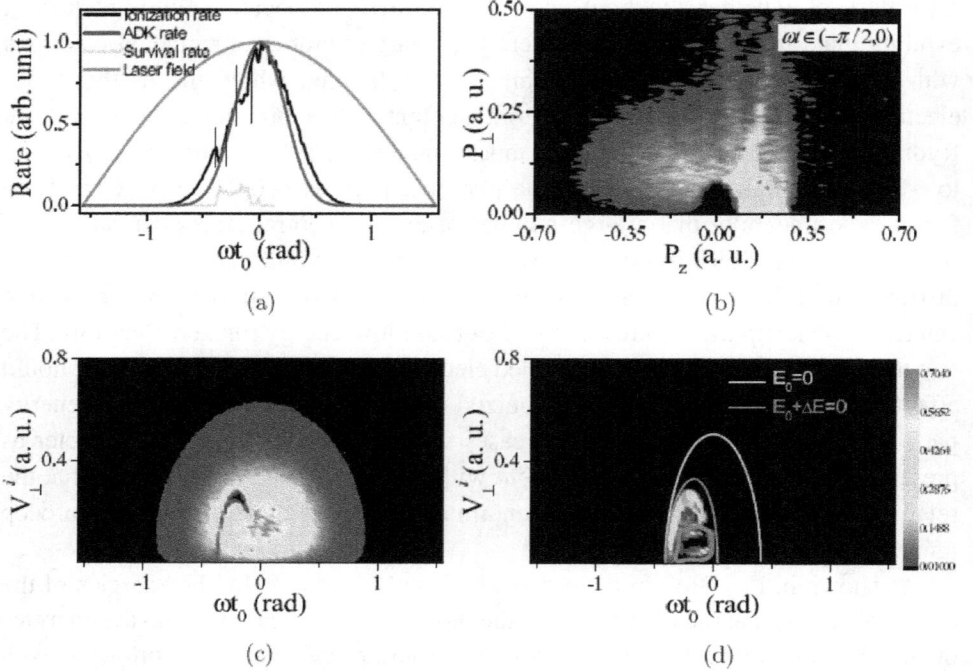

Fig. 8. (a) The ionization (or survival) rate vs. the initial phase (see the text for details). (b) Two-dimensional momentum distribution $(P_z, P_\perp)$ that is contributed from the rising front of the first half laser cycle. (c) and (d) show the initial transverse velocities of tunneled electrons with respect to the laser phase that contribute to ionization and stabilization process respectively. In (d), the solid and dashed curves indicate the theoretical predictions for the boundary for unionized window (see text for details).

to the field phases in Figs. 8(c) and 8(d) respectively. The launch window of the laser phase and initial transverse velocities for those negative energy electrons consists of a regular hippocampus-like structure [the bright regime in Fig. 8(d)], and an irregular regime [encircled in Fig. 8(d)]. Those electrons ionized slightly before the field maximum with a small transverse velocity are strongly affected by the attraction of the Coulomb field, especially moving with an initially small drift velocity along the laser field.

The physics behind the unionized window can be understood by the following approximately theoretical analysis. Considering the Coulomb potential, the final energy of the tunneled electron at the birth time of t_0 can be given by $E_0 = \frac{1}{2}(v_\perp^i \hat{e}_\perp + \frac{\varepsilon}{\omega} \sin \omega t_0 \hat{e}_z)^2 - \frac{1}{|z_0|}$, where $z_0 = (I_p + \sqrt{I_p^2 - 4\varepsilon \cos \omega t_0})/2\varepsilon \cos \omega t_0$ is the tunnel exit point. The energy value can be smaller than zero, which indicates the tunneling without ionization. The boundary expressed by $E_0 = 0$ in t_0 and $v_\perp^i$ plane is expressed by the green curve in Fig. 8d, below it the initial energy

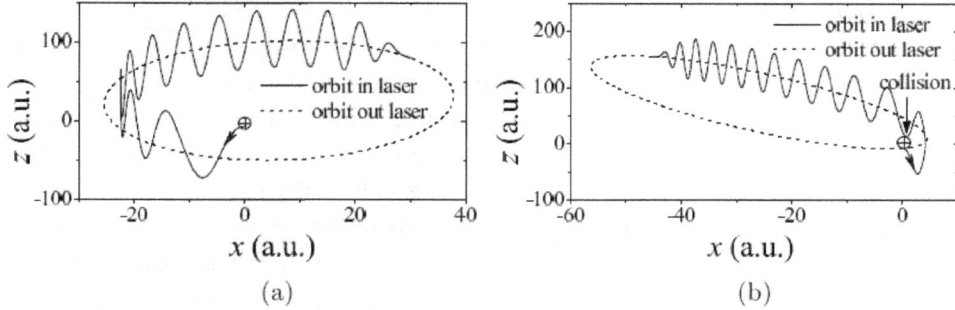

Fig. 9. (a) and (b) indicate the typical trajectories in the hippocampus-like structure regime and irregular regime (encircled with red curve) in Fig. 8(d), respectively. The first type trajectory represents the tunneled electron ejected into the elliptical orbits.

of the tunneling electron is negative. Nevertheless, the tunneled electron will be accelerated in the consequent scattering mediated by the Coulomb and light fields. The energy gain can be expressed analytically by $\Delta E = -\int_{t_0}^{\infty} \frac{z\varepsilon}{\omega r^3} \sin \omega t \, dt$. With ignoring the Coulomb focusing effect, we approximate the electron orbit to be $z = \frac{\varepsilon \sin \omega t_0}{\omega}(t - t_0) + \frac{\varepsilon}{\omega^2}(\cos \omega t - \cos \omega t_0) - z_0, r = \sqrt{[v_\perp^i (t - t_0)]^2 + z^2}$ and then the energy change can be obtained by evaluating the above integral numerically. We plot the line of $E_0 + \Delta E = 0$ in Fig. 8(d) (labeled by the purple line), which gives a good approximation for the boundary that confines the unionized window. However, since some electrons inside the *arc* region might experience multiple forward and backward scatterings, their orbits are essential chaotic [55] and cannot be described by the approximate orbits. During the chaotic scattering, those electrons can acquire additional energy and finally get ionized. This fact makes the boundary of the unionized window become irregular.

Figures 9(a) and 9(b) show the typical trajectories associated with the regular hippocampus-like structure and the irregular structure, respectively, in the specific launching time window. The electrons born with a certain field phase and transverse velocity at the rising front of a laser cycle can finally be launched into the elliptical orbits that have the negative energy. In the first optical cycle after tunneling, the electron obtains or releases energy depending on the instantaneous field phases, and then is pumped into the Rydberg elliptical orbits. The difference between those two typical trajectories is that, the electron in Fig. 9(a) is ejected directly into an elliptical orbit without collision with the atomic ion. While the tunneled electron in Fig. 9(b) experiences the hard collision with the core during its launching process. Our statistics indicate that the first type orbits constitute $\sim 90\%$ and the second type orbits contribute the residual $\sim 10\%$ of the total unionized electrons, respectively.

Those electrons with collisions prefer to move in the elliptic orbits with the smaller semi-major axes and larger eccentricities. The energy of an electron in the elliptic obit is related with the semi-major axis a, which is given by $E = -1/2a$. One can find that the distribution of semi-major axis mostly falls into the regime $1/2a \ll \omega$, i.e., the binding energies of the Rybderg states are much smaller than the photon energy. Because the classical elliptic orbit frequencies ($\sim 1/a^{3/2}$) are much smaller than the laser frequency, the fast oscillating motion driven by laser field can be safely averaged out and the electrons will finally remain on the elliptical orbits [see Figs. 9(a) and 9(b)]. This is analogous to the stabilization condition for the Rydberg atoms in the low-frequency light field [47]. However, in our case, the electrons are released from the ground state through tunneling rather than prepared in the Rydberg states directly, the stringent atomic stabilization cannot be observed.

We have investigated the partial atomic stabilization effect in strong-field tunneling ionization of atoms both experimentally and theoretically. We identify the mechanism as that, a fraction of tunneled electrons released in a certain window of the field phases and transverse velocities can be successively launched into two types of elliptical orbits and are finally stabilized there. This leads to the suppressed yield of near-zero-momentum photoelectrons, i.e., a phenomenon that is universal and is very essential in strong-field tunneling ionization. The identification provides a deep intuition for further quantum control using an attosecond VUV pulse to excite an atom at a proper time window in a strong few-cycle low-frequency light field.

5. Coulomb Asymmetry

Using elliptically polarized laser fields, it is possible to decouple the direct ionization and rescattering effect in strong-field ionization processes. The typical signature of Coulomb field effects on above-threshold ionization in an elliptically polarized field has been shown experimentally [61, 62], manifested in the lack of the fourfold symmetry of photoelectron angular distribution with respect to the both main axes of polarization ellipse that is predicted by the PPT theory (Perelomov, Popov and Terent'ev) [63] and Strong-Field-Approximation [64]. Indeed, in the formation of Coulomb asymmetry from both direct ionization and rescattering processes, the role of ionic potential should not be ignored. However, disentangling those effects from Coulomb asymmetry remains a key issue for us to understand the physics of strong-field ionization. Recently, the use of elliptically polarized laser light fields has added more dimensions to study strong laser field ionization and has attracted particular attention [61, 62, 65–70], which gives rise to more features and properties which are not accessible with a linearly polarized laser field.

For ellipticity <0.3, the formation of the asymmetric four peaks is the dominant feature in the momentum distribution for Coulomb asymmetry [70]. The measured two-dimensional photoelectron momentum distribution in the polarization plane

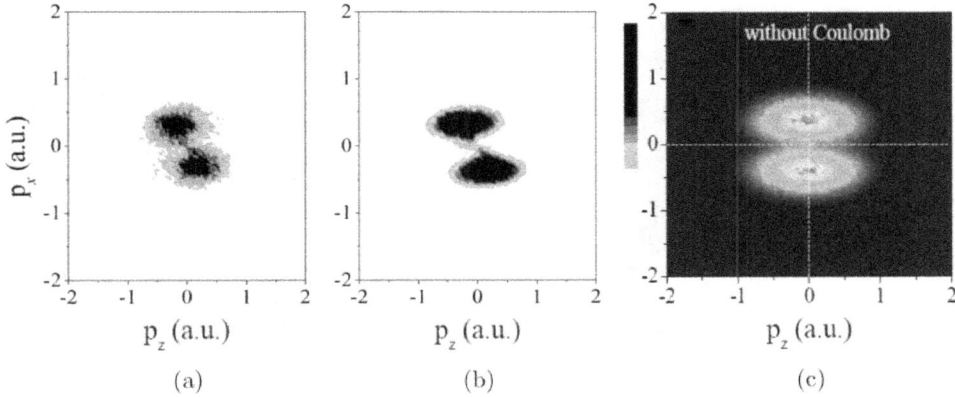

Fig. 10. The two-dimensional experimental (a) and simulated photoelectron momentum distributions with (b) and without (c) considering the Coulomb potential of Ne in an elliptically polarized laser field with the ellipticity ~ 0.25 at an intensity of 3×10^{14} W/cm^2.

for Ne at intensity of 3×10^{14} W/cm^2 with a small ellipticity ~ 0.25 is shown in Fig. 10(a). The major axis is along z direction, the minor axis is along x direction, and y is the laser propagation direction. At such a laser intensity, the Keldysh parameter γ increases from ~ 0.8 to about ~ 3.2, when the instantaneous field rotates from the major axis to the minor axis for Ne.

Using the semiclassical model described in Sec. 4, the simulated photoelectron angular distribution for Ne at intensity of 3×10^{14} W/cm^2 with a small ellipticity ~ 0.25, as illustrated in Fig. 10(b). The simulated photoelectron angular distribution agrees with the measurement nicely. Removing the Coulomb potential in the semiclassical model, the calculated angular distribution is shown in Fig. 10(c), as predicted by PPT theory. Obviously, the asymmetric distribution is due to the long range ionic potential.

We will analyze the subcycle dynamics of Coulomb asymmetry by disassembling the two-dimensional momentum distribution. To do that, we then use a half-trapezoidal pulse with constant amplitude seven cycles ramping off within three cycles, in which all of electrons tunnel in the first laser cycle. The electron momentum-integrated angular distribution with respect to the ionization time window of $(0.25T, 0.75T)$ (the instantaneous field rotates from the minor axis for a half laser cycle, here T is the laser cycle) is shown in Fig. 11(a). One can observe two groups of electrons marked with A and B respectively. The area A corresponds to the events with emission angle $\sim 30°$ in the tunneling time of $(0.25T, 0.48T)$ and the area B corresponds to the events with emission angle $\sim 150°$ in the tunneling time of $(0.48T, 0.75T)$ respectively. The separation of tunneling time is based on careful analysis on electron trajectories (details are described

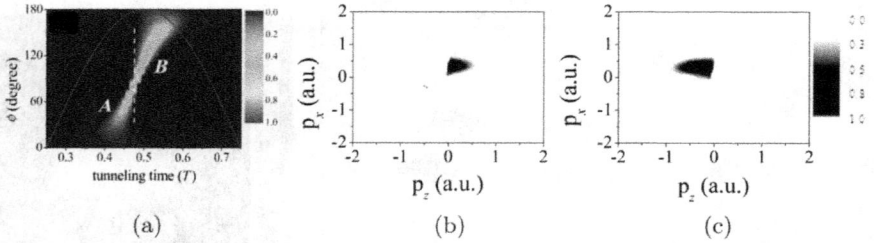

Fig. 11. (a) The momentum-integrated angular distribution with respect to the tunneling time. The solid line shows the field component along the major axis. Two groups of electrons marked A and B are mainly induced by direct ionization and multiple forward scattering, which are separated by the white dashed line. (b) and (c) show two-dimensional photoelectron momentum distributions of direct ionization and multiple forward scattering, respectively.

in [71]). We can clearly identify that the electrons of area A are mainly from direct ionization. Those electrons are ionized when the instantaneous field rotates before the major axis within a quarter of laser cycle and will be pulled away directly by the laser field. However, those electrons are ionized after $0.48T$ can be driven back towards the parent ion. The time shift from $0.5T$ (the field maximum) is due to the Coulomb potential. Depending on initial tunneling coordinates, those electrons will be scattered forward or backward by the ionic potential in the oscillating field. In an elliptically polarized field, because of the field component along the minor axis, the ionized electron will acquire a lateral drift motion with respect to the major polarization axis, and will substantially suppress the backward scattering for those rescattered electrons with small impact parameter. Accordingly, most of electrons of area B will miss the direct impaction on nucleus and will be scattered forward with a small scattering angle. Those electrons may experience forward scattering many times at their subsequent multiple returns in a long pulse. We can separate this process and direct ionization by dissecting the two-dimensional momentum distributions. The two-dimensional photoelectron momentum distributions of group A and group B producing in the first half cycle are shown in Figs. 11(b) and 11(c) respectively. By analyzing the subcycle ionization dynamics, we have decoupled the contribution of direct ionization and multiple forward scattering on Coulomb asymmetry. Obviously, the directly ionized electrons contribute to the small lobes and those multiple forward scattering electrons contribute to the main lobes of Coulomb asymmetry.

Different with linearly polarized field, the initial transverse momentum along the minor and major polarization axis is not symmetrical for an elliptically polarized field. Thus, it is very necessary to consider the role of the initial transverse velocity $\mathbf{v}_\perp$ of the tunneled electrons on the Coulomb asymmetry. In Figs. 12(a) and 12(b), we show the momentum-integrated angular distribution with respect

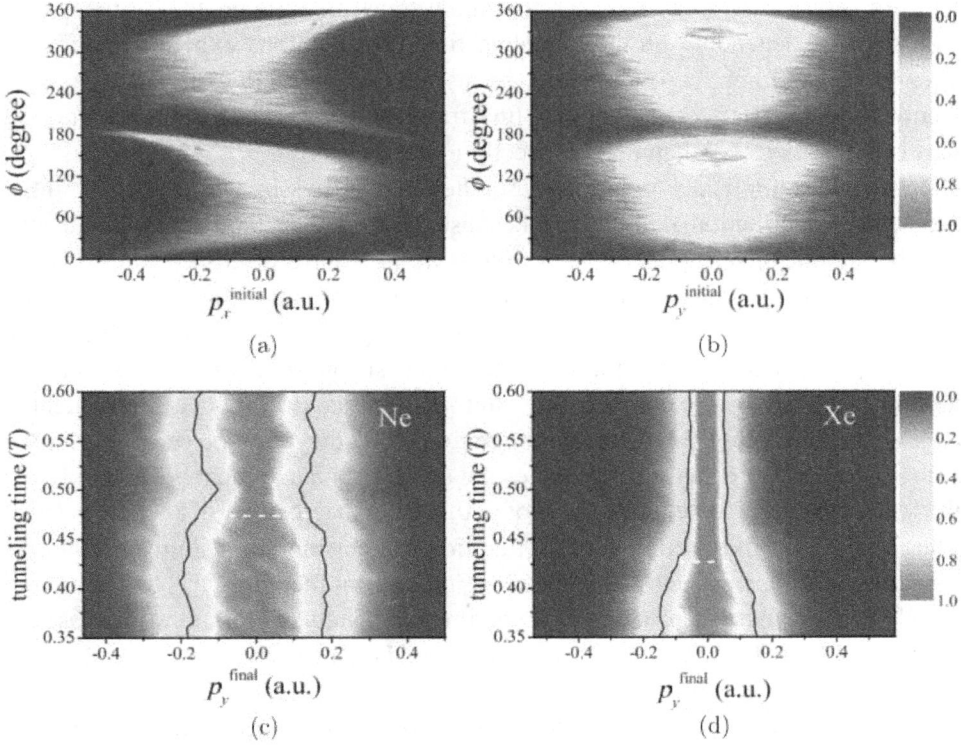

Fig. 12. (a) and (b) show the momentum integrated angular distributions with respect to the initial transverse velocities of along the minor axis and the laser propagation direction. (c) and (d) show the electron tunneling time with respect to the final transverse momentum along the laser propagation direction for Ne 3×10^{14} W/cm^2 and (d) for Xe 1.5×10^{14} W/cm^2 (the ellipticity ~ 0.25). The solid lines and dashed lines in (c) and (d) show the position of half maximum of p_y^{final} and the tunneling time when rescattering begins to occur, respectively, and the shadow scale is normalized to the maximum rate for each tunneling time.

to the initial transverse momentum along the minor axis and laser propagation direction, respectively. Because there is non-zero field component along the minor axis, the tunneled electrons with positive and negative initial transverse momentum v_x have different contributions to Coulomb asymmetry. Since there is no laser field force along the laser propagation direction, the contribution of the transverse momentum along this direction on Coulomb asymmetry is symmetric.

In order to see details of the ionic potential effect on the electrons of direct ionization and multiple forward scattering, we further show the final transverse momentum distribution along the laser propagation direction with respect to the tunneling time in Figs. 12(c) and 12(d). Both the final electron transverse momentum distributions along the laser propagation direction of direct ionization and multiple forward scattering are smaller than the initial transverse momentum at the tunnel exit. Clearly, depending on the tunneling time, the electrons experience different strength of Coulomb focusing. Since the rescattering part will experience stronger Coulomb focusing effect, the width of final transverse momentum of the rescattering part is much narrower than that of directly ionized electrons.

Using the elliptically polarized laser fields, it is possible to exact the initial momentum distribution at tunnel exit because the hard rescattering with the nucleus will be significantly suppressed. Pfeiffer *et al.* [72] have discussed the existence of the initial longitudinal momentum distribution through the comparison between the simulated and measured ellipticity-resolved angular distributions of He^+ ions. It was shown that, the longitudinal momentum spread at the tunnel exit can be around ~ 1.0 a.u. by the comparison with their semiclassical model. Subsequently, considering the elliptical segment integration [73], they further showed that the values of $\sigma_{||}^{initial}$ at different ellipticities are around 0.4 a.u., which is slightly larger than the transverse momentum spread of ~ 0.2 a.u. at the intensity of 8×10^{14} W/cm^2. The conclusion was different with the traditional understanding within the adiabatic tunneling theory, in which the initial longitudinal momentum ($p_{||}^{initial}$) is around zero [74, 75]. We can reproduce the experimental results of Helium using classical trajectory Monte Carlo (CTMC) simulation based on a similar semiclassical model with zero initial longitudinal momentum.

The experimental measurement at ellipse of 0.8 from Kr atoms is shown in Fig. 13. The tilt angle with respect to the minor axis of the polarization ellipse is due to the Coulomb interaction with the parent ion.

In order to obtain the initial longitudinal momentum at tunnel exit, we have considered the dependence of ionization probability on the longitudinal momentum

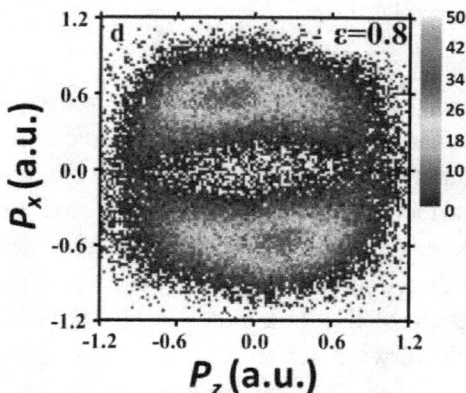

Fig. 13. The measured two-dimensional photoelectron momentum distributions of Kr in an elliptically polarized laser field with the ellipticity ~ 0.8 at an intensity of 3×10^{14} W/cm^2.

distribution. Here, we give the $W(\text{total}) = W_1 * W_2$, in which

$$W_1(p_\perp^{initial}) = p_\perp^{initial}[\sqrt{2I_p}/(\pi|E(t_1)|)]\exp[-(p_\perp^{initial})^2\sqrt{2I_p}/|E(t_1)|]$$

depends on the initial transverse momentum spread $\sigma_\perp^{initial}$ and the probability of the initial longitudinal momentum is given by,

$$W_2(p_\parallel^{initial}) = \frac{1}{\sigma_\parallel^{initial}\sqrt{2\pi}}\exp\left[-\frac{(p_\parallel^{initial})^2}{2(\sigma_\parallel^{initial})^2}\right].$$

To extract the initial longitudinal momentum spread, we have calculated the two-dimensional PADs of single ionization by varying $\sigma_\parallel^{initial}$. Figures 14 show the simulated PADs for the case of $\varepsilon = 0.8$. One can find that there is a qualitative agreement between the simulation and the measurement using a small value of $\sigma_\parallel^{initial}$. When the $\sigma_\parallel^{initial}$ is 0.4 a.u. or 0.8 a.u., the simulated PADs are strongly distorted and have a wider distribution along z-direction, which is dramatically different with the experimental data. Increasing the initial longitudinal momentum offset, the relative contribution of photoelectrons with the zero $p_\parallel^{initial}$ will decrease

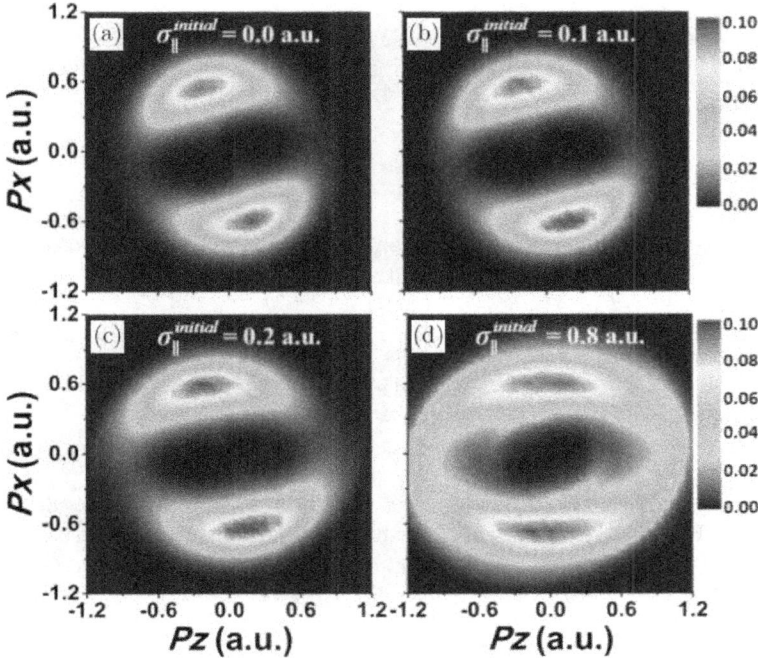

Fig. 14. The simulated two-dimensional PADs of Kr in an elliptically polarized laser field for $\varepsilon = 0.8$ with different initial longitudinal momentum spread ($\sigma_\parallel^{initial}$) of (a) 0.0 a.u., (b) 0.1 a.u., (c) 0.2 a.u. and (d) 0.8 a.u. at the intensity of 1.2×10^{14} W/cm^2.

and more electrons will emit with a larger $p_{||}^{initial}$. Ignoring the Coulomb potential effect, the electron final momentum can be given by $p_z^{final} = -\frac{1}{\varepsilon\omega}E_x(t_1) + p_{||}^{initial}(t_1)$ and, thus the final momentum distribution along z-direction becomes wider at the end of the pulse. Since the initial longitudinal momentum has less effect on the p_x, the final PADs will be strongly distorted because of the inclusion of a larger initial longitudinal momentum at the tunnel exit. The simulation indicates that the initial longitudinal momentum spread is less than 0.2 a.u.

6. Conclusion

In this chapter, we have discussed the recent study on the long-range Coulomb effect on the strong-field tunneling ionization. Implementing the high-resolution data, we are able to decouple the basic steps of direct ionization and multiple forward scattering and to resolve the role of the ionic potential on those processes using an elliptical laser field.

Acknowledgement

We acknowledge the financial support by the 973 program (No. 2013CB922403), the National Science Foundation of China (Nos. 11125416, 11134002, 11121091 and 11134002).

References

1. D. W. McCamant, P. Kukura and R. A. Mathies, *Appl. Spectrosc.* **57**, 1317 (2003).
2. J. Itatani, J. Levesque, D. Zeidler, H. Niikura, H. Pepin, J. C. Kieffer, P. B. Corkum and D. M. Villeneuve, *Nature* **432**, 867–871 (2004).
3. D. A. Long, *The Raman Effect. A Unified Treatment of the Theory of Raman Scattering by Molecules.* (John Wiley Sons, Chichester, 2002).
4. D. Lee and A. C. Albrecht, In eds. J. H. Clark, R and R. E. Hester, *Advances in Infrared and Raman Spectroscopy*, vol. 12, pp. 179–213, Wiley Heyden, London (1985).
5. S. Mukamel, *Principles of Nonlinear Optical Spectroscopy,* Oxford University Press, New York (1995).
6. M. F. Kling, S. Zherebtsov, I. Znakovskaya, T. Uphues, G. Sanone, E. Benedetti, F. Ferrari, M. Nisoli, F. Lepine, M. Swoboda, T. Remetter, A. L'Huillier, F. Kelkensberg, W. K. Siu, O. Ghafur, P. Johnsson and M. J. Vrakking, *Conference on Lasers and Electro-Optics/Quantum Electronics and Laser Science Conference and Photonic Applications Systems Technologies*, p. JFH1 (2008).
7. P. B. Corkum and F. Krausz, *Nature Phys.* **3**, 381 (2007).
8. M. Meckel, D. Comtois, D. Zeidler, A. Staudte, D. Pavičić, H. C. Bandulet, H. Pépin, J. C. Kieffer, R. Dörner, D. M. Villeneuve, and P. B. Corkum, Science **320**, 1478 (2008).
9. H. Akagi. T. Otobe, A. Staudte, A. Shiner, F. Turner , R. Dörner, D. M. Villeneuve, P. B. Corkum, *Science* **325**, 1364 (2009)

10. P. Eckle, A. N. Pfeiffer, C. Cirelli, A. Staudte, R. Dörner, H. G. Muller, M. Büttiker, and U. Keller, *Science* **322**, 1525 (2008).
11. A. N. Pfeiffer, C. Cirelli, M. Smolarski, D. Dimitrovski, M. Abu-samha, L. B. Madsen, and U. Keller, *Nature Phys.* **8**, 76 (2012).
12. M. Bashkansky, P. H. Bucksbaum, and D. W. Schumacher, *Phys. Rev. Lett.* **60**, 2458 (1988).
13. G. G. Paulus, F. Zacher, H. Walther, A. Lohr, W. Becker, and M. Kleber, *Phys. Rev. Lett.* **80**, 484 (1998).
14. S. P. Goreslavski, G. G. Paulus, S. V. Popruzhenko, and N. I. Shvetsov-Shilovski, *Phys. Rev. Lett.* **93**, 233002 (2004).
15. C. Liu and K. Z. Hatsagortsyan, *Phys. Rev. A* **85**, 023413 (2012).
16. N. I. Shvetsov-Shilovski, D. Dimitrovski, and L. B. Madsen, *Phys. Rev. A* **85**, 023428 (2012).
17. D. Shafir, B. Fabre, J. Higuet, H. Soifer, M. Dagan, D. Descamps, E. Mevel, S. Petit, H. J. Worner, B. Pons, N. Dudovich, and Y. Mairesse, *Phys. Rev. Lett.* **108**, 203001 (2012).
18. X. Wang and J. H. Eberly, *Phys. Rev. Lett.* **103**, 103007 (2009).
19. A. N. Pfeiffer, C. Cirelli, A. S. Landsman, M. Smolarski, D. Dimitrovski, L. B. Madsen, and U. Keller, *Phys. Rev. Lett.* **109**, 083002 (2012).
20. D. Comtois, D. Zeidler, H. Pépin, J. C. Kiefer, D. M. Villeneuve, and P. B. Corkum, *J. Phys. B* **38**, 1923 (2005).
21. A. Rudenko, K. Zrost, T. Ergler, A. B. Voitkiv, B. Najjari, V. L. B. de Jesus, B. Feuerstein, C. D. Schröter, R. Moshammer, and J. Ullrich, *J. Phys. B* **38**, L191 (2005).
22. Y. Liu, D. Ye, J. Liu, A. Rudenko, S. Tschuch, M. Dürr, M. Siegel, U. Morgner, Q. Gong, R. Moshammer, and J. Ullrich, *Phys. Rev. Lett.* **104**, 173002 (2010).
23. K. J. Schafer, B. Yang, L. F. DiMauro, and K. C. Kulander, *Phys. Rev. Lett.* **70**, 1599 (1993).
24. P. Corkum, *Phys. Rev. Lett.* **71**, 1994 (1993).
25. S. Basile, F. Trombetta, and G. Ferrante, *Phys. Rev. Lett.* **61**, 2435 (1988).
26. P. Krstic and M. H. Mittleman, *Phys. Rev. A* **44**, 5938 (1991).
27. H. Liu, Y. Liu, L. Fu, G. Xin, D. Ye, J. Liu, X. T. He, Y. Yang, Y. Deng, C. Wu, Q. Gong, *Phys. Rev. Lett.* **109**, 093001 (2012).
28. M. Li, Y. Liu, H. Liu, Y. Yang, J. Yuan, X. Liu, Y. Deng, C. Wu, and Q. Gong, *Phys. Rev. A* **85**, 013414 (2012).
29. J. Ullrich, R. Moshammer, A. Dorn, R. Dörner, L. Ph.H. Schmidt, and H. Schmidt-Böcking, *Rep. Prog. Phys.* **66**, 1463 (2003).
30. L. D. Landau, and E. M. Lifshitz, *Quantum Mechanics* (Pergamon, Oxford, 1977).
31. A. M. Perelomov, V. S. Popov and V. M. Teren'ev, *Zh. Eksp. Teor. Fiz.* **52**, 514 (1967) [*Sov. Phys. JETP* **25**, 336 (1967)]; M. V. Ammosov, N. B. Delone and V. P. Krainov, *Zh. Eksp. Teor. Fiz.* **91**, 2008 (1986) [*Sov. Phys. JETP* **64**, 1191 (1986)]; N. B. Delone and V. P. Krainov, *J. Opt. Soc. Am. B* **8**, 1207 (1991).
32. Linda Petzold, *Siam J. Sci. Stat. Compt.* **4**, 136 (1983).
33. P. Agostini, F. Fabre, G. Mainfray, G. Petite, and N. K. Rahman, *Phys. Rev. Lett.* **42**, 1127 (1979).
34. R. R. Freeman, P. H. Bucksbaum, H. Milchberg, S. Darack, D. Schumacher, and M. E. Geusic, *Phys. Rev. Lett.* **59**, 1092 (1987).
35. L. V. Keldysh, *Sov. Phys. JETP* **20**, 1307 (1965).
36. K. I. Dimitriou, D. G. Arbo, S. Yoshida, E. Persson, and J. Burgdorfer, *Phys. Rev. A* **70**, 061401(R) (2004).
37. Z. Chen, T. Morishita, A.-T. Le, M. Wickenhauser, X.-M.Tong, and C. D. Lin, *Phys. Rev. A* **74**, 053405 (2006).
38. L. Guo, J. Chen, J. Liu, and Y. Q. Gu, *Phys. Rev. A* **77**, 033413 (2008).
39. A. Rudenko, K. Zrost, C. D. Schro'ter, V. L. B. de Jesus, B. Feuerstein, R. Moshammer, and J. Ullrich, *J. Phys. B* **37**, L407 (2004).
40. A. S. Alnaser, C. M. Maharjan, P. Wang, and I. V. Litvinyuk, *J. Phys. B* **39**, L323 (2006).

41. F. H. M. Faisal and G. Schlegel, *J. Phys. B* **38**, L223 (2005); *J. Mod. Opt.* **53**, 207 (2006).
42. G. L. Yudin and M. Y. Ivanov, *Phys. Rev. A* **63**, 033404 (2001).
43. M. Gavrila, *J. Phys. B* **35**, R147 (2002).
44. Q. Su and J. H. Eberly, *J. Opt. Soc. Am. B* **7**, 564 (1990); J. H. Eberly and K. C. Kulander, *Science* **262**, 1229 (1993).
45. M. Dörr, R. M. Potvliege, and R. Shakeshaft, *Phys. Rev. Lett.* **64**, 2003 (1990).
46. K. C. Kulander, K. J. Schafer, and J. L. Krause, *Phys. Rev. Lett.* **66**, 2601 (1991); M. P. de Boer, J. H. Hoogenraad, R. B. Vrijen, L. D. Noordam, and H. G. Muller, *Phys. Rev. Lett.* **71**, 3263 (1993).
47. The traditional atomic stabilization requires that the laser frequency is higher than or at least comparable with the bound energy. Recently, the stabilization issue is extended to the Rydberg atoms, e.g., M. V. Fedorov and A. M. Movsesian, *J. Phys. B* **21**, L155 (1988); B. Piraux and R. M. Potvliege, *Phys. Rev. A* **57**, 5009 (1998); M. Pont and R. Shakeshaft, *Phys. Rev. A* **44**, R4110 (1991); S. Askeland, S. A. Sørngård, I. Pilskog, R. Nepstad, and M. Førre, *Phys. Rev. A* **84**, 033423 (2011).
48. M. Gavrila, I. Simbotin, and M. Stroe, *Phys. Rev. A* **78**, 033404 (2008); M. Stroe, I. Simbotin, and M. Gavrila, *Phys. Rev. A* **78**, 033405 (2008).
49. L. V. Keldysh, *Zh. Eksp. Teor. Fiz.* **47**, 1945 (1964) [*Sov. Phys. JETP* **20**, 1307 (1965)].
50. R. V. Kulyagin and V. D. Taranukhin, *Laser Phys.* **3**, 644 (1993).
51. V. P. Gavrilenk and E. Oks, *Can. J. Phys.* **89**, 849 (2011).
52. A. Rudenko, K. Zrost, C. D. Schröter, V. L. B. de Jesus, B. Feuerstein, R. Moshammer, and J. Ullrich, *J. Phys. B* **37**, L407 (2004).
53. D. G. Arbó, S. Yoshida, E. Persson, K. I. Dimitriou, and J. Burgdörfer, *Phys. Rev. Lett.* **96**, 143003 (2006).
54. Z. Chen, T. Morishita, A.-T. Le, M. Wickenhauser, X. M. Tong, and C. D. Lin, *Phys. Rev. A* **74**, 053405 (2006).
55. B. Hu, J. Liu, and S. G. Chen, *Phys. Lett. A* **236**, 533 (1997).
56. D. Ye and J. Liu, *Phys. Rev. A* **81**, 043402 (2010).
57. L. D. Landau and E. M. Lifshitz, *Quantum Mechanics* (Pergamon, New York, 1977).
58. A. M. Perelomov, V. S. Popov, and V. M. Terenév, *Zh. Eksp. Teor. Fiz.* **52**, 514 (1967) [*Sov. Phys. JETP* **25**, 336 (1967)]; M. V. Ammosov, N. B. Delone, and V. P. Krainov, *Zh. Eksp. Teor. Fiz.* **91**, 2008 (1986) [*Sov. Phys. JETP* **64**, 1191 (1986)]; N. B. Delone and V. P. Krainov, *J. Opt. Soc. Am. B* **8**, 1207 (1991).
59. T. Nubbemeyer, K. Gorling, A. Saenz, U. Eichmann, and W. Sandner, *Phys. Rev. Lett.* **101**, 233001 (2008).
60. In the simple-man model, the electron is supposed to release at origin with zero velocity and the subsequent dynamics is described classically without considering Coulomb potential. For details, refer to H. B. van Linden van den Heuvell and H. G. Muller, in *Multiphoton Processes*, edited by S. J. Smith and P. L. Knight (Cambridge University Press, Cambridge, England, 1988); T. F. Gallagher, *Phys. Rev. Lett.* **61**, 2304 (1988); P. B. Corkum, N. H. Burnett, and F. Brunel, *Phys. Rev. Lett.* **62**, 1259 (1989).
61. M. Bashkansky, P. H. Bucksbaum, and D. W. Schumacher, *Phys. Rev. Lett.* **60**, 2458 (1988).
62. S. P. Goreslavski, G. G. Paulus, S. V. Popruzhenko, and N. I. Shvetsov-Shilovski, *Phys. Rev. Lett.* **93**, 233002 (2004).
63. A. M. Perelomov, V. S. Popov, and M. V. Terent'ev, *Zh. Eksp. Teor. Fiz.* **51**, 309 (1966) [*Sov. Phys. JETP* **24**, 207 (1966)].
64. S. Basile, F. Trombetta, and G. Ferrante, *Phys. Rev. Lett.* **61**, 2435 (1988); P. Krstic and M. H. Mittleman, *Phys. Rev. A* **44**, 5938 (1991).
65. G. G. Paulus, F. Zacher, H. Walther, A. Lohr, W. Becker, and M. Kleber, *Phys. Rev. Lett.* **80**, 484 (1998).

66. C. Liu and K. Z. Hatsagortsyan, *Phys. Rev. A* **85**, 023413 (2012).
67. N. I. Shvetsov-Shilovski, D. Dimitrovski, and L. B. Madsen, *Phys. Rev. A* **85**, 023428 (2012).
68. D. Shafir, B. Fabre, J. Higuet, H. Soifer, M. Dagan, D. Descamps, E. Mevel, S. Petit, H. J. Worner, B. Pons, N. Dudovich, and Y. Mairesse, *Phys. Rev. Lett.* **108**, 203001 (2012).
69. X. Wang and J. H. Eberly, *Phys. Rev. Lett.* **103**, 103007 (2009).
70. A. N. Pfeiffer, C. Cirelli, A. S. Landsman, M. Smolarski, D. Dimitrovski, L. B. Madsen, and U. Keller, *Phys. Rev. Lett.* **109**, 083002 (2012).
71. See Supplemental Material at http://link.aps.org/supplemental/10.1103/PhysRevLett.000.000000 for brief description.
72. A. N. Pfeiffer, C. Cirelli, A. S. Landsman, M. Smolarski, D. Dimitrovski, L. B. Madsen, and U. Keller, *Phys. Rev. Lett.* **109**, 083002 (2012).
73. C. Hofmann, A. S. Landsman, C. Cirelli, A. N. Pfeiffer, and U. Keller, *J. Phys. B* **46**, 125601 (2013).
74. A. N. Pfeiffer, C. Cirelli, M. Smolarski, D. Dimitrovski, M. Abu-samha, L. B. Madsen, and U. Keller, *Nature Phys.* **8**, 76 (2012).
75. P. Eckle, A. N. Pfeiffer, C. Cirelli, A. Staudte, R. Dörner, H. G. Muller, M. Büttiker and U. Keller, *Science* **322**, 1525 (2008).

Chapter 2

Photoelectron Interference and Photoelectron Holography

Min Li*, Qihuang Gong*, and Yunquan Liu*,†

*Department of Physics and State Key Laboratory for Mesoscopic Physics,
Peking University, Beijing 100871, China
†yunquan.liu@pku.edu.cn

In strong-field physics, the electrons ionized by a strong laser field may lead to the interference effect due to the high coherence of the field-driven electron motion. A firm understanding on photoelectron interference is pre-requisite for probing ultrafast electric dynamics in atoms and molecules. In this chapter, we will review the photoelectron interference effect in strong field physics and will show the interference patterns by different interference channels. Further, we will introduce an intuitive quantum-trajectory Monte Carlo approach and will shed light on the formation of the photoelectron angular distributions in above-threshold ionization.

1. Introduction

The interference of matter wave has greatly promoted the progress of quantum mechanics in the early 20th century. When exposing atoms and molecules in strong laser pulses, an electron wave packet can be tunneled through the Coulomb barrier suppressed by the laser fields near the field maximum [1]. The liberated electron wave packet may travel along different pathways with the same final momentum, which will give rise to the interference effect in the photoelectron final momentum spectrum.

One of the most prominent interference effects in strong-field ionization is the interference of a repetitive wave-packet release at time intervals separated by the laser cycle, which will lead to periodic peak structure in the electron energy spectrum. These peaks are known as the above-threshold ionization (ATI) peaks, which were first observed by Agostini *et al.* in 1979 [2].

The liberated electron after the tunneling is accelerated by the oscillatory laser electric field and it has a possibility to be driven back toward the ion [3, 4]. The rescattering photoelectron records detailed structural and dynamic information about the atoms and molecules, and thus it has been used to image the molecular orbital and to retrieve ultrafast dynamics of molecular structure [5–7]. The rescattering wave packet may interfere with the directly emitted wave packet, allowing one to put into practice the concept of holography, a notion that was first used in optics [8]. The central idea of holography is to record the interference between a signal and a reference wave and to reconstruct the objects.

By exposing metastable Xe atom in a mid-infrared laser field at a wavelength of 7000 nm, Huismans *et al.* firstly observed the holographic structure of photoelectrons [9]. As illustrated in Fig. 1, the electron liberated by a linearly polarized laser field with nonzero initial transverse momentum will be driven directly to the detectors (reference, Path I). The electron with initial zero or very small transverse momentum may be rescattered by the nucleus and then drifts to the detectors (signal, Path II). These two channels with the same final momentum will interfere with each other and the interference patterns record the attosecond subcycle ionization dynamics. The interference between these two paths will lead to a spider-like structure in the momentum spectrum, which is a kind of forward rescattering photoelectron holography.

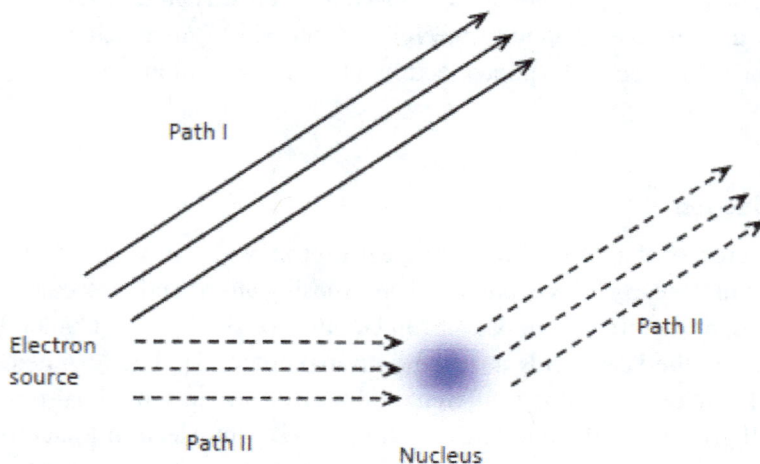

Fig. 1. The illustration of photoelectron holography. Two interfering paths with the same final momentum are indicated. Path I is a reference wave, leaving the electron source directly. Path II is a signal wave, incident on the nucleus and scattering into the same final momentum as the reference wave.

The electron may also be scattered backward at the instant of rescattering. The interference between the backward scattering wave packet and the non-scattering wave packet can be viewed as backward scattering holography. This kind of backward rescattering photoelectron holography is initially predicted by Bian *et al.* via numerically solving the time-dependent Schrödinger equation (TDSE) of molecular H_2^+ [10, 11]. The backward scattering holography can be used to reveal the spatial information of molecular structure and the birth time of the interference channels.

Generally, the study of the photoelectron interference and photoelectron holography is usually based on the strong-field approximation (SFA) and the quantum orbit theory [12, 13], in which the long-range Coulomb potential is neglected. In fact, the long-range Coulomb potential may distort the electron trajectory and modify the photoelectron interference patterns. Recently, the Coulomb-corrected SFA model [9, 14, 15] can provide more quantitative description of photoelectron interference. More accurately, one can reproduce the resonant features present in ATI spectra with astonishing precision by solving the TDSE [16]. However, because the quantum simulation is not physically transparent, no simple picture can be drawn for the mechanism leading to ATI.

In this chapter, we will review the photoelectron interference without rescattering, the forward rescattering holography, and the backward rescattering holography based on a semiclassical model with implementing interference effect. To further include the long-range Coulomb effect, we will introduce an intuitive quantum-trajectory Monte Carlo method and will study the classical-quantum correspondence of above-threshold ionization.

2. Semiclassical Model with Implementing Interference Effect

2.1. *Intercycle interference and intracycle interference*

Without considering the effect of Coulomb potential and recollision [3, 4], the electron final momentum is only determined by the vector potential at the instant of ionization. Generally, there are two types of interference in a laser pulse, known as intercycle and intracycle interference [17]. For the former, the electron wave packet emitted with time intervals separated by the laser cycles will interfere with each other, leading to the ATI peaks. For the latter, the electron wave packets are released at adjacent half-cycles, starting their motion on the opposite side of the ion. They can either be ionized directly (direct trajectory) or turn around in the laser pulse (indirect trajectory) [18]. The intracycle interference between the direct and indirect trajectories has been observed as a time double-slit experiment using few-cycle carrier-envelope phase-stabilized laser pulses [19, 20], in which the intercycle interference is suppressed.

The recollision-free photoelectron interference can be studied based on the simple-man's model with further considering the interference effect. This model is first proposed by D. Arbó *et al.* for both one-dimensional and two-dimensional cases [17, 21]. For a final momentum (p_z, p_x) (z is the laser polarization direction), the released time t_0 and the initial transverse momentum v_x of this electron satisfies (Atomic units are used throughout unless specified),

$$p_z = -A(t_0) \tag{1}$$

$$p_x = v_x \tag{2}$$

Here $A(t)$ is the vector potential of the laser field. For a monochrome field, a final momentum p_z corresponds to different ionization times t_0^j ($j = 1, 2, \ldots, 2N$) with identical field strength. So the ionization rate W is the only function of the final momentum p_z and p_x. Consequently, the momentum distributions can be written as,

$$D(p_z, p_x) = W(p_z, p_x) \left| \sum_j e^{iS(t_0^j)} \right|^2 \tag{3}$$

where S is the classical action along the trajectory, which is given by

$$S(t) = - \int_t^\infty dt' \left\{ \frac{[p_z + A(t')]^2 + p_x^2}{2} + I_p \right\} \tag{4}$$

with I_p being the ionization potential. The sum over the interfering trajectories in Eq. (3) can be decomposed into those associated with two ionization times within the same laser cycle (intracycle interference) and those associated with ionization times in different cycles (intercycle interference), i.e.,

$$\sum_{j=1}^{2N} e^{iS(t_0^j)} = \sum_{k=1}^{N} \sum_{a=1}^{2} e^{iS(t_0^{a,k})} \tag{5}$$

As a result, Eq. (3) can be rewritten as,

$$D(p_z, p_x) = 4W(p_z, p_x)F(p_z, p_x)B(p_z, p_x) \tag{6}$$

Where $F(p_z, p_x) = \cos^2(S_{\text{intra}}/2)$ is the intracycle interference and $B(p_z, p_x) = \sin^2(N S_{\text{inter}}/2)/\sin^2(S_{\text{inter}}/2)$ is the intercycle interference (here N is the laser cycle). For a monochrome laser field of $\mathbf{F}(t) = F_0 \cos(\omega t)\mathbf{z}$, the phase can be expressed as $S_{\text{inter}} = (p_z^2/2 + p_x^2/2 + I_p + U_p)T$ and $S_{\text{intra}} = (p_z^2/2 + p_x^2/2 + I_p + U_p)(t_2 - t_1) - \frac{3U_p}{2\omega}(\sin(2\omega t_2) - \sin(2\omega t_1))$, here T is the laser cycle, t_1 and t_2 are

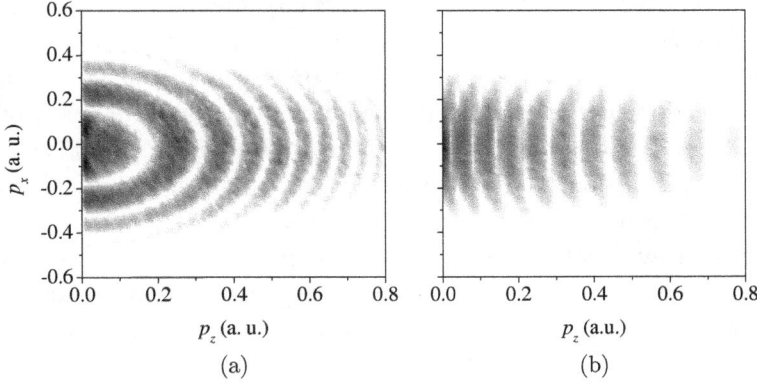

Fig. 2. The intercycleinterference (a) and the intracycle interference (b) patterns. The latter means the interference between the direct and indirect trajectories. The laser intensity is 1×10^{14} W/cm^2 and the laser wavelength is 1320 nm.

the two solutions of Eq. (1) inside each laser cycle, and U_p is the ponderomotive energy with $U_p = F_0^2/4\omega^2$.

Figures 2(a) and 2(b) show the intercycle interference and the intracycle interference patterns for a two-cycle monochrome laser field. The intercycle interference reveals ring-like patterns in the momentum spectrum, corresponding to the ATI peaks in the electron energy spectrum. The intracycle interference shows a stripe-like pattern. Note that the intracycle interference stripes are bent towards the positive p_z direction.

2.2. Forward rescattering photoelectron interference

When considering the recollision process, one can put into practice the concept of holography using photoelectrons. The recollision wave packet serves as a signal wave and the non-recollision wave packet serves as a reference wave. Because the reference wave might be the direct trajectory or the indirect trajectory, there will be two forward rescattering holographic interferences. (When the scattering angle at the instant of recollision is smaller than 90° or larger than 270°, the electron is scattered forward and the interference is dubbed forward scattering holography.)

One can perform a simulation starting with the classical rescattering model [22] with further including the interference effect [10, 23, 24] and the weight of each trajectory according to the Ammosov-Delone-Krainov (ADK) tunneling theory [25, 26]. Specifically, assuming that the laser pulse is cosine-like with $\mathbf{F}(t) = F_0 \cos(\omega t)\mathbf{z}$, where $\mathbf{z}$ is the laser polarization direction. The rescattering electrons are released at the nucleus with zero initial momentum. The electron velocity before recollision can be calculated as $v_z(t) = -F_0/\omega[\sin(\omega t) - \sin(\omega t_0)]$.

The recollision time t_c is obtained by numerically solving the classical equation of motion [22–24] $\cos(\omega t_c) - \cos(\omega t_0^{\text{sig}}) + \omega(t_c - t_0^{\text{sig}})\sin(\omega t_0^{\text{sig}}) = 0$. At t_c, the electron elastically rescatters off the nucleus with a scattering angle with respect to its impact direction. The asymptotic momentum of the rescattered electron is $p_z = F_0/\omega \sin(\omega t_c) + v_c \cos(\theta_c)$ and, $p_x = v_c \sin(\theta_c)$, where $v_c = v_z(t_c)$ is the electron momentum at recollision. The birth time (t_0^{ref}) of the direct and indirect trajectory is expressed as $[2\pi + \sin^{-1}(\omega p_z/F_0)]/\omega$ and $[\pi - \sin^{-1}(\omega p_z/F_0)]/\omega$, respectively. Those electrons will obtain the drift momentum $v_z(t) = -F_0 \sin(\omega t)/\omega$. The corresponding energy is smaller than $2U_p$. If the final drift momenta of the non-rescattering electrons (direct and indirect trajectories) and the rescattering electrons are the same, the interference will take place.

In the calculation, the phase of each trajectory is given by the classical action along the trajectory [27] $S = \int_{t_0}^{\infty}[\mathbf{v}^2(t)/2 + I_p]dt$. The phase difference between the backscattered and the non-rescattered trajectories is $\Delta S_0 = \frac{1}{2}\int_{t_0}^{t_c} v_z^2 dt - \frac{1}{2}\int_{t_0^{\text{ref}}}^{t_c} v_z^2 dt - \frac{1}{2}v_x^2(t_c - t_0^{\text{ref}}) - I_p(t_0^{\text{sig}} - t_0^{\text{ref}})$. In this model, each pair of trajectories is weighted by $W(t_0, v_0) = W_0(t_0)W_1(v_0)$, in which $W_0(t_0) \propto |(2I_p)^2/\mathbf{F}(t_0)|^{2/\sqrt{2I_p}-1}\exp[-2(2I_p)^{3/2}/|3\mathbf{F}(t_0)|]$ is the instantaneous tunnelling ionization probability, and $W_1(v_0) = \sqrt{2I_p}/|\mathbf{F}(t_0)|\exp[-\sqrt{2I_p}v_0^2/|\mathbf{F}(t_0)|]$ is the initial transverse momentum distributions [25, 26]. The electrons are sampled with the well-known Monte-Carlo method.

Figures 3(a) and 3(b) show the interference pattern of the forward rescattering trajectory with the indirect trajectory and the interference pattern of the forward scattering trajectory with the direct trajectory, respectively. The former reveals the spider-like structure, which has been observed in the experiment by Huismans *et al.* [9]. The latter shows the stripe-like pattern, which is very similar to the

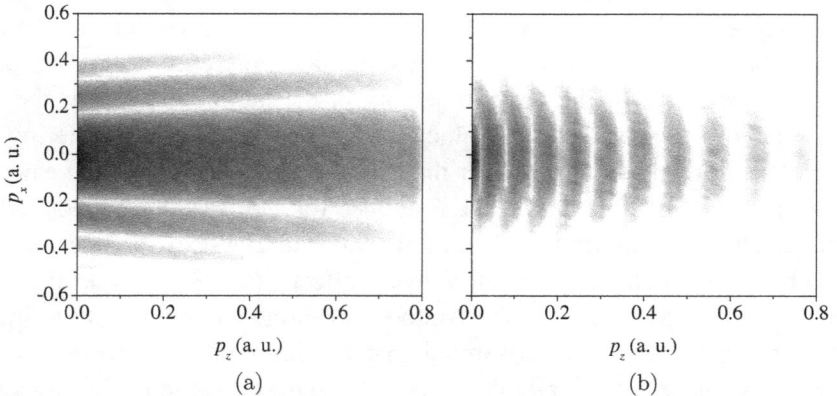

Fig. 3. The interference pattern of the forward rescatteringtrajectory with (a) indirect trajectory and (b) direct trajectory. The laser intensity is 1×10^{14} W/cm^2 and the laser wavelength is 1320 nm.

intracycle interference between the direct and indirect trajectories, as shown in Fig. 2(b). However, the bending direction of the interference stripes is different for these two cases.

2.3. *Backward rescattering photoelectron interference*

If the electron is scattered backward at the instant of recollision, one can obtain the backward rescattering photoelectron interference. As shown in Fig. 4(a), there are two kinds of trajectories in each laser cycle, i.e., the direct trajectory (t_4) and the indirect trajectory (t_3), which will obtain the same final momentum with no recollision. When the backward recollision is taken into account, there are two more trajectories that may acquire the same final momentum as that of the non-rescattered trajectory, which correspond to the long (t_1) and the short (t_2) rescattered trajectories, respectively. Here the long and the short trajectories for the rescattering photoelectrons are very similar to the trajectories in HHG [27]. For a cosinelike laser pulse, the short trajectory is released within ($17°$, $90°$) and the long trajectory within ($0°$, $17°$). Figure 4(b) and 4(c) show the typical trajectories for t_1, t_2 and t_3, t_4, respectively.

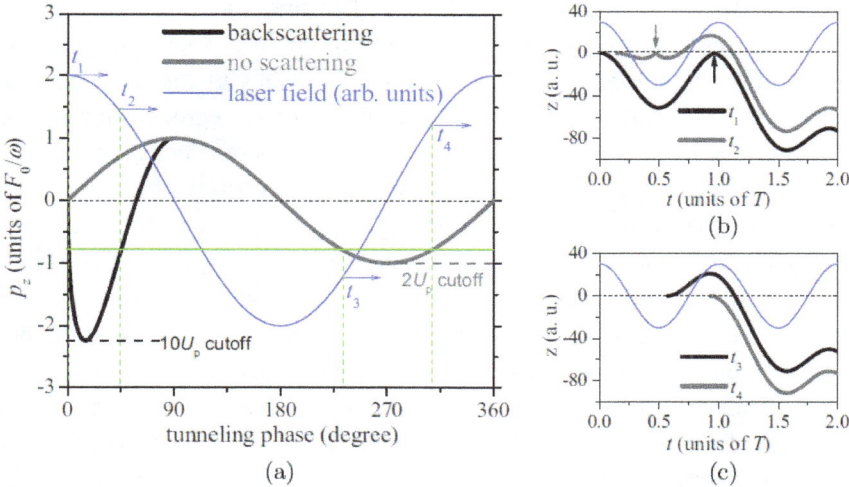

Fig. 4. (a) The electron final momentum with respect to the electron tunneling phase with (solid black curve) and without (solid red curve) recollision based on the simple-man model. The dashed black and dashed red lines correspond to the classical $10U_p$ and $2U_p$ cutoff in the photoelectron energy spectrum, respectively. For the same final momentum (green line), there are four kinds of trajectories in one laser cycle, shown as t_1 (long trajectory), t_2 (short trajectory), t_3 (indirect emission), and t_4 (direct emission), respectively. Here t_1 and t_2 result from recollision, while t_3 and t_4 are without recollision. (b) and (c) show the typical trajectories corresponding to t_1, t_2 (b) and t_3, t_4 (c) respectively. In (b), the arrows indicate the instance of recollision. Blue curves show the laser field in arbitrary units. Adapted from Ref. [23].

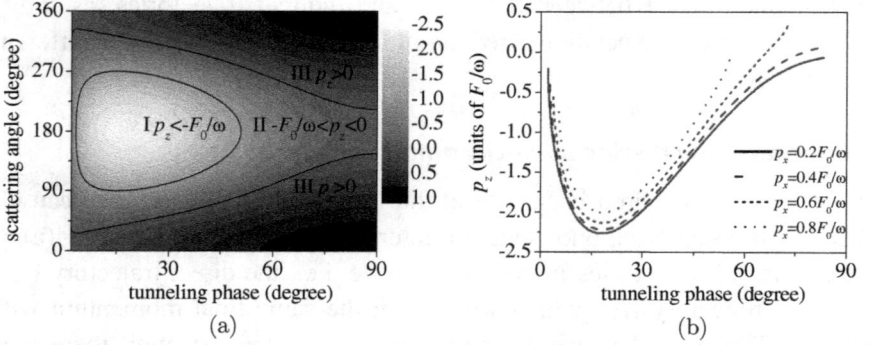

Fig. 5. (a) The final photoelectron momentum pz with respect to the tunneling phase and the scattering angle at the instant of recollision. The final momentum scales with F0/w. Three regions are dissected by the black lines. Region I shows the high-energy photoelectrons with $p_z < -F_0/\omega$, Region II shows the photoelectrons with $-F_0/\omega < p_z < 0$, and Region III with $0 < p_z < F_0/\omega$, respectively. (b) The photoelectron momentum p_z with respect to the tunneling phase with the different transverse momenta p_x. Here only the backward rescattering is calculated.

According to the method in 2.2, the final momentum of the rescattered photoelectron only depends on two variables, i.e., the tunneling phase and the scattering angle at the instant of recollision. The final electron momentum with respect to the tunneling phase and the scattering angle is shown in Fig. 5(a) for the rescattered photoelectrons. The tunneling phase lies in the interval $(0°, 90°)$ corresponding to the falling edge of the laser pulse. Three regions can be separated depending on the magnitude and direction of the final photoelectron momentum.

For region I in Fig. 5(a), the electron with the final momentum $p_z < -F_0/\omega$ corresponds to the electron final energy larger than $2U_p$, and thus they will not interfere with the non-scattering electron wave packet. For those electrons, they are released slightly later than the field maximum and are backward scattered off the nucleus. Those electrons will contribute to the plateau structure of high-order ATI [44]. In Region III, the electron energy is smaller than $2U_p$ $(0 < p_z < F_0/\omega)$. The electron final momentum is positive, which is opposite to the vector potential at the instant of electron release. Those electrons experience the forward scattering and they can interfere with the non-scattered trajectory, as shown in 2.2. For the Region II, the final energy is lower than $2U_p$ $(-F_0/\omega < p_z < 0)$ and the final momentum has the same direction as the vector potential at the instant of electron release. Therefore those electrons experience the backward scattering.

In fact, the final electron momentum can be expressed as the final electron momentum can be analytically expressed as $p_z = \frac{F_0}{\omega}\sin(\omega t_c) - \sqrt{\frac{F_0^2}{\omega^2}[\sin(\omega t_c) - \sin(\omega t_0)]^2 - p_x^2}$ under the condition of the backward scattering. In Fig. 5(b), we show the final momentum p_z with respect to the tunneling phase for different p_x. One can find that one value of p_z corresponds to two tunneling phases with a specific value of p_x. Therefore, when the scattering angle is considered, one final momentum (p_z, p_x) still corresponds to two backscattered trajectories, i.e., the long and the short trajectories.

From Fig. 5(a), one can see that some electrons can have a scattering angle smaller than 90° or larger than 270° and will also obtain negative final momentum. This results from the difference between the photoelectron emission angle $\theta_e = \cos^{-1}(p_z/\sqrt{p_z^2 + p_x^2})$ and the scattering angle θ_c. One can obtain $\cot\theta_e = \cot\theta_c - \frac{\sin(\omega t_c)}{\sin(\omega t_c) - \sin(\omega t_0)} \frac{1}{\sin\theta_c}$. When the scattering angle at instant of recollision is smaller than 90° ($\cot\theta_c > 0$), the final photoelectron emission angle might be larger than 90° and thus final p_z is still negative. Therefore, one needs to consider the full solid scattering angle θ_c at instant of recollision to study the interference between the backscattered trajectory and the non-rescattered trajectory. Strictly speaking, those electrons with scattering angles smaller than 90° or larger than 270° are not backward scattering from the view of scattering, but the final momentum direction is "backward" with respect to that of the non-recollision case. Thus those events are still regarded as backward scattering here. For those photoelectrons, one final momentum (p_z, p_x) may correspond to only one trajectory, so the long and the short trajectories means the electrons released at the tunneling phase before 17° or after 17° at this case.

For the backward rescattering holography, there will be four types of interference between the back-rescattered and non-rescattered trajectories, i.e., the long rescattered trajectories t_1 with the indirect trajectories t_3, the long rescattered trajectories t_1 with the direct trajectories t_4, the short rescattered trajectories t_2 with the indirect trajectories t_3, and the short rescattered trajectories t_2 with the direct trajectories t_4. Each pair of trajectories, i.e., the back-rescattered trajectory (signal) and the non-rescattered trajectory (reference), will have the same final momentum at the end of the pulse. Those four interference patterns are shown in Fig. 6 for a model diatomic molecular ions with an internuclear distance of 2 a.u. for the perpendicular orientation at an intensity of 1.5×10^{15} W/cm² (532 nm).[a] In those plots, the trajectory weight is not included in the calculation. Two interference patterns have comparable fringe spacing, i.e., the interference between the long rescattered trajectory and the indirect trajectory [Fig. 6(a)], and the interference between the short rescattered trajectory and the direct trajectory [Fig. 6(d)]. The fringe spacing of the other two channels [Fig. 6(b) and 6(c)] is much larger.

The fringe spacing of the interference stripes is determined by the phase differences accumulated along the trajectories, which depend sensitively on the

---

[a]In Fig. 6, we have assumed that the tunnel exit is at the position of I_p/F_0, the same as in Ref. [11]. This difference has little effect on the interference patterns.

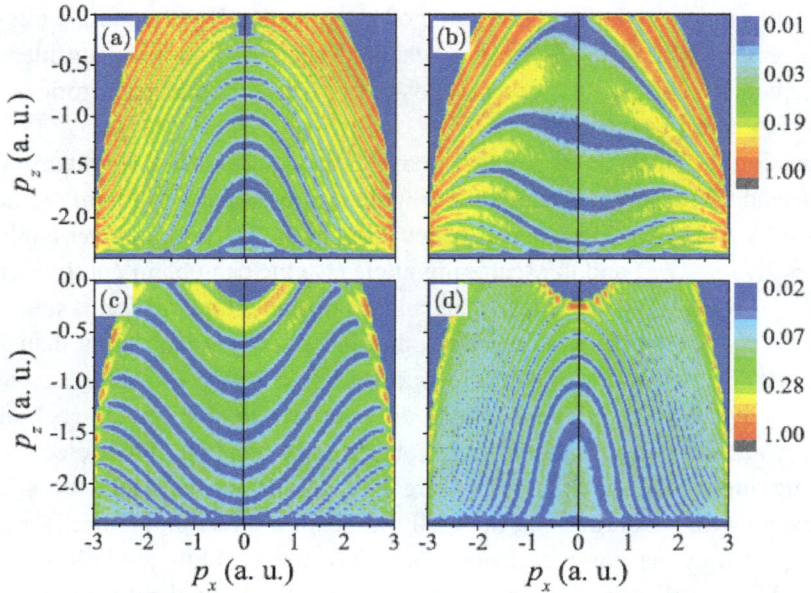

Fig. 6. The interference patterns in the two-dimensional photoelectron momentum spectrum of H2+ (a) between the long trajectory and the indirect trajectory, (b) between the long trajectory and the indirect trajectory, (c) between the short trajectory and the indirect trajectory, and (d) between the short trajectory and the direct trajectory. The molecular axis is perpendicular to the laser polarization direction. The left and right panels of each plot correspond to the photoelectron rescattered by the parent ions or the neighboring ions, respectively. The internulcrear distance is 2 a.u. Adapted from Ref. [23].

difference between the rescattering time of the rescattering trajectory and the birth time of the non-rescattering trajectory [23]. As shown in Fig. 6, because the time differences between the long rescattered trajectory and the indirect trajectory and between the short rescattered trajectory and the direct trajectory are considerably large, the fringe spacing for these two interferences is small and their interference patterns look quite similar. The much shorter time difference for the other two interference channels (the short rescattered trajectory with the indirect trajectory and the long rescattered trajectory with the direct trajectory) leads to much larger spacing of the fringes in p_z momentum distributions.

Experimentally, a linearly polarized 1320 nm radiation was generated by an optical parametric application (OPA) system that was pumped by 25 fs, 795 nm pulses from a Ti:Sa laser system with 3 kHz repetition rate, amplified pulse energy up to 0.8 mJ. The estimated pulse duration at 1320 nm was ∼30–35 fs. The two-dimensional PADs from single ionization of molecules are measured by a reaction microscope [28] with the photoelectron momentum resolution ∼0.02 a.u. along the time-of-flight direction and ∼0.05 a.u. along the transverse direction. Ions and

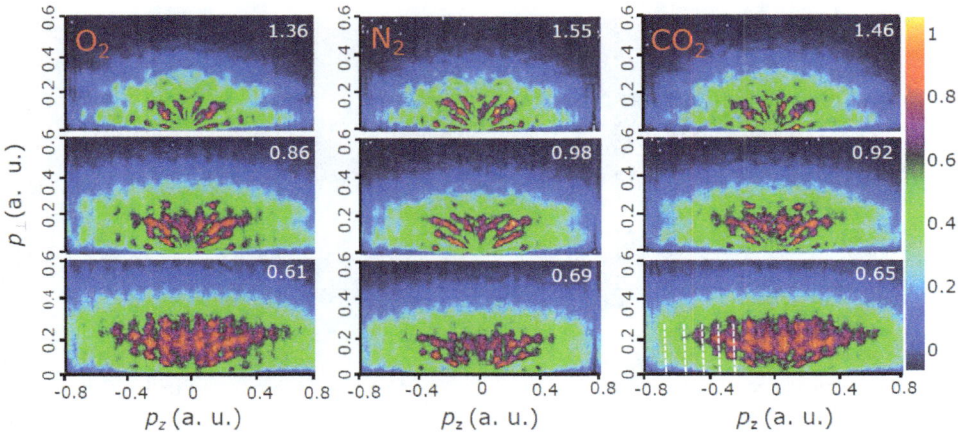

Fig. 7. Experimental two-dimensional PADs in momentum space (P_z, $P_\perp$) of O_2, N_2 and CO_2 in the intensities of $0.2{\sim}1 \cdot 10^{14}$ W/cm^2 at 1320 nm. The Keldysh parameters are labeled at the top right corner. Adapted from Ref. [40].

electrons were measured with two position-sensitive microchannel plate detectors respectively. Weak electric ($\sim$3 V/cm) and magnetic ($\sim$5 G) fields along the time-of-flight axis are applied to guide the electrons and ions to the detectors. The full momentum vectors of the particles were constructed by measuring the time of flight and the impact position on the detections. In the off-line analysis, the photoelectrons were selected in coincidence with their singly charged parent molecular ions to reduce the effect of background electrons. The laser polarization direction was along the time-of-flight axis.

Figure 7 shows the measured two-dimensional PADs in momentum space (P_z, $P_\perp$) of O_2, N_2, and CO_2 molecules at the intensities of $0.2{\sim}1 \times 10^{14}$ W/cm^2 at 1320 nm (the molecular axis is randomly oriented with respect to the laser polarization direction). The P_z and $P_\perp \left(P_\perp = \sqrt{P_x^2 + P_y^2}\right)$ represent the momentum distribution parallel and perpendicular to the laser polarization direction, respectively. At low laser intensities, i.e., in the multiphoton ionization regime, the PADs reveal ring-like multiphoton structures. The PADs of molecules reveal similar low-energy resonant structure at the wavelength of 800 nm [29–31]. At 1320 nm, the PADs of molecules (see the top panels of Fig. 7) are similar to those of atoms at low laser intensities [32]. Interesting, with increasing the laser intensity, an interference stripe structure is enhanced for strong-field ionization of molecules, which is nearly perpendicular to the laser polarization axis. The fringe spacing between adjacent interference stripes is almost 0.1 a.u., as shown by the white dashed lines in Fig. 2. This interference stripe pattern was not observed for the rare gas atoms [32, 33] and the negative molecular ions [34] at similar laser intensities and wavelengths.

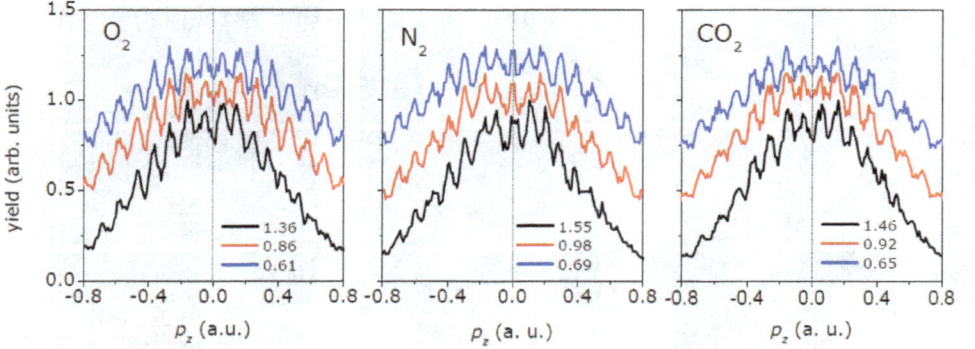

Fig. 8. The measured p_z momentum distributions of O_2, N_2 and CO_2 are shown in the intensities of 0.2–1×10^{14} W/cm^2 at 1320 nm. The Keldysh parameters are shown by the labels. One can find that it is an interference minimum for O_2 and CO_2, and a maximum for N_2 when $p_z = 0$. Adapted from Ref. [40].

By projecting the two-dimensional PADs onto the laser polarization direction, one can clearly identify the enhanced interference structure in the p_z momentum distributions with increasing the laser intensity, as shown in Fig. 8.

In fact, the rescattering cross sections for molecules are larger than those of atoms and negative molecular ions. The possibility of backward rescattering will increase for molecules because of multiple centers and strong coulomb focusing effects [11]. The averaged scattering cross section for randomly aligned molecules is given by [35] $\sigma_m(\theta_c) = 2\left(1 + \frac{\sin(sR)}{sR}\right)\sigma_a(\theta_c)$, where s is the difference between the scattered and incident wave vectors, R is the internuclear distance of molecules, θ_c is the scattering angle, and $\sigma_a(\theta_c)$ is the scattering cross section of atoms. According to the formula, the molecular scattering cross section is approximately twice as large as the atomic cross section for backward rescattering. Increasing the laser intensity, the holographic interference induced by the rescattering will become more important [36].

To study the relative contribution of the four type interference patterns to the final PADs, it is very necessary to include the ionization probability of each channel. Because the long rescattered trajectory are born near the field maximum, as seen in Fig. 4(b), the yield of the long rescattered trajectories is much higher than that of the short rescattered trajectories [23]. On the other hand, the yield of the indirect trajectories will dominate over that of the direct trajectories because of the effect of the long-range Coulomb potential [38, 39]. Thus, the interference between the long rescattered trajectories and the indirect trajectories plays a dominant role in the formation of the low-energy interference stripes. In order to see the interference pattern more clearly, one can assume that the pair of trajectories has the same weight and further calculate the interference patterns for the long rescattered trajectory

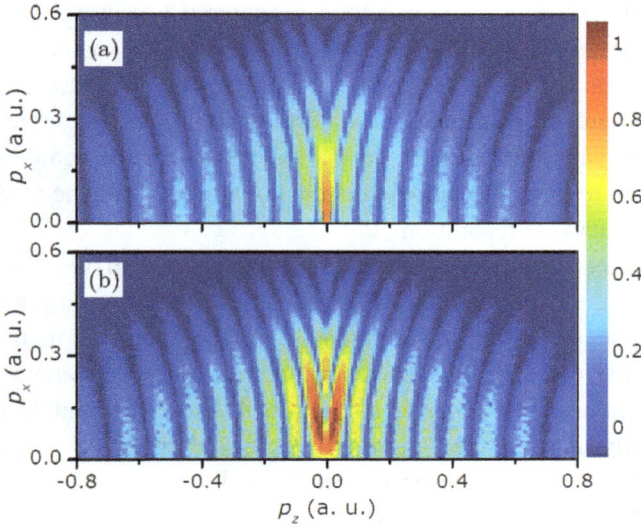

Fig. 9. The simulated interference patterns between the long trajectory and the indirect trajectory. The additional phase difference is zero for σ_g orbital (a) and π for π_g orbital (b). The intensity is 1×10^{14} W/cm^2 and the wavelength is 1320 nm. Adapted from Ref. [40].

with the indirect trajectory, as shown in Fig. 9(a). The simulated interference stripes are nearly transverse to the laser polarization direction and the fringe spacing is almost 0.1 a.u., which is similar to the experimental data. The simulated interference patterns between the long rescattered trajectory and the indirect trajectory agree qualitatively with the interference stripes observed in the measured PADs.

It has been shown that the high-energy above-threshold ionization photoelectrons depend sensitively on the molecular outermost orbital [41], which has been used to extract the electron-ion differential scattering cross section for molecules [42]. Different with the high-energy rescattering photoelectrons, the low-energy rescattering electrons are usually mixed with the non-scattering photoelectrons. We will show that the low-energy photoelectron rescattering holographic patterns is encoded with the molecular structure information, thus could also be used to probe the molecular structure. In Fig. 7, it is strikingly interesting for different molecules at zero momentum of photoelectron angular distributions. It is an interference minimum for O_2 and CO_2, while it is an interference maximum for N_2. This is more clearly seen in the momentum distribution projected onto the laser polarization direction, as shown in Fig. 8. The outermost orbitals of O_2 and CO_2 are the anti-bonding π_g geometry and that of N_2 reveals the bonding σ_g geometry. There will be another phase difference of π for the tunneling from adjacent half cycles for the π_g initial orbital [43], which will result in the destructive interference at the zero momentum, as shown in Fig. 9(b). The experimental results indicate that the interference of the long rescattered trajectories and the indirect trajectories is sensitive to the molecular structures. The low-energy rescattering photoelectron interference holography can also provide a useful probe of the molecular structure. In addition, each interference stripe corresponds to the birth time and the rescattering time of the photoelectrons in the semiclassical model. The low energy rescattering photoelectron holography has encoded the electron dynamics with subcycle resolution.

One should note that other interference channels will also have significant contributions to the experimental PADs, e.g., the ring-like multiphoton ionization structure and the spider-like interference structure. These structures have been studied previously [2, 9]. The backward rescattering interference pattern will superimpose on those characteristic structures. At lower laser intensities, the backward rescattering interference pattern will be buried below other more prominent interference patterns, e.g., the ring-like patterns. With increasing the intensity, the ring-like patterns will become blurred by the intensity averaging. As a result, the visibility of the backward rescattering photoelectron interference increases at high intensities for strong-field ionization of molecules.

The simulated backward rescattering interference pattern does not completely agree with the observed interference stripes, which may arise from the negligence of the long-range Coulomb potential after the electron ionization. The Coulomb tail has shown its importance for both the rescattering and non-rescattering photoelectrons [38, 39, 44]. Hence, the trajectories and the interference patterns will be inevitably modified by the Coulomb tail. In the next section, we will introduce another method with quantitatively including the Coulomb potential and the interference effect.

3. Classical-Quantum Correspondence of above Threshold Ionization

The measured two-dimensional PADs of multiphoton ionization of rare gas (Xe) by a Ti:sappire Laser at a wavelength of 795 nm is shown in Fig. 10(b) at the intensity of 0.75×10^{14} W/cm^2, which is dominated by a series of concentric rings centered at zero, corresponding to the characteristic ATI peaks. There are many side lobe structures inside each order ATI.

In order to quantitatively recapture the interference features observed in the experiment, we develop an intuitive Quantum Trajectory Monte Carlo (QTMC) method. Both the initial conditions for the electron propagation and the trajectory weight of each electron are the same as the CTMC method in Ref. [45]. In the QTMC model, the electron initial condition after tunneling is derived from cthe Laudau's effective potential theory [46] and ADK theory [25, 26]. After tunneling, the electron motion in the combined laser and Coulomb fields is governed by the Newtonian equations, $\ddot{\mathbf{r}} = -\mathbf{r}/r^3 - \mathbf{E}(t)$, where r is the distance between electron

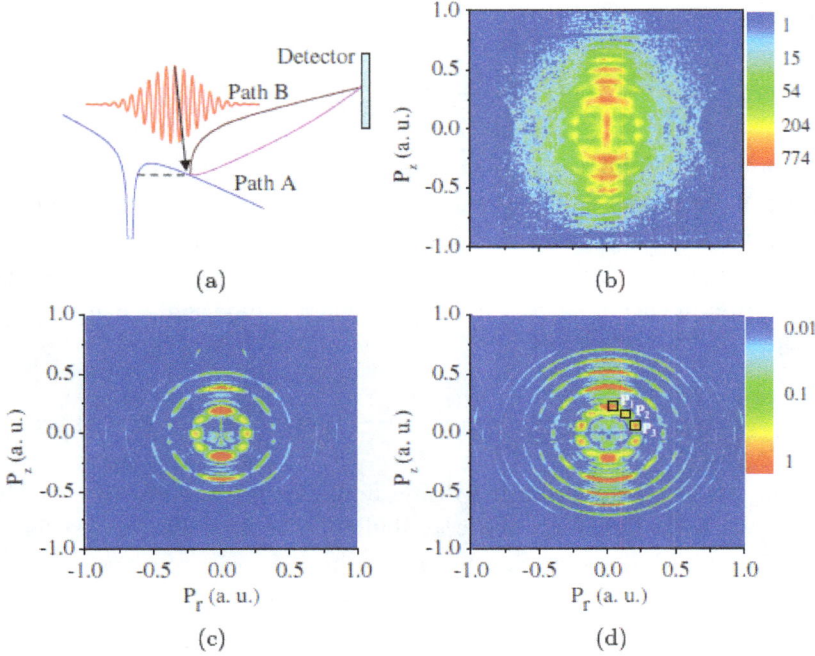

Fig. 10. (a) Illustration of the QTMC model: exposing an atom in a strong linearly polarized laser field, the atomic potential barrier is suppressed by the laser field allowing an electron to tunnel through the narrowed barrier. The initial momentum distributions at the tunnel exit are predicted by ADK tunneling theory. The subsequent electron motion in the combined laser field and Coulomb potential is governed by Newtonian equation. The phase of each electron trajectory is calculated with the phase equation within Feynman's path integral approach. (b) The experimental photoelectron angular distribution of Xe at the intensity of 0.75×10^{14} W/cm^2 (795 nm). (c) The simulated photoelectron angular distribution of Xe at the intensity of 0.5×10^{14} W/cm^2 (795 nm). (d) The simulated photoelectron angular distribution of Xe at the intensity of 0.75×10^{14} W/cm^2 (795 nm). Adapted from Ref. [39].

and nucleus. At the same time, each trajectory will acquire a phase of S according to Feynman path-integral approach [47].

In Feynman's path integral approach [47], the propagator of the electron defined by the probability transition amplitude between two space-time points (r_0, t_0) [at tunnel exit] and (r_1, t_1) [at virtual detector] in the classical limit can be expressed as

$$U(\mathbf{r}_1, t_1; \mathbf{r}_0, t_0) = C(t_1; t_0) \sum_{\text{classical trajectory}} e^{iS(\mathbf{r}_1, t_1; \mathbf{r}_0, t_0)} \tag{7}$$

where $C(t_1; t_0)$ is a normalization constant, which is independent on any individual path, and $S(\mathbf{r}_1, t_1; \mathbf{r}_0, t_0)$ is the action of the electron along the trajectory given by $S(\mathbf{r}_1, t_1; \mathbf{r}_0, t_0) = \int_{t_0}^{t_1} L dt$, here L is the Lagrangian of the electron. In this case, the

Lagrangian in the length gauge, is given by

$$L = \frac{\mathbf{v}^2(t)}{2} + \mathbf{E}(t) \cdot \mathbf{r} + \frac{1}{|\mathbf{r}|} \tag{8}$$

The Coulomb potential in Eq. (8) can be treated as a perturbation when the electron is released at the tunnel exit. We will first calculate the zeroth-order perturbative action, i.e., the Coulomb-free action [neglecting the last term of Eq. (8)], and then include the Coulomb effect as a first-order perturbation. For the zeroth-order case, the contribution of each trajectory to the propagator can be analytically represented as [47, 48],

$$U_j(\mathbf{r}_1, t_1; \mathbf{r}_0, t_0) \approx \frac{\theta(t_1 - t_0)}{[2\pi i (t_1 - t_0)]^{3/2}} \exp\{i S(\mathbf{r}_1, t_1; \mathbf{r}_0, t_0)\} \tag{9}$$

In the final state, we are interested in the final electron momentum distribution. Therefore, we need transform the contribution of each trajectory to the mixed position-momentum representation by performing a Fourier transformation,

$$U_j(\mathbf{p}_1, t_1; \mathbf{r}_0, t_0) \approx (2\pi)^{-3/2} \exp\{i [S(\mathbf{p}_1, t_1; \mathbf{r}_0, t_0) - \mathbf{p}_1 \cdot \mathbf{r}_1]\} \tag{10}$$

where $S(\mathbf{p}_1, t_1; \mathbf{r}_0, t_0) = \int_{t_0}^{t_1} [\frac{\mathbf{v}^2(t)}{2} + \mathbf{E}(t) \cdot \mathbf{r}] dt$ is the classical Coulomb-free action. Since $-\mathbf{p}_1 \cdot \mathbf{r}_1$ is the upper boundary term for the integral $-\int_{t_0}^{t_1} \frac{d}{dt}(\mathbf{v}(t) \cdot \mathbf{r}(t)) dt [\mathbf{A}(t_1) = 0]$, whose lower boundary is $\mathbf{v}(t_0) \cdot \mathbf{r}(t_0) = 0$ according to the above initial tunnelling condition. Equation (10) can be expressed as,

$$U_j(\mathbf{p}_1, t_1; \mathbf{r}_0, t_0) \approx (2\pi)^{-3/2} \exp\left\{i \left[\int_{t_0}^{t_1} \left[\frac{\mathbf{v}^2(t)}{2} + \frac{d\mathbf{v}(t)}{dt} \cdot \mathbf{r}\right] dt \right.\right.$$

$$\left.\left. - \int_{t_0}^{t_1} \frac{d}{dt}(\mathbf{v}(t) \cdot \mathbf{r}(t)) dt \right]\right\}$$

$$= (2\pi)^{-3/2} \exp\left\{-i \int_{t_0}^{t_1} \frac{\mathbf{v}^2(t)}{2} dt\right\}$$

$$= (2\pi)^{-3/2} \exp\left\{-i \int_{t_0}^{t_1} H_0(t) dt\right\} \tag{11}$$

Where $H_0(t)$ is the Coulomb-free Hamiltonian of the electron in the laser field. Depending on the tunnelling phase, different electron trajectory will include an additional phase related with the ionization potential, which is given by $\exp(i I_p t_0)$. Thus the effective propagator is $U_j(\mathbf{p}_1, t_1; \mathbf{r}_0, t_0) \cdot \exp(i I_p t_0) = (2\pi)^{-3/2} \exp(i I_p t_1) \exp\left\{-i \int_{t_0}^{t_1} \left(\frac{\mathbf{v}^2(t)}{2} + I_p\right) dt\right\}$. When electrons reach the virtual detector ($t_1 \to \infty$), the phase is given by,

$$\Phi_j^{(0)} = \int_{t_0}^{\infty} \{\mathbf{v}^2(t)/2 + I_p\}dt \tag{12}$$

Now we consider the Coulomb potential effect on the phase in the first-order perturbation theory. The Coulomb potential has two importance effects. First, it directly introduces the Coulomb-corrected phase due to the variation of the Hamiltonian,

$$\Phi_j^{(1)} = \int_{t_0}^{\infty} -\frac{1}{|\mathbf{r}|}dt \tag{13}$$

Second, the Coulomb potential could change the trajectory of the electron, so it introduces another effect on the term of $\int_{t_0}^{\infty} \{\mathbf{v}^2(t)/2\}dt$. The total phase of each trajectory is the sum of Eq. (12) and Eq. (13), given by

$$\Phi_j = \int_{t_0}^{\infty} \left\{\mathbf{v}^2(t)/2 + I_p - \frac{1}{|\mathbf{r}|}\right\} dt \tag{14}$$

We should note that in Eq. (14) $\mathbf{v}(t)$ is the velocity of electron in the combined Coulomb and laser fields and is different with that in the Coulomb-free case.

The phase and the asymptotic momenta for each trajectory are stored in a table after numerically calculating the Newtonian equation and the phase equation [Eq. (14)]. Then the asymptotic momenta of all trajectories are assigned one-by-one into an interval of $[-2, 2] \times [-1, 1](p_z \times p_r)$ with 1601×801 bins. It is sufficient to sample the electron in a two-dimensional $p_z \times p_r$ phase plane due to the cylindrical symmetry about the z axis, where z is the laser polarization direction and r direction is perpendicular to that. The trajectories will interfere with each other when their asymptotic momenta are in the same bin. Accordingly, the probability of each bin is determined by adding coherently the trajectories in that bin

$$|A|^2_{\text{bin}} = \left|\sum_j \sqrt{W(t_0, v_\perp^j)} \exp(-i\Phi_j)\right|^2$$

$$= \left|\sum_j \sqrt{W(t_0, v_\perp^j)} \cos(\Phi_j)\right|^2 + \left|\sum_j \sqrt{W(t_0, v_\perp^j)} \sin(\Phi_j)\right|^2 \tag{15}$$

where j is jth electron trajectory in this bin. The final two-dimensional momentum distribution is obtained after the probability of all the bins is calculated. Supposing the phase is randomly distributed, the interference term will disappear. The Eq. (14) can transfer back to $|A|_{\text{bin}}^2 = \sum_j W(t_0, v_\perp^j)$, which is the exact result of the CTMC simulation.

Our model is different with the method of CC-SFA [14, 15] and TDSE [16]. The CC-SFA method must introduce the concept of "complex time" or "imaginary time method" and apply the saddle-point approximation. Our model is based on both ADK tunnelling theory and Feynman's path integral approach. All the trajectories propagate in the real (not complex) spatial-temporal space. We can extract the initial tunnelling coordinates at the tunnel exit of ATI photoelectrons. The QTMC model provides more direct and transparent physical picture.

According to Eq. (13), the phase is contributed by three terms, i.e., the trajectory phase related with the motion in laser field, $\Phi^L = \int_{t_0}^{\infty} \mathbf{v}(t)^2/2dt$, the constant phase related with the ionization potential $\Phi^I = \int_{t_0}^{\infty} I_p dt$, and the Coulomb-corrected phase $\Phi^C = \int_{t_0}^{\infty} -1/|\mathbf{r}(t)|dt$. Note that the classical action S is integrated with time together with the Newtonian equations, and thus the Coulomb effect is fully considered in the QTMC model.

The simulated two-dimensional PAD of Xe at the intensity of 0.75×10^{14} W/cm^2 [Fig. 10(d)] agrees well with the experiment [Fig. 10(b)]. The calculated PADs of Xe at the intensity of 0.5×10^{14} W/cm^2 is also shown in Fig. 10(c). One can find that the dominant angular momentum of the first-order ATI is $L = 4$, which may attribute to the absorption one more photon from f Rydberg states. At the intensity of 0.75×10^{14} W/cm^2, the dominant angular momentum of the first-order ATI is $L = 5$ and reveals the character of absorption one more photons from g Rydberg states. One can clearly observe the multiphoton channel-switching effect [49, 50], which corresponds to 10-photon and 11-photon channel at those laser intensities, respectively.

The Coulomb potential is very important in the strong-field atomic ionization. However, this effect can not be directly unraveled by the quantum simulation. As seen in the QTMC model, the Coulomb effect plays essential roles in the two aspects, (i) the electron motion in the Newtonian equation; (ii) both the Coulomb-corrected trajectory and phase in the phase equation. One can separate the Coulomb effect in both the Newtonian equation and the phase equation [Eq. (14)] by the QTMC model. The calculated PAD within the Simpleman model is shown in Fig. 11(a). Compared the PAD that calculated by the CTMC model in Fig. 11(b), it is distorted by the strong scattering in the presence of Coulomb potential. However, the results of both the Simpleman and CTMC models deviate very much from the experimental measurement if the interference effect is not included.

When the Coulomb effect is not included in both the Newtonian equations and the phase equation after tunneling in the QTMC model, the characteristic ATI rings

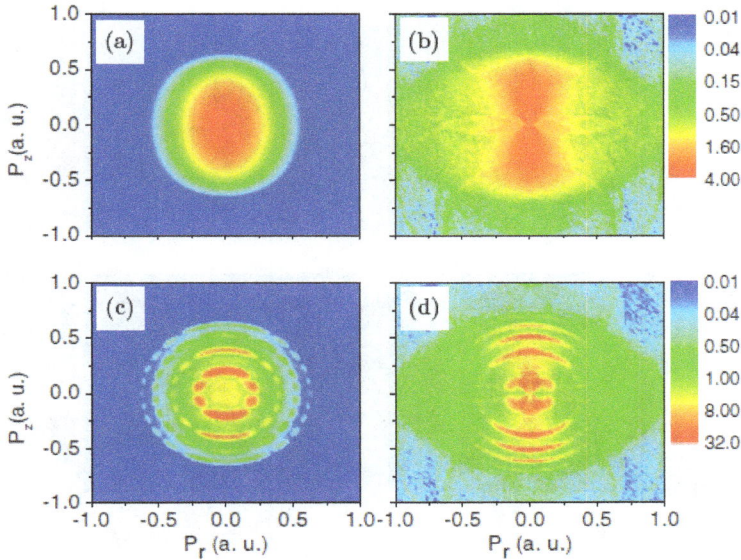

Fig. 11. (a) The simulated photoelectron angular distribution of Xe at the intensity of 0.75×10^{14} W/cm^2 (795 nm) without the Coulomb potential in the Newtonian equation and without solving the phase equation (strong-field approximation). (b) The simulated photoelectron angular distribution of Xe at the intensity of 0.75×10^{14} W/cm^2 (795 nm) with the Coulomb effect in the Newtonian equation and without solving the phase equation (CTMC model). (c) The simulated photoelectron angular distribution of Xe at the intensity of 0.75×10^{14} W/cm^2 (795 nm) without consideration of Coulomb effect in both the Newtonian equation and phase equation. (d) The simulated photoelectron angular distribution of Xe at the intensity of 0.75×10^{14} W/cm^2 (795 nm) with Coulomb effect in the Newtonian equation and without Coulomb-corrected phase. Only the Newtonian equation is solved in (a) (without Coulomb potential) and (b) (with Coulomb potential). The Newtonian and phase equations are solved simultaneously in (c) and (d). In (c), Coulomb potential is not considered in both equations. In (d), Coulomb potential is considered in both equations, but without the Coulomb-corrected phase. Adapted from Ref. [39].

will appear [Fig. 11(c)]. The simulated PAD by the QTMC model reveals the so-called "interference carpet" structure [51]. When the Coulomb potential is fully included, the relative contribution of each ATI spot is redistributed and most of events are shifted along the laser polarization direction because of the Coulomb focusing effect [52], as seen in Fig. 10(d). In order to achieve deep understanding on the Coulomb effect on PAD of ATI, we illustrate the calculated PADs with the inclusion of the Coulomb effect in the Newtonian equations, but not included for the phase equation in the QTMC model in Fig. 12(d). Interestingly, without the Coulomb-corrected phase in the action, one can only observe the ATI along the longitudinal direction (parallel to the laser polarization). This indicates that the Coulomb-corrected phase plays an essential role in the formation of ATI spots transverse to the laser polarization, i.e., "interference carpet". To shed more light

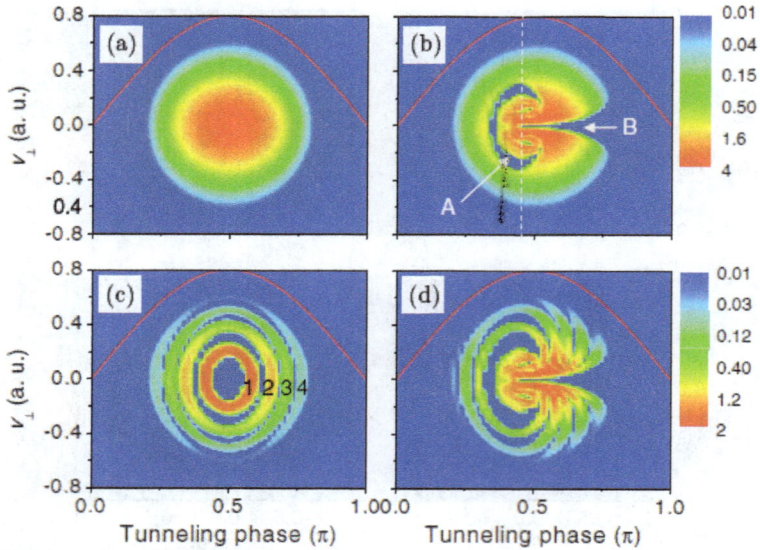

Fig. 12. (a) The initial transverse momentum distribution with respect to the tunneling phase of tunneled electrons without the Coulomb potential in the Newtonian equation and without solving the phase equation (Simplemam model). (b) The initial transverse momentum distribution with respect to the tunneling phase of tunneled electrons with consideration of Coulomb potential in the Newtonian equation and without solving the phase equation (CTMC model). (c) The initial transverse momentum distribution with respect to the tunneling phase of tunneled electrons without consideration of Coulomb potential in both the Newtonian and phase equation in QTMC model. (d) The initial transverse momentum distribution with respect to the tunneling phase with consideration of Coulomb effect in both Newtonian and phase equation with QTMC model. Adapted from Ref. [39].

on that, one must study the tunneling coordinates of ATI and consider the Coulomb effect on both the trajectory and phase of the tunneled electrons.

The Coulomb potential has shown its importance in the formation of ATI. In the Feynman's Path integral picture, the trajectories can contribute to the destructive or constructive interference that depends on the classical action along their paths. Now we investigate the classical origin of ATI patterns by tracing back the initial coordinates (tunneling phase and tunneling momentum) of electrons at the tunnel exit.

Classically, there are two kinds of tunneled electrons, i.e., direct electrons and rescattered electrons. In the Simpleman model, the electron ionized before the peak of the electric field within half an optical cycle will be pulled away from the ion directly. The electron released after the peak of the electric field, will be driven back to the parent ion. Direct electrons experience a small Coulomb attraction immediately after the tunnel exit, while the rescattered electrons will be further influenced by the parent ion's potential upon their subsequent returns. In order to find the signatures of the Coulomb effect on electron trajectories and quantum interference, we trace back the initial coordinates (the initial transverse momentum with respect to the tunneling phase) of all of tunneled electrons in a half laser cycle that contribute to the final PAD with different models.

Without consideration of the Coulomb effect on the trajectory and phase, all of tunneled electrons will contribute to the final PAD, as seen in Fig. 12(a), where the initial coordinates of tunneled electrons in a half laser cycle are predicted by ADK tunneling theory. Including the Coulomb effect, not all of tunneled electrons will contribute to the final PAD, as seen in Fig. 12(b). Compared with Fig. 12(a), the Coulomb potential has three important effects on the final PAD without considering the interference: (i) a fraction of tunneled electrons can be recaptured by the ionic potential during electron recollision; (ii) some tunneled electrons are scattered into high energy. The initial transverse momentum distribution and the Coulomb effect manifest their importance in both the recapture and backscattering process. As shown in Fig. 12(b), the missing area *A* is due to the trapping of tunneled electrons by the parent ions and those electrons can be stabilized in the Rydberg states [32, 53]. The missing electrons in the area *B* will contribute to high energy electrons in the plateau because of strong backward scattering; and (iii) the tunneling time boundary of direct electrons and rescattered electrons is shifted ahead of the field maximum due to the Coulomb effect, indicated by the white dashed line in Fig. 12(b) [38].

If the intercycle interference is considered, depending on the initial tunneling coordinates, some electrons will contribute to the constructive or destructive interference. Not all of tunneled electrons will contribute to the ATI peaks in energy domain. With the back analysis of PADs of the low-order ATI, one can identify the initial tunneling coordinates of all of tunneled electrons. If the Coulomb effect is not considered, the initial momentum distributions for those low-order ATI peaks also reveal rings, as shown in Fig. 12(c) (the corresponding ATI orders are also marked). When the effects of Coulomb potential and interference are fully considered in the QTMC model, the initial momentum distributions of tunneled electrons that contribute to the final PADs are strongly distorted [Fig. 12(d)]. Physically, the back analysis of initial tunneling coordinates of tunneled electrons with consideration of the long-range Coulomb potential provides more sophisticated information.

In fact, both intracycle and intercycle interference will come into effect. Each order ATI is composed of several spots, e.g., there are the destructive or constructive interference patterns within each order of ATI. The dominant angular momentum of the first ATI ring is determined by both the Coulomb-corrected trajectory and Coulomb-corrected action. Employing the advantage of the QTMC model, we can further trace back the classical origin of each ATI spot. As shown in Figs. 13(a)–(c), the initial transverse momentum and the tunneling phase for the tunneled electrons

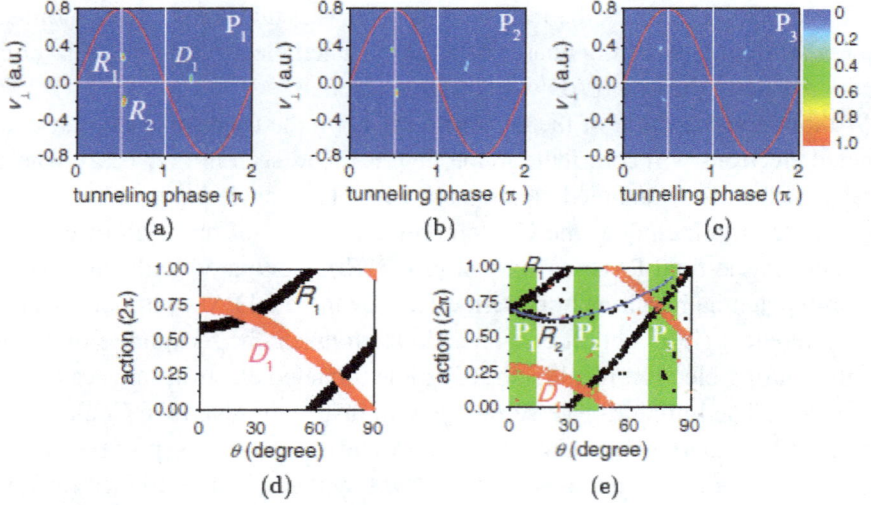

Fig. 13. (a–c) show the initial transverse momentum with respect to the tunneling time of P1, P2, P3 of the first-order ATI in a laser cycle. The red solid lines show the laser field. P_1–P_3, represent the three ATI spots in first quadrant ($p_z > 0$ and $p_x > 0$) of Fig. 1(d). In (a), $R1$ and $R2$ indicate the group of the rescattered electrons with a larger and smaller transverse momentum respectively. $D1$ indicates the direct electrons. (d)–(e) The electron action with respect to the emission angle $\theta = \cos^{-1}\left(p_z/\sqrt{p_z^2 + p_x^2}\right)$ without Coulomb potential (d) and with Coulomb potential (e) for the first order ATI peaks in the quadrant I (supplementary information). Black (red) dots show the electrons released at the first (second) half cycle of the laser fields in (a)–(c). In (d), there are only the group of $R1$ and $D1$. The trajectory weights for three ATI spots are not included in (d) and (e). In (e), the blue solid line indicates the group of $R2$ electrons. The green regions corresponds the three spots of P1-P3. The action values are normalized into an interval of $[0, 2\pi]$. The simulation parameters are the same as Fig. 1(d). Adapted from Ref. [39].

for three spots with positive p_z and p_r of the first-order ATI [P_1–P_3 in Fig. 10(d)] are shown. Obviously, there are two types of intracycle interference to form those spots, (i) the interference between rescattered electrons (marked with R_1 and R_2) in the same field direction, R_1 and R_2 represent the groups of tunneled electrons with large positive and small negative initial transverse momentum distributions, respectively; (ii) the interference of direct electrons (D_1) with the rescattered electrons (R_1 or R_2). The typical trajectories of D_1, R_1 and R_2 are shown in Fig. 14. If the Coulomb effect is not considered, only the groups of R_1 and D_1 with the positive transverse momentum contribute to the final interference patters. In the presence of Coulomb field, another group of rescattered electrons with small negative transverse momenta (R_2) will significantly contribute to the ATI because of its higher ionization rate. The final momenta of the groups R_1, R_2 and D_1 of tunneled electrons are very similar and will show constructive or destructive interference patterns within the ATI rings. Indeed, all of intercycle and intracycle interference [17] of direct electrons and

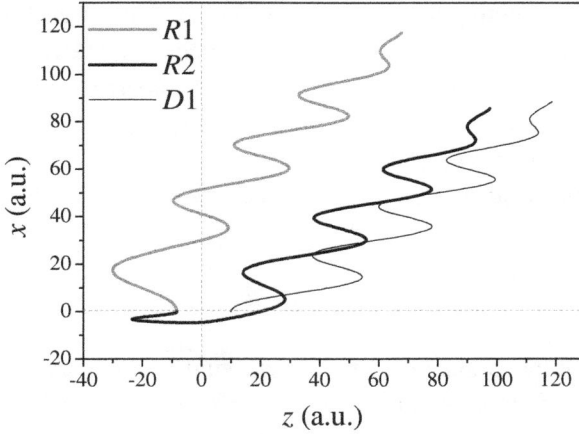

Fig. 14. Typical trajectories for three kinds of contributions to the ATI side lobes with considering the Coulomb potential. The initial coordinates (tunnelling phase, initial transverse momentum) of these three trajectories are $R1$ (0.24 T, 0.380 a.u.), $R2$ (0.263 T, −0.083 a.u.) and $D1$ (0.664 T, 0.234 a.u.), respectively. Their final momentum distributions (p_z, p_x) are nearly the same. Because of Coulomb potential, the transverse momenta of $R2$ have changed the sign from negative to positive. In fact, $R1$ corresponds to the indirect trajectory and $R2$ the forward rescattering trajectory in Section 2.

rescattered electrons from a long laser pulse contributes to the final PADs. The higher order ATI is mainly contributed from the interference among the rescattered electrons.

To reveal how the intracycle interferences take place, one can plot the action of tunneled electrons from the positive field and negative field in a laser cycle with respect to the emission angle in the range of 0~90° for the first order ATI, as shown in Figs. 13(d) and 13(e). The action difference ΔS of those groups of electrons can reflect the constructive or destructive interference of the tunneled electrons. Without including the Coulomb effect in both the electron trajectory and phase, the normalized action of two groups of electrons intersects near the angle of 0° and 75° (where ΔS is $2n\pi$, n = 0, 1, 2...), i.e., ATI spots are formed by the constructive interference between R_1 and D_1 group of electrons [Fig. 13(d)]. Figure 13(f) illustrates the results with the inclusion of both the Coulomb-corrected trajectory and the Coulomb-corrected phase in the range of 0~90° for the first-order ATI. One can find that the spot P_1 is dominantly formed by the constructive interference between R_1 and R_2, P_2 is mainly formed by R_1 and D_1, and P_3 is contributed by all of them. Because the group of electrons of R_2 has smaller initial transverse momenta, the Coulomb potential has much stronger influence on R_2 than R_1 and D_1. The Coulomb-corrected phase will evidently slow down the increase of the action of R_2 with respect to the emission angle. Accordingly, the interference between R_1 and R_2 can give rise to more destructive patterns within an ATI ring than that in the case of neglecting the Coulomb-corrected phase [Fig. 11(d)], in which the action difference between R_1 and R_2 is nearly independent on the emission angle. The relative contribution of those trajectories will change when varying the laser intensity. The observed interference patterns of each order ATI record the underlying electron dynamics of multiphoton ionization on a subcycle time scale, enabling the photoelectron spectroscopy of the excited states created by the strong laser fields.

4. Conclusion

In this chapter, we have discussed the recent study on photoelectron interference effect of strong-field tunneling of atoms. We have discussed the interference patterns for different channels and photoelectron holography. In the end, we introduce an intuitive quantum-trajectory Monte Carlo method and shed some light on the formation of the photoelectron angular distributions of ATI.

References

1. L. V. Keldysh, *Sov. Phys. JETP* **20**, 1307–1314 (1965).
2. P. Agostini, F. Fabre, G. Mainfray, G. Petite, and N. K. Rahman, *Phys. Rev. Lett.* **42**, 1127 (1979).
3. P. B. Corkum, *Phys. Rev. Lett.* **71**, 1994 (1993).
4. K. J. Schafer, B. Yang, L. F. DiMauro, and K. C. Kulander, *Phys. Rev. Lett.* **70**, 1599 (1993).
5. J. Itatani, J. Levesque, D. Zeidler, H. Niikura, H. Pépin, J. C. Kieffer, P. B. Corkum, and D. M. Villeneuve, *Nature* **432**, 867–871 (2004).
6. M. Meckel, D. Comtois, D. Zeidler, A. Staudte, D. Pavičić, H. C. Bandulet, H. Pépin, J. C. Kieffer, R. Dörner, D. M. Villeneuve, and P. B. Corkum, *Science* **320**, 1478 (2008).
7. C. I. Blaga, J. Xu, A. D. Dichiara, E. Sistrunk, K. Zhang, P. Agostini, T. A. Miller, L. F. DiMauro, and C. D. Lin, *Nature* **483**, 194 (2012).
8. D. Gabor, *Nature* **161**, 777 (1948).
9. Y. Huismans, A. Rouzée, A. Gijsbertsen, J. H. Jungmann, A. S. Smolkowska, P. S. W. M. Logman, F. Lépine, C. Cauchy, S. Zamith, T. Marchenko, J. M. Bakker, G. Berden, B. Redlich, A. F. G. Van der Meer, H. G. Muller, W. Vermin, K. J. Schafer, M. Spanner, M. Yu. Ivanov, O. Smirnova, D. Bauer, S. V. Popruzhenko, and M. J. J. Vrakking, *Science* **331**, 61 (2011).
10. X. B. Bian, Y. Huismans, O. Smirnova, K. J. Yuan, M. J. J. Vrakking, and A. D. Bandrauk, *Phys. Rev. A* **84**, 043420 (2011).
11. X. B. Bian, and A. D. Bandrauk, *Phys. Rev. Lett.* **108**, 263003 (2012).
12. W. Becker, F. Grasbon, R. Kopold, D. B. Milošević, G. G. Paulus, and H. Walther, *Adv. At. Mol. Opt. Phys.* **48**, 35 (2002).
13. P. Salières, B. Carré, L. Le Déroff, F. Grasbon, G. G. Paulus, H. Walther, R. Kopold, W. Becker, D. B. Milošević, A. Sanpera, and M. Lewenstein, *Science* **292**, 902 (2001).
14. T. Yan, S. V. Popruzhenko, M. J. J. Vrakking, and D. Bauer, *Phys. Rev. Lett.* **105**, 253002 (2010).
15. T. Yan and D. Bauer, *Phys. Rev. A* **86**, 053403 (2012).
16. T. Morishita, Z. Chen, S. Watanabe, and C. D. Lin, *Phys. Rev. A* **75**, 023407 (2007).
17. D. G. Arbó, K. L. Ishikawa, K. Schiessl, E. Persson, and J. Burgdörfer, *Phys. Rev. A* **81**, 021403(R) (2010).
18. D. G. Arbó, E. Persson, and J. Burgdörfer, *Phys. Rev. A* **74**, 063407 (2006).

19. F. Lindner, M. G. Schatzel, H. Walther, A. Baltuska, E. Goulielmakis, F. Krausz, D. B.Milosevic, D. Bauer, W. Becker, and G. G. Paulus, *Phys. Rev. Lett.* **95**, 040401 (2005).
20. R. Gopal, K. Simeonidis, R. Moshammer, Th. Ergler, M. Dürr, M. Kurka, K.-U. Kühnel, S. Tschuch, C.-D. Schröter, D. Bauer, J. Ullrich, A. Rudenko, O. Herrwerth, Th. Uphues, M. Schultze, E. Goulielmakis, M. Uiberacker, M. Lezius, and M. F. Kling, *Phys. Rev. Lett.* **103**, 053001 (2009).
21. D. G. Arbó, K. L. Ishikawa, E. Persson, and J. Bergdörfer, *Nucl. Instr. Meth. Phys. Res. B* **279**, 24–30 (2012).
22. G. G. Paulus, W. Becker, W. Nicklich, and H. Walther, *J. Phys. B* **27**, L703 (1994).
23. M. Li, J. Yuan, X. Sun, J. Yu, Q. Gong, and Y. Liu, *Phys. Rev. A* **89**, 033425 (2014).
24. M. Lein, J. P. Marangos, and P. L. Knight, *Phys. Rev. A* **66**, 051404 (2002).
25. M. V. Ammosov, N. B. Delone and V. P. Krainov, *Sov. Phys. JETP* **64**, 1191(1986).
26. N. B. Delone and V. P. Krainov, *J. Opt. Soc. Am. B* **8**, 1207–1211 (1991).
27. M. Lewenstein, Ph. Balcou, M. Yu. Ivanov, A. L'Huillier, and P. B. Corkum, *Phys. Rev. A* **49**, 2117 (1994).
28. J. Ullrich, R. Moshammer, A. Dorn, R. Dörner, L. Ph. H. Schmidt, and H. Schmidt-Böcking, *Rep. Prog. Phys.* **66**, 1463–1545 (2003).
29. A. Rudenko, K. Zrost, C. D. Schröter, V. L. B. De Jesus, B. Feuerstein, R. Moshammer, and J. Ullrich, *J. Phys. B- At. Mol. Opt. Phys.* **37**, L407–L413 (2004).
30. C. M. Maharjan, A. S. Alnaser, I. Litvinyuk, P. Ranitovic, and C. L. Cocke, *J. Phys. B- At. Mol. Opt. Phys.* **39**, 1955–1964 (2006).
31. Y. Deng, Y. Liu, X. Liu, H. Liu, Y. Yang, C. Wu, and Q. Gong, *Phys. Rev. A* **84**, 065405 (2011).
32. H. Liu, Y. Liu, L. Fu, G. Xin, D. Ye, J. Liu, X. T. He, Y. Yang, X. Liu, Y. Deng, C. Wu, and Q. Gong, *Phys. Rev. Lett.* **109**, 093001 (2012).
33. D. D. Hickestein, P. Ranitovic, S. Witte, X.-M. Tong, Y. Huismans, P. Arpin, X. Zhou, K. Keister, C. W. Hogle, B. Zhang, C. Ding, P. Johnsson, N. Toshima, M. J. J. Vrakking, M. M. Murnane, and H. C. Kapteyn, *Phys. Rev. Lett.* **109**, 073004 (2012).
34. H. Hultgren and I. Yu. Kiyan, *Phys. Rev. A* **84**, 015401 (2011).
35. C. Cornaggia, *Phys. Rev. A* **78**, 041401(R) (2008).
36. T. Marchenko, Y. Huismans, K. J. Schafer and M. J. J. Vrakking, *Phys. Rev. A* **84**, 053427 (2011).
37. M. Lewenstein, P. Salieres, and A. L'Huillier, *Phys. Rev. A* **52**, 4747–4754 (1995).
38. M. Li, Y. Liu, H. Liu, Q. Ning, L. Fu, J. Liu, Y. Deng, C. Wu, L. Peng and Q. Gong, *Phys. Rev. Lett.* **111**, 023006 (2013).
39. M. Li, J. Geng, H. Liu, Y. Deng, C. Wu, L. Peng, Y. Liu and Q. Gong, *Phys. Rev. Lett.* **112**, 113002 (2014).
40. M. Li, X. Sun, X. Xie, Y. Shao, Y. Deng, C. Wu, Q. Gong, and Y. Liu, *Sci. Rep.* **5**, 8519 (2015).
41. M. Okunishi, R. Itaya, K. Shimada, G. Prümper, K. Ueda, M. Busuladžić, A. Gazibegović-Busuladžić, D. B. Milošević, and W. Becker, *Phys. Rev. Lett.* **103**, 043001 (2009).
42. M. Okunishi, H. Niikura, R. R. Lucchese, T. Morishita, and K. Ueda, *Phys. Rev. Lett.* **106**, 063001 (2011).
43. J. Müth-Bohm, A. Becker, and F. H. M. Faisal, *Phys. Rev. Lett.* **85**, 2280 (2000).
44. J. Yuan, M. Li, X. Sun, Q. Gong, and Y. Liu, *J. Phys. B- At. Mol. Opt. Phys.* **47**, 015003 (2014).
45. B. Hu, J. Liu, and S. G. Chen, *Phys. Lett. A* **236**, 533 (1997).
46. L. D. Landau and E. M. Lifschitz, *Quantum Mechanics: Non-relativistic Theory* (Oxford Univ. Press, 1958).
47. R. P. Feynman, *Rev. Mod. Phys.* **20**, 367 (1948).
48. V. S. Popov, *Phys. At. Nu.* **68**, 686–708 (2005).
49. V. Schyja, T. Lang, and H. Helm, *Phy. Rev. A* **57**, 3692 (1998).
50. M. J. Nandor, M. A. Walker, L. D. Van Woerkom, and H. G. Muller, *Phys. Rev. A* **60**, R1771 (1999).

51. Ph. A. Korneev, S. V. Popruzhenko, S. P. Goreslavkski, T.-M. Yan, D. Bauer, W. Becker, M. Kübel, M. F. Kling, C. Rödel, M. Wünsche, and G. G. Paulus, *Phys. Rev. Lett.* **108**, 223601 (2012).
52. T. Brabec, M. Yu. Ivanov, and P. B. Corkum, *Phys. Rev. A* **54**, R2551 (1996).
53. T. Nubbemeyer, K. Gorling, A. Saenz, U. Eichmann, and W. Sandner, *Phys. Rev. Lett.* **101**, 233001 (2008).

Chapter 3

Dissociation of Hydrogen Molecular Ions in Strong Laser Fields

Feng He

Key Laboratory for Laser Plasmas (Ministry of Education)
and Department of Physics and Astronomy,
Shanghai Jiaotong University, Shanghai 200240, China
IFSA Collaborative Innovation Center,
Shanghai Jiao Tong University, Shanghai 200240, China
fhe@sjtu.edu.cn

The dissociation of H_2^+ and its isotopes in strong laser fields has attracted a lot of interests in past decades. In this chapter, we review recent theoretical progress, including the control of electron localization, tunneling dissociation, rescattering dissociation, and the laser intensity effect on dissociation, cooling down of the nuclear vibrational states.

1. Introduction

When molecules are exposed to strong laser fields, one of the most basic processes is dissociation [1]. The molecular dissociation is important because it directly relates to the production of chemical reactions. Molecular dissociation happens in the timescale of femtoseconds or even attoseconds. It has been a longstanding goal for physical chemists to visualize and further control the dissociation process [2]. This dream has not been realized until the advent of ultrashort laser pulses, i.e., phase stabilized few-cycle infrared pulses [3] and isolated attosecond pulses [4, 5].

As the simplest molecule, the hydrogen molecular ion (H_2^+), and its isotopes, play very important roles for understanding molecular dissociation [6]. Though most experiments start from a neutral H_2 or D_2, some techniques are already developed to generate a beam of H_2^+ [7–9], which allows a comprehensive comparison with ab initio numerical simulations.

In past decades, many experiments and numerical simulations have explored a series of mechanisms about molecular dissociation, to say a few of them, bond softening [10], above threshold dissociation [11, 12], vibrational trapping [13], rescattering dissociation [14]. Some strategies have been raised to control the electron localization after the molecular dissociation [15]. With the pump-probe technique, people are able to trace the nuclear movement [16]. The studies of H_2^+ and its isotopes offer blueprints for observing, then understanding, and ultimately controlling complex chemical reactions.

In this chapter, we review recent theoretical progresses by numerically solving the time-dependent Schrödinger equation (TDSE). In Sec. 2, we briefly introduce the numerical model. In Sec. 3, which includes several subsections, we present several simulation results. In the first subsection, we discuss the control of electron localization beyond the two lowest electronic states, i.e., $1s\sigma_g$ and $2p\sigma_u$. In the second subsection, we discuss the tunneling dissociation of H_2^+ and its isotopes in laser fields with very long wavelengths. The systematically calculations about the laser-intensity effect on dissociation pathways are presented in the followed subsection. In the other two subsections, we present the results how to design an optimal laser pulse to cool down nuclear vibrational states, and rescattering dissociation of H_2^+ and its isotopes. We end this section by discussing how to control a two-dimensional nuclear wavepacket and achieve an angle-resolved electron localization. A summary is given in Sec. 4.

2. Numerical Methods

A hydrogen molecular ion exposed to ultrashort laser fields works as a perfect example for studying ultrafast molecular dynamics. The perturbation theory breaks down when the laser intensity is more than 10^{12} W/cm^2 [17]. To theoretically explore the dissociation of H_2^+ in strong laser fields, one numerically solves the time-dependent Schrödinger equation (atomic units are used unless indicated otherwise) [18].

$$i\frac{\partial}{\partial t}\Psi(r, R; t) = [H_0 + H_I]\Psi(r, R; t), \tag{1}$$

where H_0 is the field-free Hamiltonian, and H_I describes the laser-molecule interaction. r is the electron coordinate, and R denotes the internuclear displacement. Equation (1) can be changed into the following form

$$\Psi(r, R; t + \Delta t) = e^{-i(H_0 + H_I)\Delta t}\Psi(r, R; t). \tag{2}$$

With Cayley's form, $e^{-i(H_0+H_I)\Delta t}$ is replaced by $\frac{1-i(H_0+H_I)\Delta t}{1+i(H_0+H_I)\Delta t}$ [19]. Therefore, Eq. (2) is equivalent to

$$[1 + i(H_0 + H_I)\Delta t]\Psi(r, R; t + \Delta t)$$
$$= [1 - i(H_0 + H_I)\Delta t]\Psi(r, R; t). \qquad (3)$$

By discretizing the wave function in space, Eq. (3) becomes a matrix equation, which can be solved with the LU decomposition method [19]. For multiple dimensional problems, one may decompose several dimensions and solve Eq. (3) in each dimension orderly. After iterating enough times, one may propagate the wave function until all observed quantities are converged.

To fully precisely solve Eq. (1) is a very time-consuming project. Considering concrete physical problems, Eq. (1) can be simplified. For example, when an external laser field is only a few tens of femtosecond, and the molecule is well oriented, then the molecular rotation can be neglected and R changes into a scalar expressing the internuclear distance. If the external laser field is linearly polarized and the polarization axis is along the molecular axis, the interacting system persists the rotational symmetry, and the electron dynamics can be fully described by a two dimensional cylindrical coordinate. Further simplification, for example, restricting the electron movement along the molecular axis, may significantly save the calculation time and qualitatively describe physical problems.

The initial state is obtained by imaginary time propagation [20]. The spatial and time step should be small enough that the numerical grids can resolve the maximum momentum and maximum energy in a physical problem. To count on dissociation probabilities, one may define different districts and integral the probabilities in according areas [21]. When the electron and two nuclei are far away from each other, and each particle can be described by plane waves, then one may Fourier transform the molecular wave function into momentum representation, and obtain the energy spectra [22]. Alternatively, one may project the ultimate wave packet into eigenstates and obtain the probabilities for different components [23].

In the study of molecular dissociation, one widely accepted method is the Born-Oppenheimer approximation [24]. In this approximation, because the electron and nuclei masses differ by orders of magnitude, the electron movement and nuclear movement can be decoupled. In this case, the molecular wavefunction is written as

$$\Psi(R, r, t) = \sum_m \chi_m(R, t)\Phi_m(R, r), \qquad (4)$$

and χ_m is the nuclear wave function and Φ_m is the corresponding electronic wave function. Inserting Eq. (4) into Eq. (1) yields

$$
i\frac{\partial}{\partial t}
\begin{pmatrix}
\chi_1(R,t) \\
\chi_2(R,t) \\
\vdots \\
\chi_m(R,t) \\
\vdots
\end{pmatrix}
=
\begin{pmatrix}
T+V_1 & D_{1,2} & \cdots & D_{1,m} & \cdots \\
D_{2,1} & T+V_2 & \cdots & D_{2,m} & \cdots \\
\vdots & \vdots & \vdots & \vdots & \vdots \\
D_{m,1} & D_{m,2} & \cdots & T+V_m & \cdots \\
\vdots & \vdots & \vdots & \vdots & \vdots
\end{pmatrix}
\begin{pmatrix}
\chi_1(R,t) \\
\chi_2(R,t) \\
\vdots \\
\chi_m(R,t) \\
\vdots
\end{pmatrix},
\tag{5}
$$

where the kinetic operator $T = -\frac{1}{M}\frac{\partial^2}{\partial R^2}$ with M the nuclear mass, V_m is the potential energy surface corresponding to the electronic state $\Phi_m(R,r)$ and the dipole transition matrix element $D_{m,n}$ is written as

$$
D_{m,n} = -\langle r_{m,n}(R)\rangle \bullet E(t)
\tag{6}
$$

with the dipole

$$
\langle r_{m,n}(R)\rangle = \langle \Phi_m(R,r)|r|\Phi_n(R,r)\rangle
\tag{7}
$$

and $E(t)$ the electric field. Compared to Eq. (1), Eq. (5) reduces the simulation time significantly. We use MOLPRO [25] to calculate electron orbits, potential energy surfaces and dipole transition moments. The basis we use is augcc-pvtz [26], and the calculation is based on the theory MCSCF [27, 28]. The eight lowest electronic states are shown in Fig. 1 [29]. The molecular axis is along the horizontal axis (x-axis), and two nuclei are represented by two white dots in the middle. The blue and red colors indicate opposite phases. As the orbital names imply, the gerade (a,c,e,g) and ungerade (b,d,f,h) states are symmetric and antisymmetric under reflection through an inversion center. Only the mixture of gerade and ungerade states with the same azimuthal quantum number will cause the electron asymmetric localization. Figure 2 shows several potential surfaces of H_2^+. Because of the symmetry and antisymmetry of orbits, the dipole transition only happens between gerade and ungerade states. The relative angle between the laser polarization axis and the molecular axis results in different preferred transitions. The transitions between σ and π states are through y or z axis, while those between σ and σ or π and π states are though x axis. The MOLPRO calculation can give very similar results with those in Refs. [30, 31]. In many cases, only a few electronic states play roles for the dissociation, and many other electronic states are not involved. Hence, Eq. (5) can be significantly reduced. For example, if only the two lowest electronic states $1s\sigma_g$ and $2p\sigma_u$ are related to the dynamics, the Hamiltonian in Eq. (5) can be reduced to a 2×2 matrix [32, 33].

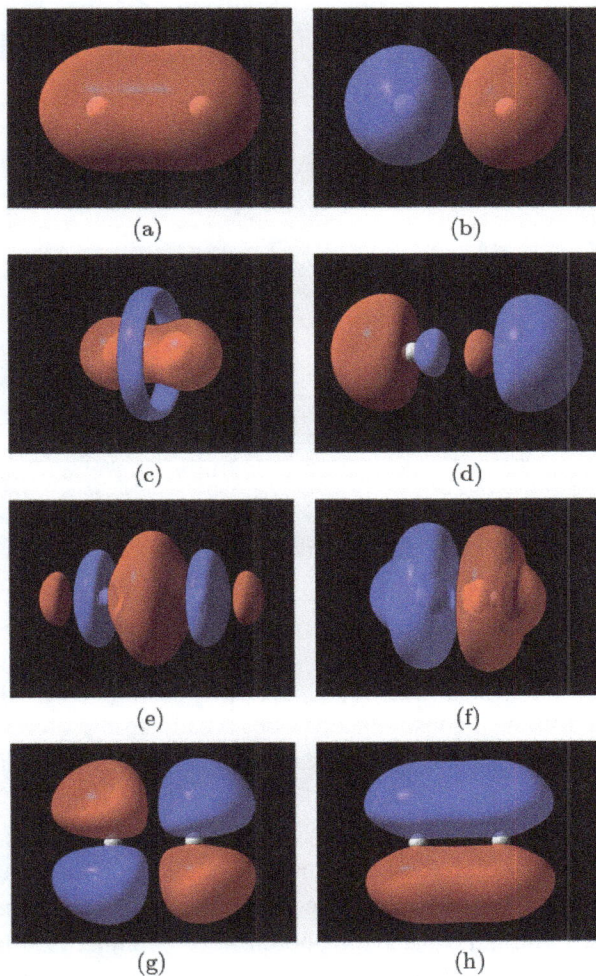

Fig. 1. (Color online) Electron orbits for different electronic states. The internuclear distance is fixed at 5 a.u. and isovalue (electron density) = 0.02. The molecular axis is along the horizontal axis. The nuclei are represented by two white dots. (a) $1s\sigma_g$, (b) $2p\sigma_u$, (c) $3d\sigma_g$, (d) $3p\sigma_u$, (e) $2s\sigma_g$, (f) $4f\sigma_u$, (g) $3d\pi_g$, (h) $2p\pi_u$. (adapted from [29]).

3. Dissociation in Strong Laser Fields

The ground state of H_2^+ is stable. When H_2^+ is exposed to strong laser fields, the electron is excited from $1s\sigma_g$ to other excitation states, and dissociation happens. This dissociation depends on nuclear masses [34, 35], laser parameters, and the orientation of molecules in laser fields. In this section, we show some recent examples by numerically solving the TDSE.

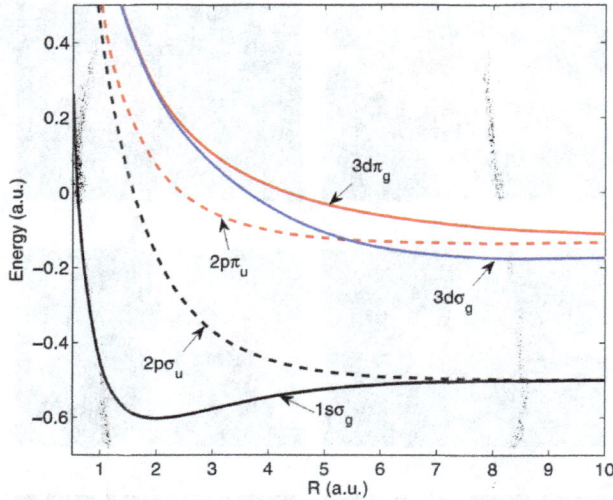

Fig. 2. (Color online) Several potential-energy surfaces of H_2^+ (adapted from [29]).

A. *Electron localization*

After the dissociation of H_2^+, the unique electron in the system will attach on one of the nuclei. The followed questions are: Does the electron preferentially locate on one of the nuclei? Is it possible to highly control the electron localization? In past decades, these questions have attracted a lot of interests, and quite a lot of strategies have been raised to control the electron localization. For example, by varying the relative phase of a two-color ($\omega - 3\omega$) laser field with ω the fundamental angular frequency, the target molecule may be constructively or destructively excited by simultaneously absorbing the ω and 3ω photons [36]. Another example is to change the time delay between two laser pulses so that the molecule can be first pumped to an intermediate state, then evolves and accumulates the time-dependent phases, and later be dumped to a different final state, thereby changing the production of a chemical reaction [37]. Furthermore, specific tailoring of the laser field may dictate a complex chemical reaction to follow one particular channel and stay away from all the others, achieving a selective terminal state [38]. With the development of phase-stabilized few-cycle pulses [3], people are able to control the electron localization by tuning the carrier envelope phase [14, 39–41]. In these control methods, the molecular excitation and electron steering are performed by the same pulse, therefore the time delay between excitation and electron steering are locked. To control the electron localization with higher degrees, He *et al.* [18] proposed to decouple these two processes via a pump-probe strategy, in which H_2^+ is firstly resonantly excited by a single attosecond pulse, then a time-delayed IR field is used to steer the electron movement between two nuclei. In this new method, one may realize an unprecedented asymmetry degree. This proposal has been realized in experiment by Sansone *et al.* [42]

In all previous studies, the laser polarization axis is parallel to the molecular axis, therefore only σ orbitals can be excited. On the contrary, we proposed another control strategy [43]: First, an attosecond pulse, whose polarization axis is perpendicular to the molecular axis, excites H_2^+ to a π orbital, then a time-delayed laser pulse is introduced to mix H_2^+ in some excited electronic states, resulting in the electron asymmetric localization. In such a calculation, the electron movement is restricted in the plane constructed by laser polarization and molecular axes. The wavefunction is in a three spatial spaces x-z-R, where z is the electron coordinate along the molecular axis, while x is the electron coordinate in the direction perpendicular to the molecular axis. By doing Fourier transformation, the wavefunction distribution in the phase space $(x; z; p_R; t)$ is obtained. To characterize the electron asymmetry, an KER-dependent asymmetry parameter is defined as

$$A(E_{KER}) = \frac{P_-(E_{KER}) - P_+(E_{KER})}{P_-(E_{KER}) + (E_{KER})}, \tag{8}$$

where $P_-(E_{KER})$ and $P_+(E_{KER})$ are the probabilities of directional localization of the electron along the positive and negative z-axis as the corresponding proton energy is E_{KER}, respectively.

Figures 3(a) and (b) show the wavefucntion distribution in z-R and x-R spaces, respectively, where the third dimension has been integrated. The snapshots are taken when the nuclear momentum and asymmetry parameters are converged. The node in the vicinities of $x = 0$ in (x, R) space confirms that the electron is in the π orbital. The dissociative states in the (z, R) space present some interference patterns along the R axis. Further analysis confirms that the electron localization is due to the superposition of $2p\sigma_u$ and $3d\pi_g$, which have opposite parities.

The electron localization can be controlled by changing the time delay between the UV and IR pulses, or by changing the carrier envelope phase of the IR pulse. Figure 4(a) shows the dependence of the asymmetry parameter as functions of the proton energy KER and the carrier envelope phase when the time delay is fixed at $t = 500$ a.u. Similarly, by fixing the carrier envelope phase at 0 while tuning the time delay, one may obtain the dependence of the asymmetry parameter as functions of the proton energy and the time delay, as shown in Fig. 4(b). The simulation results show the control degree is up to 90% for the proper laser parameters.

In this control strategy, the main dynamics happens among states $1s\sigma_g$ and $2p\pi_u$ and $3d\pi_g$, therefore, alternatively, one may simulate Eq. (5) by only picking up these

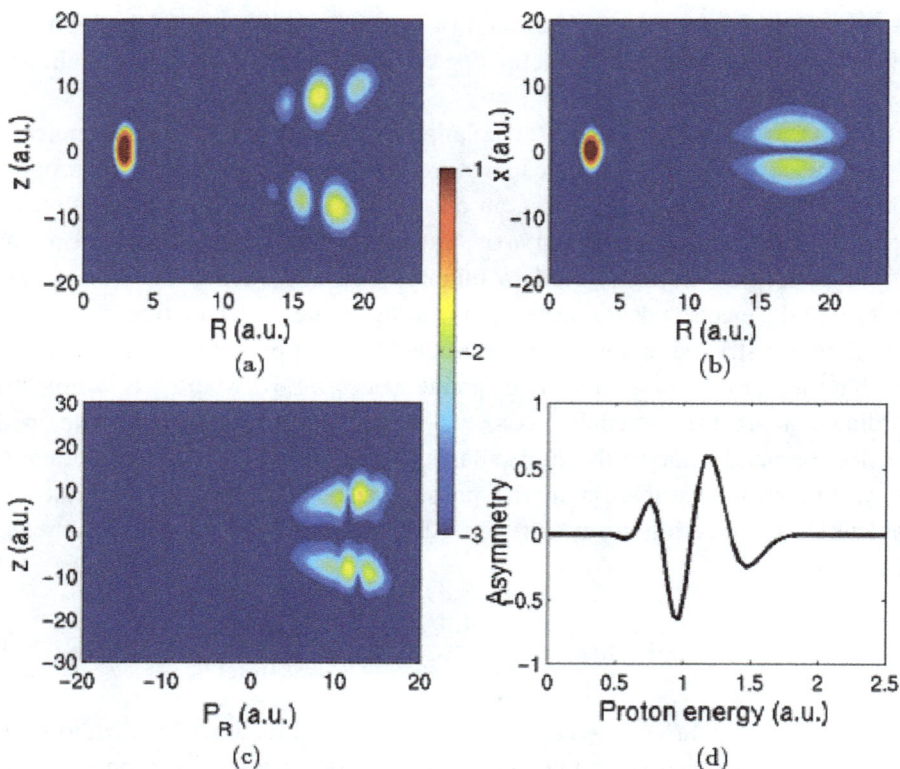

Fig. 3. (Color online) The logarithmic distribution of the wave function in the representations (a) (R, z), (b) (R, x), and (c) (P_R, z). (d) The asymmetry parameter as a function of the proton energy. The time delay is $t = 500$ and the CEP is $\theta = 0$. The snapshots are taken at $t = 1350$ a.u. (adapted from [43]).

related electronic states. In Ref. [29], we simulate a three-channel TDSE, and the KER-dependent electron localization as functions of time delay and carrier envelope phase are shown in Fig. 4(c) and (d). Similarly as Fig. 4(a) and (b), the asymmetry parameters vary periodically with Δt, θ, and KER, as shown by the stripes in each panel.

The potential curves and the coupling matrix in the above two numerical models are slightly different, leading to the quantitative difference between left and right panels in Fig. 4. For the given laser parameters in this subject, the three-channel model is more convincible because the energy curves are more accurate. However, in the numerical model used in Ref. [43], the electron dynamics is confined in a two-dimensional plane, therefore the description is not completed. We conclude if the involved electronic states are clearly known, it is better to use the multi-channel model to calculate the dissociation.

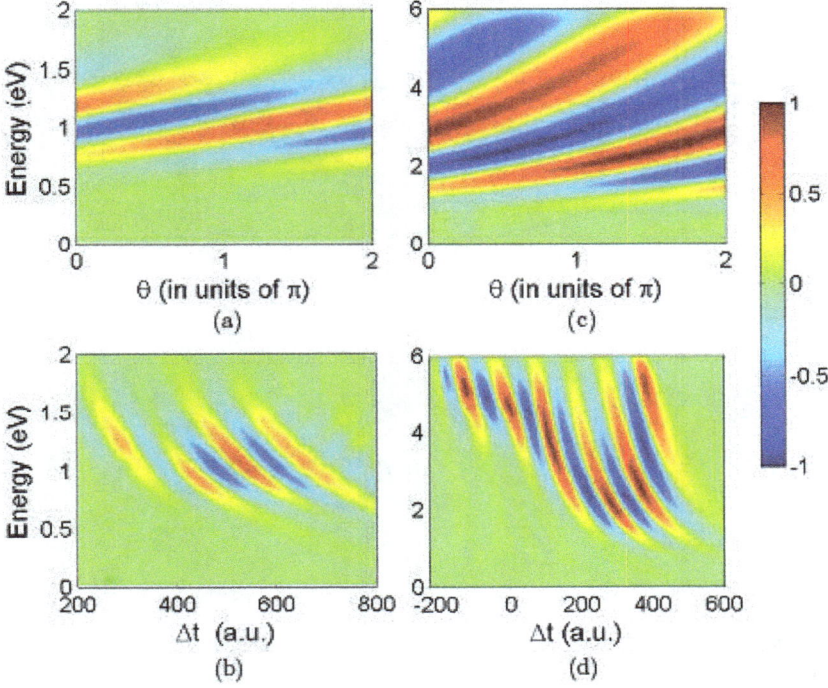

Fig. 4. (Color online) (a) The asymmetry parameter as functions of the CEP and proton energy when the time delay is fixed at $t = 500$ a.u. (b) The asymmetry parameter as functions of the time delay and proton energy when the CEP is fixed at $\theta = 0$. (c) and (d) are similar to (a) and (b) but calculated with the three-channel model. (adapted from [29, 43]).

Recently, Wang *et al.* [44] proposed to use two ultraviolet pulses to mix two highly excited electronic states and achieved the asymmetric electron localization with very high degrees.

B. *Dissociation in long wavelengths and short wavelengths*

According to Keldysh parameters [45], ionization can be generally distinguished as tunneling and multiphoton dissociation. We graft the same idea to molecular dissociation, and introduce the nuclear Keldysh parameter $\gamma_n = \sqrt{2D_p/U_{pn}}$ [46–48], where D_p is the dissociation potential and U_{pn} is the nuclear ponderomotive energy. We expect H_2^+ and its isotopes will undergo tunneling dissociation in the case $\gamma_n < 1$.

To make $\gamma_n < 1$ and avoid ionization, a THz field is applied. In Ref. [49], a laser field with wavelength of 0.1 mm (frequency 3 THz), and laser intensities 10^{13} W/cm^2 and 2×10^{13} W/cm^2 are chosen. The electric field is expressed as $E = E_0 \cos(\omega t) \sin^2(\pi t/\tau)$, $0 < t < \tau$, where the pulse duration is $\tau = 4T$,

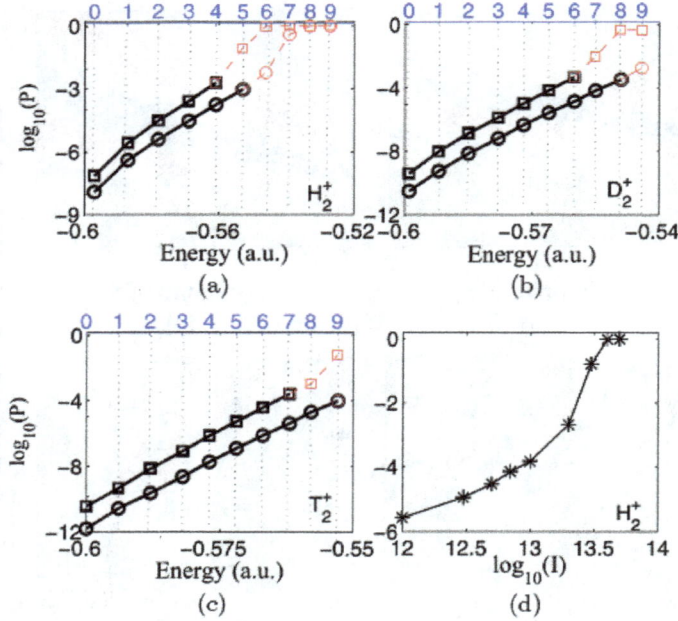

Fig. 5. (Color online) The dissociation probability in logarithmic scale of (a) H_2^+, (b) D_2^+, and (c) T_2^+ initially in different vibrational states when the laser intensity is 10^{13} W/cm^2 (circles) and 2×10^{13} W/cm^2 (squares). The abscissas on tops of panels indicate the indices of nuclear vibrational levels. (d) The dissociation probability as a function of laser intensity for H_2^+ initially in the vibrational state $v = 4$. (adapted from [49]).

and the laser period is $T = 13779$ a.u. Figures 5(a), (b) and (c) show the final dissociation probabilities for molecules in different initial vibrational states when the field intensity is 10^{13} W/cm^2 (circles) and 2×10^{13} W/cm^2 (squares). Figure 5(d) shows the dissociation probability as a function of the field intensity when H_2^+ is initially in the vibrational state $v = 4$. In simulations, the initial vibrational states are obtained by diagonalizing the Hamiltonian. Each curve in panels (a), (b), and (c) is divided in two different parts marked as black bold and light red (gray) lines. These two parts have two distinct slopes, indicating two different dissociation mechanisms: tunneling and overbarrier dissociation. For H_2^+, the inflexion points of these two dissociation mechanisms are found at the vibrational states $v = 6$ and $v = 5$ when laser intensities are 10^{13} W/cm^2 and 2×10^{13} W/cm^2, respectively. Isotopes D_2^+ and T_2^+ have similar behaviors although inflexion points occur at other vibrational states. When comparing the same vibrational states for different isotopic molecules, lighter molecules have smaller dissociation potentials D_p and larger nuclear ponderomotive energies U_{pn}, leading to smaller γ_n. Hence, lighter molecules are easier to tunneling dissociate. This remark also works for heteronuclear hydrogen molecular ions, such as HD$^+$. In Fig. 5(d), the dissociation probability increases very quickly as the laser

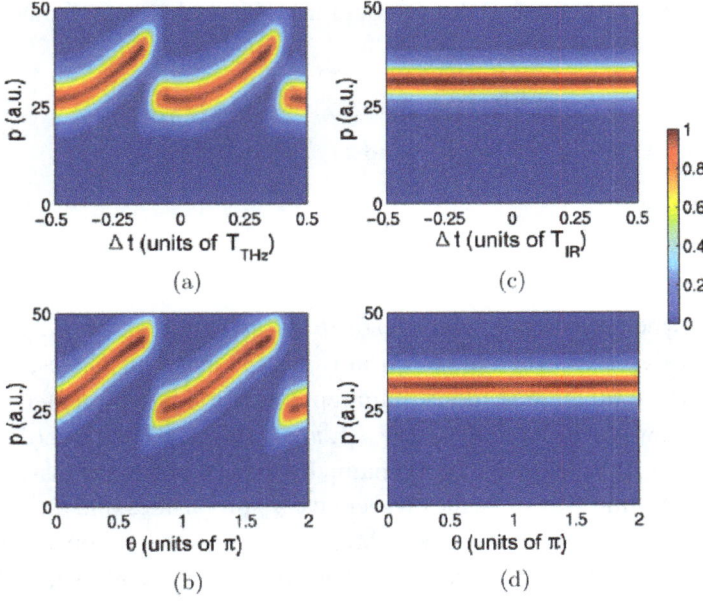

Fig. 6. The nuclear momentum as functions of the time delay at $\theta = 0$ (a) and the CEP θ of the THz pulse at $\Delta t = 0$(b). (c) Same as (b) but substituting the THz pulse by a Ti:spphire pulse with the intensity 4×10^{13} W/cm^2 and pulse duration 26 fs. (adapted from [52]).

intensity increases to 2×10^{13} W/cm^2. This intensity is the critical intensity where the transition between tunneling and overbarrier dissociation occurs.

The molecular dissociation in the long- and short-wavelength fields are distinct. To see this phenomena very clearly, the dissociation of H$_2^+$ is triggered resonantly by an UV pulse, then a time-delayed THz field is applied to modulate the dissociation. Figure 6 shows the dependence of the converged nuclear momenta on (a) the time delay Δt and (b) the CEP θ of the THz pulse. For reference, Fig. 6(c) and (d) show the same with Fig. 6(a) and (b) except that the THz wave field is replaced by a Ti:sapphire light pulse (800nm, with a similar intensity and pulse shape), respectively. One may see the nuclear momentum is periodically modulated, and the period is half of the THz light period. The modulation depth is equal to the THz laser vector potential. For some time delays or carrier envelope phases, the nuclear momentum has two distinct peaks, separated by the THz vector potential. However, for an 800nm laser pulse, the nuclear momentum is perturbed invisibly.

The phase-adiabatic presentation [50, 51] is helpful to explain such wavelength dependence. In the phase-adiabatic presentation, the field-dressed potentials become

$$V_{\pm} = \frac{V_g(R) + V_u(R)}{2} \pm \sqrt{\frac{(V_g(R) - V_u(R))^2}{4} + v_{gu}^{THz}(R, t)^2}.$$

Here, v_{gu}^{THz} is the THz-induced dipole coupling. The nuclear wavefunctions χ_+ and χ_- evolving on $V_\pm$ are

$$\chi_-(R, t) = \cos(\alpha)\chi_g + \sin(\alpha)\chi_u, \qquad (9)$$

$$\chi_+(R, t) = -\sin(\alpha)\chi_g + \cos(\alpha)\chi_u, \qquad (10)$$

where α satisfies

$$\tan(2\alpha) = -2V_{gu}^{THz}(R, t)/(V_u(R) - V_g(R)), \qquad (11)$$

α trends to 0 when R is small, which leads to $\chi_+ = \chi_u$ and $\chi_- = \chi_g$. χ_g and χ_u are the nuclear wavepacket when the electron is in $1s\sigma_g$ and $2p\sigma_u$ states, respectively. The nuclear wave function in momentum representation is represented by $\tilde{\chi}$.

Figure 6 shows (a) $|\tilde{\chi}_g(p, t)|^2$, (b) $|\tilde{\chi}_u(p, t)|^2$, (c) $|\tilde{\chi}_-(p, t)|^2$, and (d) $|\tilde{\chi}_+(p, t)|^2$ when $\theta = 0$ and $\Delta t = 0$. Just after the pump by the UV pulse, the dissociative state is only the χ_u. Within 6 to 10 fs, the dissociative χ_u propagates into the area $R > 10$, and the NM increases to about 35 a.u. Meanwhile, V_+ and V_- open up gradually, and χ_u changes adiabatically into χ_+. When the temporary electric field changes the direction, α swaps between $\pm \pi/4$, therefore χ_+ and χ_- swap each other.

Figure 8 shows the same with Fig. 6 except $\theta = 0.77\pi$. As one can clearly see, very distinct from Fig. 6, each panel shows two stripes, ending with two momenta

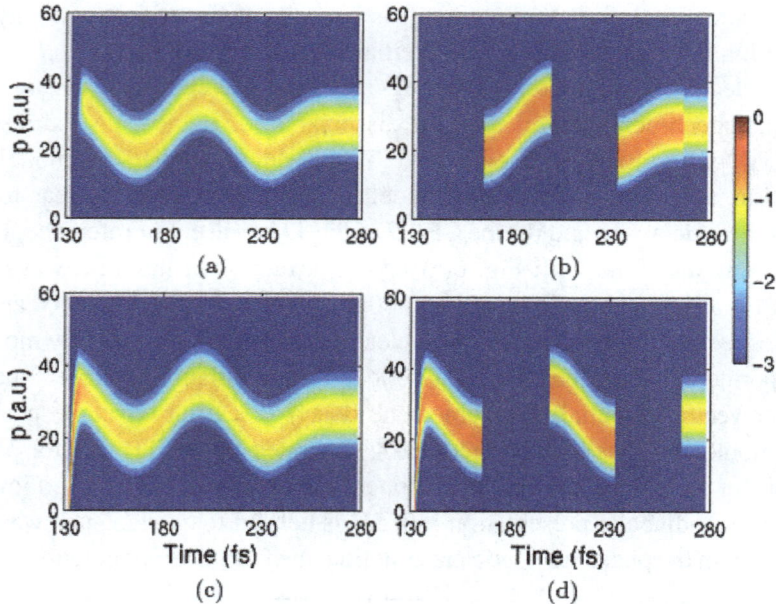

Fig. 7. (a) $|\tilde{\chi}_g(p, t)|^2$, (b) $|\tilde{\chi}_u(p, t)|^2$, (c) $|\tilde{\chi}_-(p, t)|^2$, (d) $|\tilde{\chi}_+(p, t)|^2$ in the logarithmic scale when $\theta = 0$ and $\Delta t = 0$.

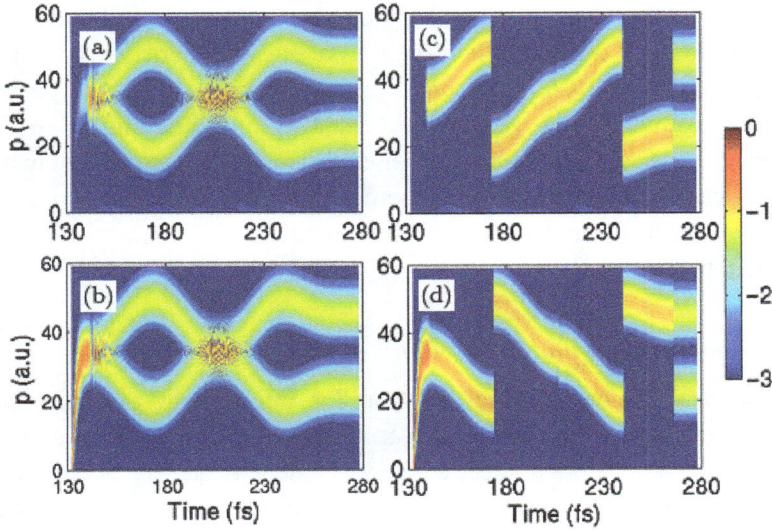

Fig. 8. (Color online) Same as Fig. 6 but for $\theta = 0.77\pi$ and $\Delta t = 0$. (adapted from [52]).

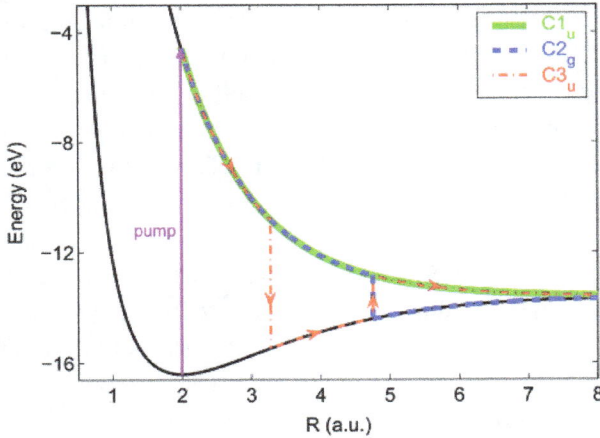

Fig. 9. (Color online) Potential curves for H_2^+ in two lowest electronic states. The UV pump pulse resonantly excites H_2^+ from $1s\sigma_g$ to $2p\sigma_u$, followed by the dissociation controlled by the time-delayed IR laser pulse. Several possible dissociation pathways are sketched. (adapted from [35]).

separated by the THz vector potential. This difference can be explained as follows. For Fig. 8, after the UV excitation, the dissociative χ_+ (or χ_u) sees a falling electric field, which makes V_+ gradually close up to V_u. When the internuclear distance increases to around 10 a.u. where V_g and V_u degenerate, V_- and V_+ also degenerate. This synchronization of the two degenerations leads to $\chi_u = \chi_+$ even when R is around 10 a.u. After that, the temporary electric field starts to increase, and V_- and V_+ deviate each other. Mathematically, α quickly increases from 0 to $\pi/4$. According to Eq. (9) and 10, χ_u becomes the superposition of χ_- and χ_+ who have almost the same probabilities. $\chi_\pm$ will subsequently propagate along $V_\pm$ respectively, and experience the acceleration or deceleration oppositely, resulting in two distinct momentum peaks. However, for the parameters in Fig. 6, the molecule is in pure χ_+ when the nuclear wave packet propagates to $R = 10$ a.u. where V_g and V_u degenerate, but V_+ and V_- do not.

C. *Dissociation in short and intense laser fields*

During the dissociation of H_2^+, the time for internuclear distance increasing from 2 a.u. the equilibrium internuclear distance of H_2^+ to 10 a.u. where the energies of the two lowest electronic states degenerate, is about 10 fs. We define "long" and "short" wavelengths according to their periods. If the period is much shorter than the timescale for the nuclear movement, which is about 10 fs, both the period and the wavelength are short. In this case, the dissociating molecule will experience a fast oscillation of the laser field, and the photon concept works. However, if the laser period is significantly longer than the timescale of the nuclear movement, the dissociating molecule only experiences part of the field, hence, the whole dissociation is not dressed by photons, but a temporary electric field. In this case, tunneling dissociation happens. In this subsection, we present the dissociation of H_2^+ in short and intense laser fields [35]. According to the above analysis, both UV and Ti:sapphire laser pulse have short wavelengths. In these fields, H_2^+ will absorb certain photons, depending on the laser intensity.

To see the dissociation dressed by laser fields with short wavelengths, we first use an UV pulse resonantly excited H_2^+ from $1s\sigma_g$ to $2p\sigma_u$, then use a strong infrared pulse to dress the dissociative process [52]. The dissociation process happened in this UV-pump-IR-probe system are depicted in Fig. 9. The attosecond UV pulse resonantly excites H_2^+ from $1s\sigma_g$ to $2p\sigma_u$, generating a dissociating wave packet. The time delayed IR pulse guides the following dissociation channels: (1) The wave packet directly dissociates through $2p\sigma_u$ (channel 1, abbreviated as $C1_u$); (2) The molecule emits one photon and is dumped to $1s\sigma_g$, then dissociates along $1s\sigma_g$ ($C2_g$); (3) The molecule is dumped to $1s\sigma_g$ by emitting three IR photons, then is pumped to $2p\sigma_u$ by absorbing one IR photon, and finally dissociates through $2p\sigma_u$ ($C3_u$); (4) The molecule emits three IR photons, and dissociates through $1s\sigma_g$ ($C4_g$, not shown in Fig. 9). The one-photon and three-photon interactions mainly occur when $R = 4.75$ and 3.28 a.u. (for 800 nm pulses), where the energy gaps between $1s\sigma_g$ and $2p\sigma_u$ equal to ω_{IR} and $3\omega_{IR}$, respectively. Physically it is possible to

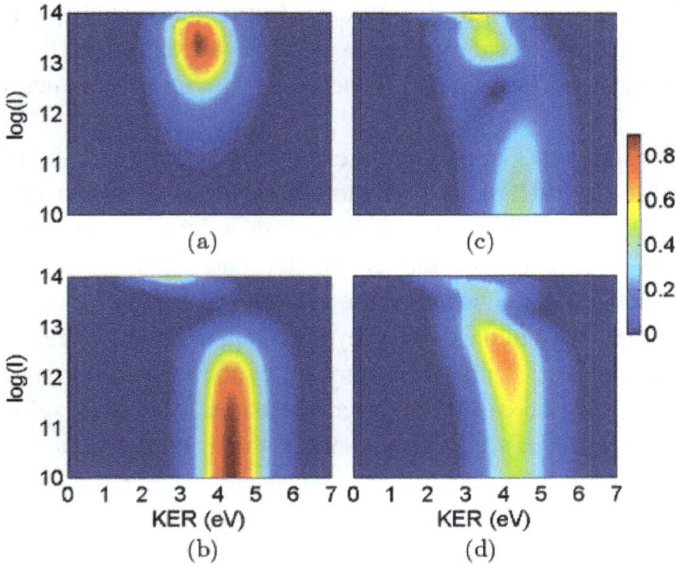

Fig. 10. (Color online) Intensity-dependent KER distribution for H_2^+ dissociating along (a) $1s\sigma_g$, (b) $2p\sigma_u$, or the states that the electron locates on (c) the left nucleus and (d) the right nucleus. The time delay is fixed at $\tau = -1.2T_{800}$, the carrier envelope phase $\theta = 0$. (adapted from [35]).

absorb 5 or even 7 photons with much smaller probabilities. Laser intensities and pulse durations decide the dissociation probabilities along these several dissociation channels.

In this project, a two-channel TDSE is solved. Figure 10 displays (a) $|\tilde{\chi}_g(I, KER)|^2$, (b) $|\tilde{\chi}_u(I, KER)|^2$, (c) $|\tilde{\chi}_l(I, KER)|^2$, and (d) $|\tilde{\chi}_r(I, KER)|^2$, where $\tilde{\chi}_{l/r} = (\tilde{\chi}_g \pm \tilde{\chi}_u)/\sqrt{2}$. Note that the bound vibrational states have been filtered out. Figures 10(a) and (b) clearly show different dissociation behaviors when different IR intensities are applied. As depicted in Fig. 9, the UV pulse resonantly excites H_2^+ to $2p\sigma_u$. When the IR intensity is very weak (around 10^{10} W/cm^2), the molecule mainly directly dissociates along $2p\sigma_u$ (C1$_u$), with little probabilities to interact with the IR field during its dissociation. The KER is simply converted from the molecular potential. In this case, the electron distributes on two nuclei equally. With the increasing of the laser intensity ($10^{12} \sim 10^{13}$ W/cm^2), more and more populations on $2p\sigma_u$ are transferred to $1s\sigma_g$. When the IR intensity is about $10^{12.5}$ W/cm^2, the dissociation probabilities along C1$_u$ and C2$_g$ are almost equal, and the superposition of these two channels makes the electron almost locate all on the right nucleus, as shown in Fig. 10(c) and (d). When the intensity is $10^{13} \sim 6 \times 10^{13}$ W/cm^2, χ_u is very likely dumped to χ_g, i.e. C2$_g$ is the dominant dissociation channel. The further increase of the IR intensity ($10^{13.8} \sim 10^{14}$ W/cm^2) opens C3$_u$. The superposition of C2$_g$ and C3$_u$ makes the electron locate on the left nucleus with larger probabilities when the IR intensity is about 10^{14} W/cm^2. In the above analysis, the net three-photon channel C4$_g$ is absent. McKenna *et al.* have experimentally demonstrated that the enough short laser pulse is necessary to observe a net three-photon process [53].

Figure 11(a) shows the dependence of electron localization on UV-IR time delays and IR intensities. For delays $\tau < 0$, the UV pulse precedes the IR pulse. When delays vary in $[-4T_{800}, T_{800}]$, the IR field with a constant intensity acts on the dissociating H$_2^+$, consequently, the asymmetry parameter oscillates periodically with the time delay. On the contrary, the asymmetry parameter deviates from the periodical oscillation when the time delay is around $-5T_{800}$ or $2T_{800}$, which is due to the fact that the laser fields of the turn-on or the turn-off parts have different intensities. With the increasing of laser intensities, the stripes for asymmetry parameters tilt rightwards. Along each stripe, the asymmetry parameters change

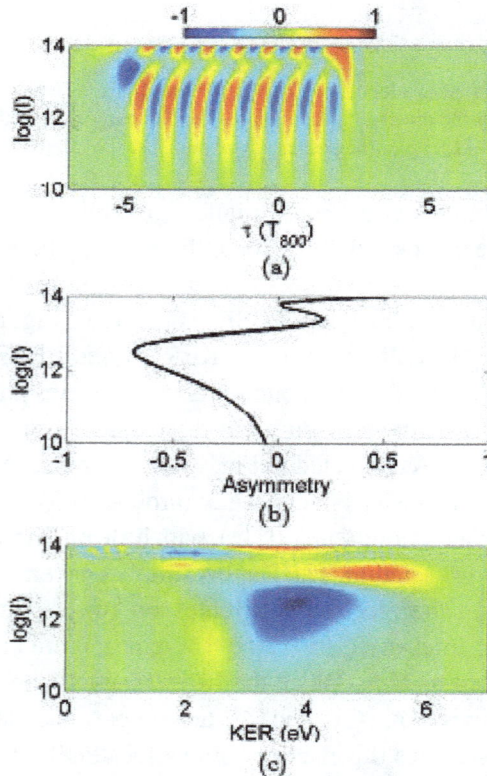

Fig. 11. (Color online) (a) The two dimensional map of left-right asymmetry parameters as a function of the time delay and the IR intensity. (b) The left-right asymmetry as a function of IR intensities when the time delay is $-1.2T_{800}$. This curve is taken from (a). (c) The KER-dependent asymmetry when the time-delay is $-1.2T_{800}$. (adapted from [35]).

signs when the IR intensity is around $10^{13.6}$ W/cm^2. For a fixed time delay (for example $\tau = -1.2T_{800}$), the asymmetry parameters may be positive or negative, as shown in Fig. 11(b), meaning that the expected electron moving directions can follow or oppose the laser electric force. This is because the main dissociation pathways are different when the driving laser intensities are different. Figure 11(c) shows the KER-dependent left-right asymmetry, which can explain why the laser pulse with same carrier envelope phases and same time delays but different intensities will steer the electron along different directions. The decomposed left-right asymmetry is a complementary to the conception of "momentum gate" [54].

If the laser intensity is very strong, especially in the case that the photon energy is very large, the electron might be pumped to highly excited states. In that case, more electronic states must be included in the simulation. In the example in Ref. [29], two orthogonally polarized UV pulses are introduced to interact with H$_2^+$. The polarization axis of the first and second UV pulses (abbreviated as UV1 and UV2) are parallel and perpendicular to the molecular axis, respectively. UV1 and UV2 have the same sin2 profiles. UV1 (UV2) has the frequency 0.43 a.u. (0.65 a.u.) and the intensity 10^{14} W/cm^2 (10^{14} W/cm^2). For these laser parameters, the eight lowest electronic states have been included. Two UV pulses pump H$_2^+$ to several electronic states simultaneously, whose populations depend on laser intensities and pulse durations. Figure 12 shows the time-dependent probabilities for eight electronic states. UV1 and UV2 resonantly excite H$_2^+$ to $2p\sigma_u$ and $2p\pi_u$, who therefore have large populations. The large population in $3d\sigma_g$ can be explained based on the resonant transition between $2p\sigma_u$ to $3d\pi_g$, since the energy gap between $2p\sigma_u$ and $3d\sigma_g$ at the equilibrium internuclear distance is close to the UV1 photon energy. These time-dependent state populations show Rabi oscillations. With time evolutions, UV1 and UV2 keep pumping H$_2^+$ to $2p\sigma_u$ and $2p\pi_u$, respectively. Once the population of $2p\sigma_u$ is relatively large, the molecule is further resonantly excited from $2p\sigma_u$ to $3d\sigma_g$ by absorbing another UV1 photon. For this reason, a decrease in $2p\sigma_u$ in the time interval $-30 < t < 30$ a.u. is shown, while the population of $3d\sigma_g$ increases rapidly. When $30 < t < 70$ a.u. the states $3d\sigma_g$ and $1s\sigma_g$ are both transferring to $2p\sigma_u$. It is clear if only two or three states are included, the numerical model will not give a reasonable physical result.

D. *Control vibrational states*

The singly ionization of H$_2$ is generally approximated with Frank-Condon approximation [24], which assumes the nuclear wavepacket of H$_2^+$ just after the singly ionization of H$_2$ is same as that of H$_2$. Such a nuclear wavepacket can be regarded as the superposition of a series of vibrational states. Once H$_2^+$ is produced, it starts

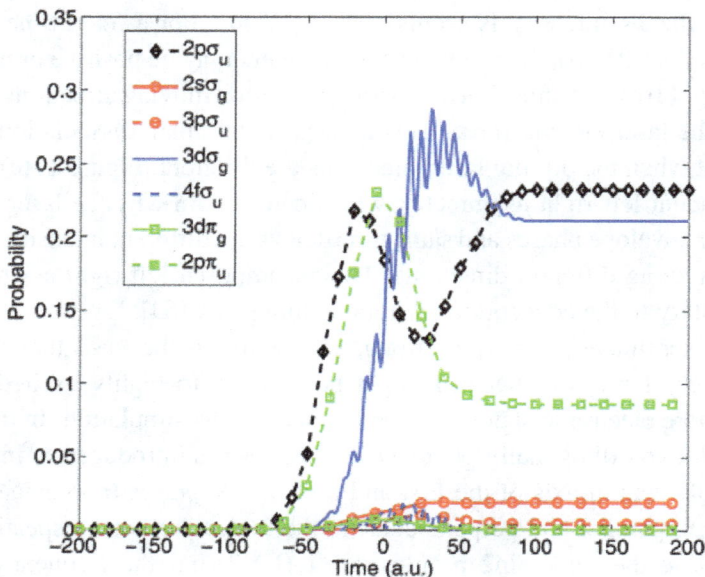

Fig. 12. (Color online) The time-dependent probabilities for different states when both UV pulses have the intensity 10^{14} W/cm^2 and pulse duration 5 fs. (adapted from [29]).

to stretch, or the superimposed nuclear wave packet will propagate in the potential well. The nuclear wave packet can be written as

$$\chi(R, t) = \sum_v a_v |v> e_v^{-iE_v t} + \int b_E |E> e^{-iEt} dE. \tag{12}$$

The first and second term in the right hand side of Eq. (12) are bound vibrational states and free dissociative states, whereas the probability of the latter one is less than 3% and is neglected in many studies. a_v is the Frank-Condon factor and $|a_v|^2$ indicates the probability of different vibrational states. Figure 13(a) shows the probabilities of different vibrational states under the Frank-Condon approximation. During the transition from H_2 to H_2^+, the $v = 2$ state has the largest probability. The Frank-Condon coefficients depend on nuclear masses. Figure 13(b) shows the different vibrational wavefunction, which presents some similar characters with the eigenstates of harmonic oscillators but of course are different. In the following free propagation, the free dissociative states will quickly propagate to large R areas, i.e. the molecule is broken. However, the bound vibrational states evolve in the potential well, and the wave packet collapse after a few tens of femtosecond. The wavepacket revivals around 300 fs later, as shown in Fig. 13(c).

The superposition of the nuclear wave packets make the followed dissociation complex if it is exposed to strong laser fields. Thus, it would be useful to design

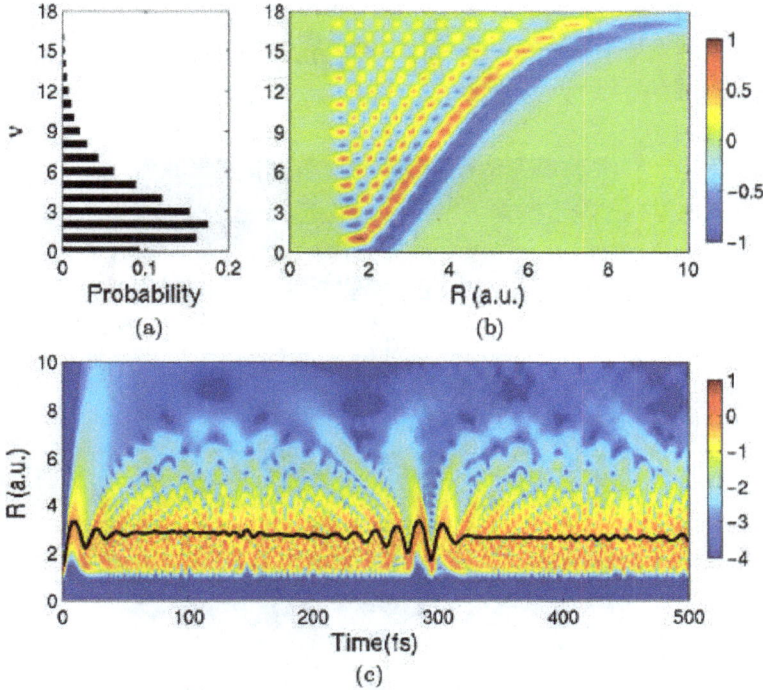

Fig. 13. (Color online) (a) Franck-Condon coefficients of the nuclear wavepacket of H_2^+. (b) Wave functions for the 19 vibrational states (in linear scale). (c) The free propagation of the Franck-Condon nuclear wavepacket of H_2^+ in the $1s\sigma_g$ potential curve (in logarithmic scale). (adapted from [55]).

an optimal laser pulse which can transfer the superimposed vibrational states into a selective stationary state, especially the ground vibrational state $\chi_{v=0}$. We formulate the design of a laser field E to realize the nuclear wavepacket (NWP) transfer of $\chi_{v=0}$ as a minimax problem, and then apply a sequential linear programming algorithm to solve it [55].

By restricting the laser amplitude within 0.01 and pulse duration a few tens of femtosecond, the numerically designed optimal laser field is shown in Fig. 14(a). Figures 14(b), (c), (d) show the optimal laser pulse induced NWP evolution for $|\psi_g(R,t)|^2$, $|\psi_u(R,t)|^2$, and $|\psi_g(R,t)|^2 + |\psi_u(R,t)|^2$, respectively.

Figure 14(a) reveals an interesting physical story. First of all, the optimal laser field is a pulse train. After the inception of H_2^+, nearly no electric field is introduced until $t = 11$ fs. In this period, the NWP propagates freely to the outer turning point and then turns back, as shown in Fig. 13(b). The main pulse appears at around $t = 11$ fs, at which time the NWP is moving inward instead of outward. This is important because if the laser field starts interacting with the NWP when it is moving outward, the part of the wave packets promoted to $2p\sigma_u$ surface will directly dissociate and

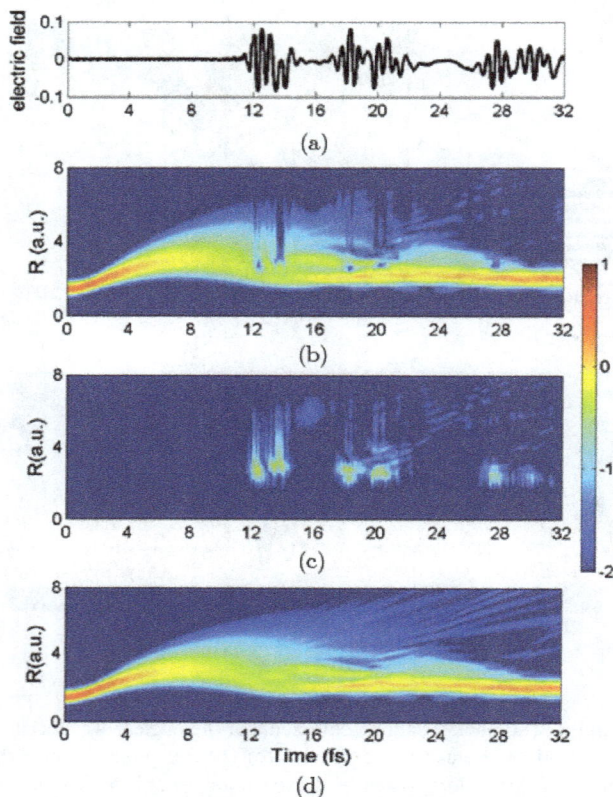

Fig. 14. (Color online) (a) The designed optimal laser pulse train. The evolution of $|\psi_g(R, t)|^2$ (b), $|\psi_u(R, t)|^2$ (c), $|\psi_g(R, t)|^2 + |\psi_u(R, t)|^2$ (d) (all in logarithmic scale). (adapted from Ref. [55]).

the subsequent laser pulse has little chance to pull them back to the bound states [33]. To transfer $\psi_g(R, t)$ to $\psi_g^{v=0}(R)$, $\psi_u(R, t)$ must be an intermediate state. From a closer look at Fig. 13(b) and (c), one may find that within each oscillation of the electric field, part of ψ_g and ψ_u are exchanged. The wave function $\psi_u(R, t)$ mainly distributes close to the range $R = 3$. The quantity $|\psi_g(R, t)|^2 + |\psi_u(R, t)|^2$ gives a smooth evolution of the NWP, as depicted in Fig. 13(d). At the terminal time, the NWP has been transferred to the ground vibrational state.

E. *Rescattering dissociation*

Rescattering is one of the central processes in strong field physics [56]. When H_2 is exposed to strong laser fields, one electron first tunnels out the laser-dressed Coulomb barrier, then backs to and rescatters with its parent ion once the laser electric field changes direction. During this rescattering, the bound electron may be excited to $2p\sigma_u$, forming a dissociative H_2^+. This process has been discussed in experiment [14]. Several theories have also been developed to confirm it though different theories show conflict explanations [57–59]. Recently, an *ab initio* non-Born-Oppenheimer simulation for H_2 has been performed [60]. The governed TDSE is

$$i\frac{\partial}{\partial t}\Psi(R, x_1, x_2; t) = [T + V(R, x_1, x_2)]\Psi(R, x_1, x_2; t), \tag{13}$$

where the kinetic operator

$$T = \frac{p_R^2}{2\mu} + \frac{[p_1 + A(t)]^2}{2} + \frac{[p_2 + A(t)]^2}{2}, \tag{14}$$

the potential

$$V(R, x_1, x_2) = \frac{1}{R} + \frac{1}{\sqrt{(x_1 - x_2)^2 + \alpha(R)}}$$

$$- \sum_{s=\pm 1}\sum_{i=1}^{2} \frac{1}{\sqrt{(x_i + sR/2)^2 + \frac{\beta(R)^2}{25} + \frac{1}{\beta(R)} - \frac{\beta(R)}{5}}}. \tag{15}$$

In Eq. (14), μ is the reduced nuclear mass, and $A(t)$ is the laser vector potential and $A(t) = -\int_0^t E(t')dt'$ with $E(t)$ being the electric field. p_R, p_1 and p_2 are the relative nuclear momentum operator, the first and second electron momentum operators, respectively. By adjusting the soft-core parameters $\alpha(R)$ and $\beta(R)$ [61], this reduced-dimensional model may generate the potential curves which are very close to the actual ones [62, 63], as shown in Fig. 15(a). Therefore, this model can well describe the ionization and dissociation of H_2. Figure 15(b) sketches the ionization and rescattering process.

Following the wavefunction propagation, one may analyze all physical processes triggered by the laser pulse. Figure 16 shows the snapshot of wavefunction distribution in $x_1 - R$ space at $t = 324$ a.u. i.e. $W(x_1, R) = \int dx_2 |\Psi(x_1, x_2, R, t = 324)|^2$. The release of the first electron in H_2, marked by the dashed rectangle, gives birth to H_2^+ with bound nuclei vibrational states, which will stay in the area where R is small. The electron wave packets ejected around two optical peaks superimpose, resulting in the intracycle interference [64–66]. In the single ionization of H_2, the photon energy is shared by the photoelectron and the nuclear vibration [67]. Differently, some other wave packets, marked by the dotted rectangle, gradually propagate towards larger R areas with time evolution and finally clearly separates from the bound H_2 and H_2^+. This part corresponds to molecular dissociation. By watching the wavefunction propagation movie, we may identify that the electron released at around $t = 115$ a.u. slams its parent ion H_2^+ at around $t = 175$ a.u.

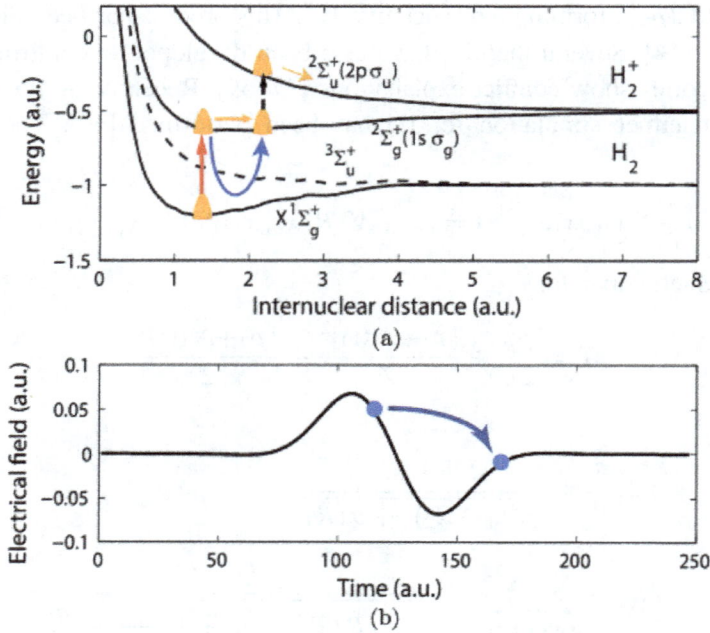

Fig. 15. (a) Calculated potential energy surfaces of H_2 and H_2^+. The single ionization of H_2 gives birth to H_2^+, which will stretch and then be excited to $2p\sigma_u$ by rescattering electron wave packets, followed by molecular dissociation. (b) The laser field used in the simulations. (adapted from [60]).

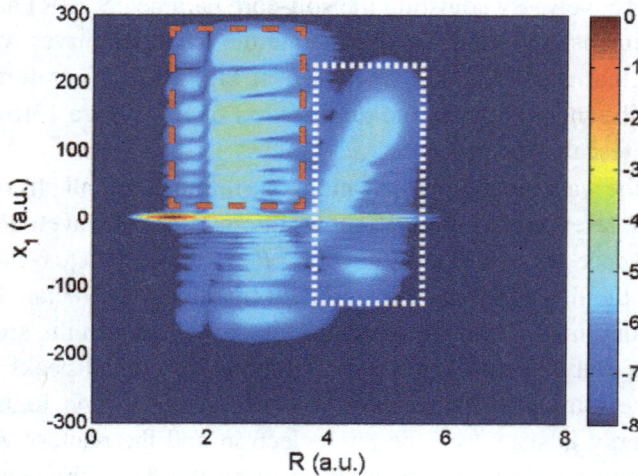

Fig. 16. The snapshot of wave packet distribution in $x_1 - R$ space at $t = 324$ a.u. in logarithm scale. The single ionization with bound H_2^+ and rescattering dissociation are marked by dashed and dotted rectangles. (adapted from [60]).

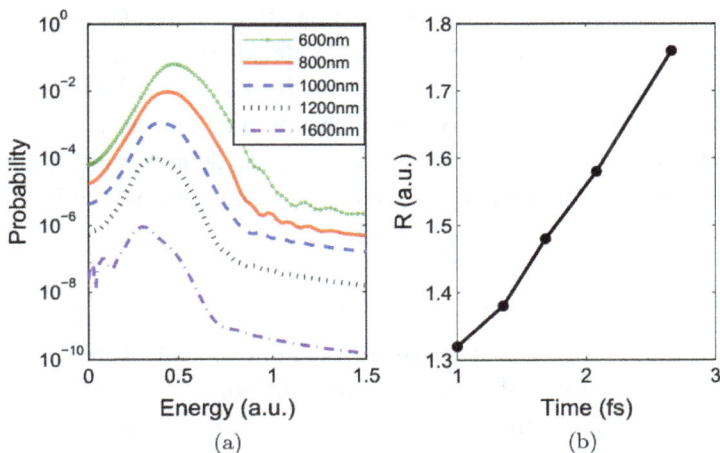

Fig. 17. (a) The KER spectra when laser fields with different wavelengths are applied. (b) The calibrated time-resolved internuclear distance. (adapted from [60]).

leading to rescattering dissociation. Most rescattering electron finally propagates along $+x_{1,2}$ axis, which is in coincidence with classical expectation. A very small part of the rescattering dissociation displayed in the lower half space is attributed to the back rescattering [68].

Once H_2 is tunneling ionized, H_2^+ starts to relax and the internuclear distance increases. The time interval between the tunneling ionization and rescattering directly relates to laser periods or laser wavelengths. For longer wavelengths, H_2^+ has longer time to relax before it is excited to $2p\sigma_u$, therefore, the rescattering dissociation will start from a larger internuclear distance and finally dissociative fragments gain smaller KER. Figure 17(a) shows KER spectra when driving laser fields with different wavelengths are applied. The intensities for 600 nm, 800 nm, 1000 nm, 1200 nm, 1600 nm pulses are 7.5×10^{14} W/cm^2, 4×10^{14} W/cm^2, 2.5×10^{14} W/cm^2, 1.7×10^{14} W/cm^2, 1×10^{14} W/cm^2, respectively, for ensuring the corresponding maximum rescattering energies are all 0.6 a.u. As expected, the peaks of KER spectra shift towards lower energies as driving laser wavelengths are longer. Since the laser pulses with longer wavelengths trigger smaller ionization probabilities, the total rescattering dissociation probabilities decline when wavelengths increase. Even when the rescattering energy is quite small, considerable rescattering dissociation may still happen. This simulation support the explanation in [59] that the first rescattering has been used to read the molecular dynamics in attosecond timescales.

KER spectra can be used to calibrate the internuclear distance at the moment of rescattering, considering the fact that $2E_{KER} = E_{1s\sigma_g}(1.3) - E_{1s\sigma_g}(R) + E_{2p\sigma_u}(R) - E_{2p\sigma_u}(\infty)$ (it is assumed that the dissociation along $2p\sigma_u$ is dominated), where R

is the internuclear distance where dissociation starts, $E_{1s\sigma_g}(R)$ and $E_{2p\sigma_u}(R)$ are potential energies of $1s\sigma_g$ and $2p\sigma_u$ states at R. Recall that 1.3 a.u. is the equilibrium internuclear distance of H_2. Figure 17(b) plots the calibrated internuclear distance as a function of time intervals between ionization and rescattering. Here, the time intervals are two-thirds of laser periods. As already demonstrated experimentally by Niikura *et al.* [59], the numerical simulation also confirms that it is possible to read the time-resolved internuclear distance.

F. *Two dimensional wave packet propagation*

Almost all of the foregoing directional dissociation controls were demonstrated in one-dimension stage along the direction of field polarization [9, 18, 21, 35, 39, 40, 69–84]. The control of the directional bond breaking rapidly diminishes for the molecule orientating away from the polarization axis of the dissociation field [9, 39]. Recently, a two-dimensional directional dissociative ionization of H_2 by using an intense phase-controlled orthogonally polarized two-color (OTC) laser pulse has been demonstrated [85]. It stands for a two-dimensional pathway interference of the nuclear wave packets with opposite parities. The numerical model for this project can be written as follows

$$
i\frac{\partial}{\partial t}\begin{pmatrix}\chi_g(y,z;t)\\\chi_u(y,z;t)\end{pmatrix}
$$

$$
=\begin{pmatrix}-\dfrac{1}{2\mu}\dfrac{\partial^2}{\partial y^2}-\dfrac{1}{2\mu}\dfrac{\partial^2}{\partial z^2}+V_g(y,z) & \mathbf{E}(t)\cdot\mathbf{D}(y,z)\\[2mm] \mathbf{E}(t)\cdot\mathbf{D}(y,z) & -\dfrac{1}{2\mu}\dfrac{\partial^2}{\partial y^2}-\dfrac{1}{2\mu}\dfrac{\partial^2}{\partial z^2}+V_u(y,z)\end{pmatrix}
$$

$$
\times\begin{pmatrix}\chi_g(y,z;t)\\\chi_u(y,z;t)\end{pmatrix},
$$

where μ is the reduced nuclear mass, χ_g and χ_u are the two dimensional nuclear wave packets corresponding to the electron in $1s\sigma_g$ and $2p\sigma_u$ state whose potential energy surfaces are V_g and V_u, respectively. Two nuclei sit on (y, z) and $(-y, -z)$ symmetrically, and $\mathbf{D}(y, z)$ is the dipole coupling matrix. The OTC field is written as $\mathbf{E}(t) = [E_{FW}\cos(\omega_{FW}t + \phi_{FW})\mathbf{e}_z + E_{SH}\cos(2\omega_{SH}t + 2\phi_{FW}+\phi_L)\mathbf{e}_y]\sin^2(\pi t/\tau), 0 < t < \tau$, where ϕ_{FW} is the carrier-envelope phase of the FW pulse, and ϕ_L is the relative phase between the orthogonal FW and SH waves. The pulse duration τ equals 14 FW optical cycles. The asymmetry parameter is defined as Eq. (8). The simulation results shown in Fig. 18(b) can well reproduce the experimental measurement as shown in Fig. 18(a).

The measured asymmetry parameter can be fitted by using $A_0\sin(\phi_L + \phi_{A0})$ where A_0 is the asymmetry amplitude and ϕ_{A0} is the phase. Figures 19(a) and (b)

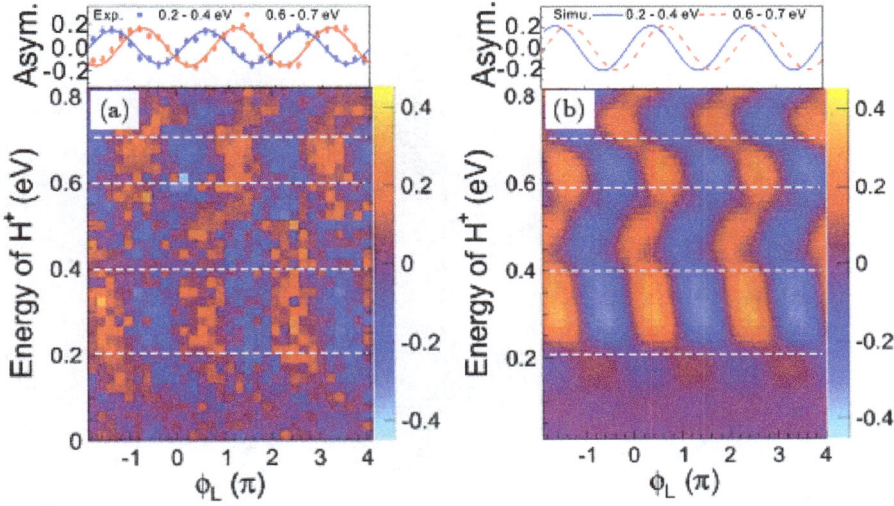

Fig. 18. The (a) experimentally measured and (b) numerically simulated two-dimensional maps of the asymmetry parameter as a function of ϕ_L and E_k for H^+ emitting to $45° < \phi_{H+} < 65°$. The top panel shows the asymmetry parameter of the directional emission of H^+ versus ϕ_L at low-E_k and high-E_k regions as indicated between the dashed lines, fit to sinusoidal curves in (a). (adapted from [85]).

plot the amplitude A_0 and phase ϕ_{A0} of the asymmetry versus the H^+ emission direction ϕ_{H+} for the low-E_k and high-E_k regions, respectively. As compared to the one-dimensional stage where the asymmetry is along the polarization axis of the employed laser field [3–24], here significant asymmetric emission of H^+ appears in the direction between the polarization axes of the FW and SH waves of the OTC pulse. As shown in Figs. 19(a) and (b), there is an angle shift of the ϕ_{H+}-dependences of the asymmetry amplitude A_0 between the low-E_k and high-E_k regions. Rather than at $\phi_{H+} = \pm45°$ or $\pm135°$, the asymmetry is favored when H^+ emits slightly close to the SH polarization ($\phi_{H+} = 0°$ or $\pm180°$) and FW polarization ($\phi_{H+} = \pm90°$) for the low-E_k and high-E_k regions, respectively. The π-shift of the phase ϕ_{A0} of the asymmetry for H^+ emitting to ϕ_{H+} and $\phi_{H+}+\pi$ validates our analysis and directly shows the directional dissociative ionization of H_2 in the phase-controlled OTC pulse. The numerically simulated A_0 and ϕ_{A0} as a function of ϕ_{H+} for the low-E_k and high-E_k regions are presented in Figs. 19(c) and (d), which are well comparable with the experiments.

The asymmetric directional emission of H^+ requires the coherent superposition of the nuclear wave packets with same kinetic energies but opposite parities. For the low-E_k region, χ_u comes from the $1\omega_{FW}$ pathway on the $2p\sigma_u$ curve, χ_g comes either from the $1\omega_{SH} - 1\omega_{FW}$ pathway (propagation on the $1s\sigma_g$ curve undergoes

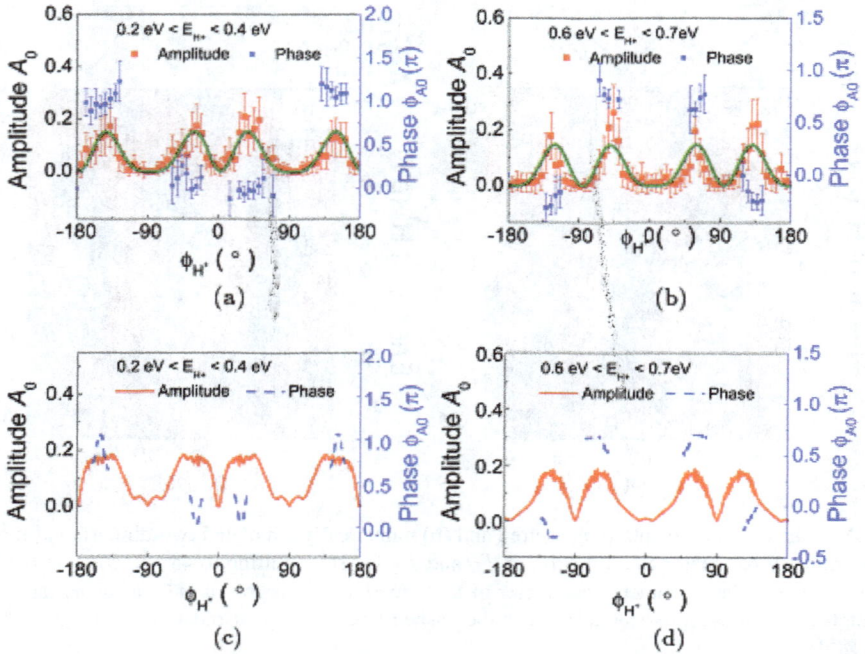

Fig. 19. (Color online) The (a), (b) experimentally measured and (c), (d) numerically simulated asymmetry amplitude A_0 phase and phase ϕ_{A0} as a function ϕ_{H+} at (a), (c) low-E_k and (b), (d) high-E_k regions. The solid curves in (a), (b) are the fits of the angle-resolved asymmetry by assuming dipole-allowed photon transitions parallel and perpendicular to the molecular axis. (adapted from [85]).

one-ω_{SH}-photon transition to the $2p\sigma_u$ curve, followed by propagation on the $2p\sigma_u$ curve and coupling back to the $1s\sigma_g$ curve by emitting one ω_{FW} photon, followed by dissociation along the $1s\sigma_g$ curve) or from the $3\omega_{FW} - 1\omega_{SH}$ pathway (propagation on the $1s\sigma_g$ and curve undergoes three-ω_{FW}-photon transition to the $2p\sigma_u$ curve, followed by propagation on the $2p\sigma_u$ curve coupling back to the $1s\sigma_g$ curve by emitting one ω_{SH} photon, followed by dissociation along the $1s\sigma_g$ curve). As compared to the $1\omega_{SH}$-$1\omega_{FW}$ pathway, the $3\omega_{FW}$-$1\omega_{SH}$ pathway is energetically unfavored since four photons are required.

Conclusion

We have reviewed recent theoretical progress about the dissociation of H_2^+ and its isotopes in strong laser fields. To understand the dissociation process very precisely, we suggest to use an isolated attosecond pulse to trigger the dissociation, which will be controlled by a time-delayed pulse. Such a pump-probe strategy will make the dissociative signal clear and easy to be analyzed. For example, with this control method, one can localize the electron in a selective proton by mixing selective electronic states with opposite parities. Molecules may undergo tunneling dissociation and rescattering dissociation when proper external laser conditions are given. The molecular dissociation can be controlled by adjusting a series of laser parameters, such as wavelength, intensity, polarization. The studies of the dissociation of H_2^+ and its isotopes pave the way to control more complex chemical reactions.

Acknowledgment

We appreciate discussions with Tian-Yu Xu, Xin-Yuan You, Zhi-Chao Li, Jun Zhang, Pei-Lun He, Xiao-Chun Gong and Jian Wu. We further acknowledge financial support by NSF of China (Grant No. 11104180, 11175120, 11121504, 11322438), and the Fok Ying-Tong Education Foundation for Young Teachers in the Higher Education Institutions of China (Grant No. 131010).

References

1. F. Krausz and M. Ivanov, *Rev. Mod. Phys.* **81**, 163 (2009).
2. A. H. Zewail, *J. Phys. Chem. A* **104**, 5660 (2000).
3. A. Baltuška, Th. Udem, M. Uiberacker, M. Hentschel, E. Goulielmakis, Ch. Gohle, R. Holzwarth, V. S. Yakovlev, A. Scrinzi, T. W. Hänsch and F. Krausz, *Nature* **421**, 611 (2003).
4. M. Hentschel, R. Kienberger, Ch. Spielmann, G. A. Reider, N. Milosevic, T. Brabec, P. Corkum, U. Heinzmann, M. Drescher and F. Krausz, *Nature* **414**, 509 (2001).
5. E. Goulielmakis, M. Uiberacker, R. Kienberger, A. Baltuska, V. Yakovlev, A. Scrinzi, Th. Westerwalbesloh, U. Kleineberg, U. Heinzmann, M. Drescher, F. Krausz, *Science* **305**, 1267 (2004).
6. A. Giusti-Suzor, F. H. Mies, L. F. DiMauro, E. Charron and B. Yang, *J. Phys. B* **28**, 309 (1995).
7. M. Odenweller, N. Takemoto, A. Vredenborg, K. Cole, K. Pahl, J. Titze, L. Ph. H. Schmidt, T. Jahnke, R. Dörner, and A. Becker, *Phys. Rev. Lett.* **107**, 143004 (2011).
8. T. Rathje, A. M. Sayler, S. Zeng, P. Wustelt, H. Figger, B. D. Esry, and G. G. Paulus, *Phys. Rev. Lett.* **111**, 093002 (2013).
9. N. G. Kling, K. J. Betsch, M. Zohrabi, S. Zeng, F. Anis, U. Ablikim, B. Jochim, Z. Wang, M. Kübel, M. F. Kling, K. D. Carnes, B. D. Esry, and I. Ben-Itzhak, *Phys. Rev. Lett.* **111**, 163004 (2013).
10. G. A. Suzor, X. He, O. Atabek, F. H. Mies, *Phys. Rev. Lett.* **64**, 515 (1990).
11. A. D. Bandrauk, M. L. Sink, *J. Chem. Phys.* **74**, 1110 (1981).
12. P. H. Bucksbaum, A. Zavriyev, H. G. Muller, D. W. Schumacher, *Phys. Rev. Lett.* **64**, 1883 (1990).
13. L. J. Frasinski, J. H. Posthumus, J. Plumridge, K. Codling, *Phys. Rev. Lett.* **83**, 3625 (1999).
14. M. F. Kling, Ch. Siedschlag, A. J. Verhoef, J. I. Khan, M. Schultze, Th. Uphues, Y. Ni, M. Uiberacker, M. Drescher, F. Krausz, and M. J. J. Vrakking, *Science* **312**, 246 (2006).
15. A. Becker, F. He, A. Picon, C. Ruiz, N. Takemoto and A. Jaron-Becker, in *Attosecond Physics: Attosecond Measurements and Control of Physical Systems*, eds. L. Plaja, R. Torres and A. Zair, Springer Series in Optical Sciences, Vol. 177 (Springer, Berlin-Heidelberg, 2013) p. 207–229.

16. Th. Ergler, A. Rudenko, B. Feuerstein, K. Zrost, C. D. Schröter, R. Moshammer, and J. Ullrich, *Phys. Rev. Lett.* **97**, 193001 (2006).
17. F. H. M. Faisal, *Theory of Multiphoton Processes* (New York, Plenum, 1987).
18. F. He, C. Ruiz, and A. Becker, *Phys. Rev. Lett.* **99**, 083002 (2007).
19. W. H. Press, S. A. Teukolsky, W. T. Vetterling, and B. P. Flannery, *Numerical Recipes* (Cambridge University Press, Cambridge, 1992).
20. L. Lehtovaara, J. Toivanen, J. Eloranta, *J. Comp. Phys.* **221**, 148 (2007).
21. V. Roudnev and B. D. Esry, *Phys. Rev. Lett.* **99**, 220406 (2007).
22. F. He and U. Thumm, *Phys. Rev. A* **81**, 053413 (2010).
23. B. Abeln, J. V. Hernandez, F. Anis, and B. D. Esry, *J. Phys. B* **43**, 155005 (2010).
24. B. H. Bransden, C. J. Joachain, *Physics of Atoms and Molecules* (Longman Pub Group, 1982).
25. H.-J. Werner, P. J. Knowles, F. R. Manby, M. Schütz, P. Celani, G. Knizia, T. Korona, R. Lindh, A. Mitrushenkov, G. Rauhut, T. B. Adler, R. D. Amos, A. Bernhardsson, A. Berning, D. L. Cooper, M. J. O. Deegan, A. J. Dobbyn, F. Eckert, E. Goll, C. Hampel, A. Hesselmann, G. Hetzer, T. Hrenar, G. Jansen, C. Köppl, Y. Liu, A. W. Lloyd, R. A. Mata, A. May, S. J. McNicholas, W. Meyer, M. E. Mura, A. Nicklaß, P. Palmieri, K. Pflüger, R. Pitzer, M. Reiher, T. Shiozaki, H. Stoll, A. J. Stone, R. Tarroni, T. Thorsteinsson, M. Wang and A. Wolf, MOLPRO, version 2010.1.
26. T. H. Dunning Jr., *J. Chem. Phys.* **90**, 1007 (1989).
27. H.-J. Werner and P. J. Knowles, *J. Chem. Phys.* **82**, 5053 (1985).
28. P. J. Knowles and H.-J. Werner, *Chem. Phys. Lett.* **115**, 259 (1985).
29. Xin-Yuan You and Feng He, *Phys. Rev. A* **89**, 063405 (2014).
30. T. E. Sharp, *At. Data. Nucl. Data Tables* 2, 119 (1970).
31. D. E. Ramaker, and J. M. Peek, *At. Data. Nucl. Data Tables* 5, 167 (1973).
32. E. Charron, A. Giusti-Suzor, and F. H. Meis, *J. Chem. Phys.* **103**, 7359 (1995).
33. H. Niikura, D. M. Villeneuve, and P. B. Corkum, *Phys. Rev. Lett.* **92**, 133002 (2004).
34. J. J. Hua and B. D. Esry, *J. Phys. B* 42, 085601 (2009).
35. T. Y. Xu and F. He, *Phys. Rev. A* **90**, 053401 (2014).
36. M. Shapiro and P. Brumer, *Quantum Control of Molecular Processes*, (Wiley-VCH, New York, 2011).
37. D. J. Tannor and S. A. Rice, "Control of selectivity of chemical reaction via control of wave packet evolution," *J. Chem. Phys.* **83**, 5013–5026 (1985).
38. S. Shi, A. Woody and H. Rabitz, "Optimal control of selective vibrational excitation in harmonic linear chain molecules," *J. Chem. Phys.* **88**, 6870–6873 (1988).
39. M. Kremer, B. Fischer, B. Feuerstein, V. L. B. de Jesus, V. Sharma, C. Hofrichter, A. Rudenko, U. Thumm, C. D. Schröter, R. Moshammer, and J. Ullrich, *Phys. Rev. Lett.* **103**, 213003 (2009).
40. B. Fischer, M. Kremer, T. Pfeifer, B. Feuerstein, V. Sharma, U. Thumm, C. D. Schröter, R. Moshammer, and J. Ullrich, *Phys. Rev. Lett.* **105**, 223001 (2010).
41. V. Roudnev, B. D. Esry, and I. Ben-Itzhak, "Controlling HD^+ and H_2^+ Dissociation with the Carrier-Envelope Phase Difference of an Intense Ultrashort Laser Pulse," *Phys. Rev. Lett.* **93**, 163601 (2004).
42. G. Sansone, F. Kelkensberg, J. F. Perez-Torres, F. Morales, M. F. Kling, W. Siu, O. Ghafur, P. Johnsson, M. Swoboda, E. Benedetti, F. Ferrari, F. Lepine, J. L. Sanz-Vicario, S. Zherebtsov, I. Znakovskaya, A. L'Huillier, M. Yu. Ivanov, M. Nisoli, F. Martin and M. J. J. Vrakking, "Electron localization following attosecond molecular photoionization," *Nature (London)* **465**, 763–766 (2010).
43. F. He, *Phys. Rev. A* **86**, 063415 (2012).
44. Z. Wang, K. Liu, P. Lan, P. Lu, *J. Phys. B* **48**, 015601 (2015).
45. L.V. Keldysh, *Sov. Phys. JETP* **20**, 1307 (1965).
46. J. T. Paci and D. M. Wardlaw, *J. Chem. Phys.* **119**, 7824 (2003).

47. S. Chelkowski, A. D. Bandrauk, and P. B. Corkum, *Phys. Rev. Lett.* **93**, 083602 (2004).
48. D. Ursrey, F. Anis, and B. D. Esry, *Phys. Rev. A* **85**, 023429 (2012).
49. Z. C. Li and F. He, *Phys. Rev. A* **90**, 033421 (2014).
50. I. Kawata, H. Kono and Y. Fujimura, *J. Chem. Phys.* **110**, 11152 (1999).
51. F. Kelkensberg, G. Sansone, M. Y. Ivanov and M. Vrakking, *Phys. Chem. Chem. Phys.* **13**, 8647 (2011).
52. T. Y. Xu and F. He, *Phys. Rev. A* **88**, 043426 (2013).
53. J. McKenna, A. M. Sayler, F. Anis, B. Gaire, Nora G. Johnson, E. Parke, J. J. Hua, H. Mashiko, C. M. Nakamura, E. Moon, Z. Chang, K. D. Carnes, B. D. Esry, and I. Ben-Itzhak, *Phys. Rev. Lett.* **100**, 133001 (2008).
54. F. He, A. Becker and U. Thumm, *Phys. Rev. Lett.* **101**, 213002 (2008).
55. J. Zhang, G. Q. He and F. He, *Mol. Phys.* **112**, 1929 (2014).
56. P. B. Corkum, *Phys. Today* **64**, 36 (2011).
57. X. M. Tong, Z. X. Zhao, and C. D. Lin, *Phys. Rev. Lett.* **97**, 049301 (2006).
58. Jie Hu, Ke-Li Han, and Guo-Zhong He, *Phys. Rev. Lett.* **97**, 049302 (2006).
59. H. Niikura, F. Legare, R. Hasbani, M. Yu. Ivanov, D. M. Villeneuve, and P. B. Corkum, *Nature (London)* **421**, 826 (2003).
60. Z. C. Li and F. He, *Phys. Rev. A* **90**, 053423 (2014).
61. B. Feuerstein and U. Thumm, *Phys. Rev. A* **67**, 043405 (2003).
62. F. Martín, J. Fernández, T. Havermeier, L. Foucar, Th. Weber, K. Kreidi, M. Schöffler, L. Schmidt, T. Jahnke, O. Jagutzki, A. Czasch, E. P. Benis, T. Osipov, A. L. Landers, A. Belkacem, M. H. Prior, H. Schmidt-Böcking, C. L. Cocke, and R. Dörner, *Science* **315**, 629 (2007).
63. T. E. Sharp, *Atomic Data* **2**, 119 (1971).
64. R. Gopal, K. Simeonidis, R. Moshammer, Th. Ergler, M. Dürr, M. Kurka, K.-U. Kühnel, S. Tschuch, C.-D. Schröter, D. Bauer, J. Ullrich, A. Rudenko, O. Herrwerth, Th. Uphues, M. Schultze, E. Goulielmakis, M. Uiberacker, M. Lezius, and M. F. Kling, *Phys. Rev. Lett.* **103**, 053001 (2009).
65. D. G. Arbo, K. L. Ishikawa, K. Schiessl, E. Persson, and J. Burgdörfer, *Phys. Rev. A* **81**, 021403 (R) (2010).
66. M.-H. Xu, L.-Y. Peng, Z. Zhang, Q.u. Gong, X.-M. Tong, E. A. Pronin, and A. F. Starace, *Phys. Rev. Lett.* **107**, 183001 (2011).
67. C. B. Madsen, F. Anis, L. B. Madsen, and B. D. Esry, *Phys. Rev. Lett.* **109**, 163003 (2012).
68. T. Morishita, A. T. Le, Z. Chen, and C. D. Lin, *Phys. Rev. Lett.* **100**, 013903 (2008).
69. M. F. Kling, Ch. Siedschlag, A. J. Verhoef, J. I. Khan, M. Schultze, Th. Uphues, Y. Ni, M. Uiberacker, M. Drescher, F. Krausz, and M. J. J. Vrakking, *Science* **312**, 246 (2006).
70. J. McKenna, F. Anis, A. M. Sayler, B. Gaire, N. G. Johnson, E. Parke, K. D. Carnes, B. D. Esry, and I. Ben-Itzhak, *Phys. Rev. A* **85**, 023405 (2012).
71. E. Charron, A. Giusti-Suzor, and F. H. Mies, *Phys. Rev. Lett.* **71**, 692 (1993).
72. B. Sheehy, B. Walker, and L. F. DiMauro, *Phys. Rev. Lett.* **74**, 4799 (1995).
73. H. Ohmura, N. Saito, and M. Tachiya, *Phys. Rev. Lett.* **96**, 173001 (2006).
74. D. Ray, F. He, S. De, W. Cao, H. Mashiko, P. Ranitovic, K. P. Singh, I. Znakovskaya, U. Thumm, G. G. Paulus, M. F. Kling, I. V. Litvinyuk, and C. L. Cocke, *Phys. Rev. Lett.* **103**, 223201 (2009).
75. K. J. Betsch, D. W. Pinkham, and R. R. Jones, *Phys. Rev. Lett.* **105**, 223002 (2010).
76. J. Wu, A. Vredenborg, L. Ph. H. Schmidt, T. Jahnke, A. Czasch, and R. Dörner, *Phys. Rev. A* **87**, 023406 (2013).
77. X. M. Tong and C. D. Lin, *Phys. Rev. Lett.* **98**, 123002 (2007).
78. J. J. Hua, and B. D. Esry, *J. Phys. B* **42**, 085601 (2009).
79. F. Kelkensberg, G. Sansone, M. Y. Ivanov, and M. Vrakking, *Phys. Chem. Chem. Phys.* **13**, 8647 (2011).
80. A. D. Bandrauk, S. Chelkowski, and H. S. Nguyen, *Int. J. Quant. Chem.* **100**, 834 (2004).

81. K. P. Singh, F. He, P. Ranitovic, W. Cao, S. De, D. Ray, S. Chen, U. Thumm, A. Becker, M. M. Murnane, H. C. Kapteyn, I. V. Litvinyuk, and C. L. Cocke, *Phys. Rev. Lett.* **104**, 023001 (2010).

82. G. Sansone, F. Kelkensberg, J. F. Pérez-Torres, F. Morales, M. F. Kling, W. Siu, O. Ghafur, P. Johnsson, M. Swoboda, E. Benedetti, F. Ferrari, F. Lépine, J. L. Sanz-Vicario, S. Zherebtsov, I. Znakovskaya, A. L'Huillier, M. Yu. Ivanov, M. Nisoli, F. Martín, and M. J. J. Vrakking, *Nature* **465**, 763 (2010).

83. J. Wu, M. Magrakvelidze, L. P. H. Schmidt, M. Kunitski, T. Pfeifer, M. Schöffler, M. Pitzer, M. Richter, S. Voss, H. Sann, H. Kim, J. Lower, T. Jahnke, A. Czasch, U. Thumm, and R. Dörner, *Nat. Commun.* **4**, 2177 (2013).

84. T. Rathje, A. M. Sayler, S. Zeng, P. Wustelt, H. Figger, B. D. Esry, and G. G. Paulus, *Phys. Rev. Lett.* **111**, 093002 (2013).

85. X. Gong, P. He, Q. Song, Q. Ji, H. Pan, J. Ding, F. He, H. Zeng, and J. Wu, *Phys. Rev. Lett.* **113**, 203001 (2014).

Chapter 4

Nonsequential Double Ionization of Atoms in Strong Laser Field: Identifying the Mechanisms behind the Correlated-Electron Momentum Spectra

Difa Ye[*], Libin Fu[*] and Jie Liu[*,†]

*Institute of Applied Physics and Computational Mathematics,
Beijing 100088, China
†CAEP-Center for Fusion Energy Science and Technology,
Beijing 100088, China
†liu_jie@iapcm.ac.cn

Within the strong-field physics community, there has been increasing interest on nonsequential double ionization (NSDI) induced by electron-electron (*e-e*) correlation. A large variety of novel phenomena has been revealed in experiments during the past decades. However, the theoretical understanding and interpretation of this process is still far from being complete. The most accurate simulation, i.e. the exact solution of the time-dependent Schrödinger equation (TDSE) for two electrons in a laser field is computationally expensive. In order to overcome the difficulty, we proposed a feasible semiclassical model, in which we treat the tunneling ionization of the outmost electron quantum mechanically according to the ADK theory, sample the inner electron from microcanonical distribution and then evolve the two electrons with Newton's equations. With this model, we have successfully explained various NSDI phenomena, including the excessive DI yield, the energy spectra and angular distribution of photoelectrons. Very recently, it is adopted to reveal the physical mechanisms behind the fingerlike structure in the correlated electron momentum spectra, the unexpected correlation-anticorrelation transition close to the recollision threshold, and the anomalous NSDI of alkaline-earth-metal atoms in circularly polarized field. The obvious advantage of our model is that it gives time-resolved insights into the complex dynamics of NSDI, from the turn-on of the laser field to the final escape of the electrons, thus allowing us to disentangle and thoroughly analyze the underlying physical mechanisms.

1. Introduction

NSDI of atoms in strong laser field has been studied for almost four decades since its first observation in experiment [1]. Until recently, it has been the consensus that rescattering is the dominant mechanism for NSDI. In this three-step mechanism, the first electron is freed by a quasistatic tunneling ionization, and is driven back to its parent ion and imparts part of its energy to dislodge a second electron [2]. The electron recollision picture as a cornerstone of strong field physics successfully explain various experimental observations of NSDI, e.g., the knee structure in the double- over single-ion yield curve [3], the double hump structure in the ion momentum spectrum [4, 5] and the signature of e-e correlation in the joint electron momentum spectrum parallel to the field polarization [6]. It also inspires further investigations that achieve insight into the microscopic dynamics of the ionization process on the time scale of subfemtosecond or even attosecond [7–10].

Despite the great progress, novel experimental data still constantly emerge and challenge our existing knowledge. For instance, with the help of sophisticated Cold Target Recoil Ion Momentum Spectroscopy (COLTRIMS), combined with high-repetition-rate lasers, it is now possible to record differential information on strong-field few-electron reactions [11]. Plotting these data in the parallel momentum plane of $(p_1^{\parallel}, p_2^{\parallel})$, two experimental groups independently found a surprising fingerlike (or V-shaped) structure in the first and third quadrants for helium NSDI that was not observed before [12, 13], giving rise to an extensive discussion of its underlying physical mechanisms. Moreover, the pattern is found to depend sensitively on the laser intensity and to vary with the atomic species. With a slight decrease in the intensity, from $0.09\,\mathrm{PW/cm^2}$ to $0.07\,\mathrm{PW/cm^2}$, it was found that a transition from correlation (dominant population in the first and third quadrants) to anticorrelation (dominant population in the second and fourth quadrants) emerged for argon NSDI [14]. This kind of transition, however, has not been observed for neon, even though the laser intensity was decreased down to the regime where no NSDI event was detected over weeks of data collection [15]. More recently, measurements on strong-field double ionization of Xe reveal dramatic differences as compared with other low-Z rare-gas targets, e.g., He, Ne, and Ar [16]. The characteristic double-hump structure in the longitudinal momentum distributions of doubly charged ions of low-Z atomic targets disappears for double ionization of Xe. Moreover, the joint electron momentum spectra reveal the remarkable noncorrelation behavior, i.e. no dominance of the side-by-side or back-to-back emission is observed. These fresh experiments indicate that the common picture of NSDI is far from complete and thus have triggered a new surge of investigations of NSDI.

From the theoretical side, different approaches, e.g. classical or semiclassical Monte Carlo simulations, S-matrix theories based on strong-field approximation, as well as exact solution of the full-dimensional two-electron time-dependent Schrödinger equation (TDSE), were developed to account for the experimental observations, see Ref. [17] for a recent review. In this chapter, we focus on the semiclassical model, which successfully combines the ADK tunneling theory with the classical trajectory Monte Carlo (CTMC) simulation. The advantages of this model include: (i) it significantly reduces the budget of computer speed and memory, as compared with TDSE simulation, which usually requires a large-scale high-performance supercomputer and can only be run under laser wavelengths shorter than those commonly used in experiments; (ii) it naturally avoids the nonphysical autoionization as compared with purely classical simulations, thus the electron-ion potentials can be exactly taken as Coulombic, not necessary to introduce the soften parameters; and (iii) more importantly, it offers the opportunity to back analyzing the NSDI trajectories of interest, from the turn-on of the laser field to the final escape of the electrons, thus allowing us to disentangle the underlying physical mechanisms and to explain the experiments qualitatively or even sometimes quantitatively.

2. Semiclassical Model

The development of our semiclassical model dates back to 1995 and is roughly divided into four stages: (i) we first calculated the ionization rate of atoms in strong laser field by representing the atoms with microcanonical ensembles and found good agreement with quantum simulation, thus confirming the validity of classical description of atoms [18, 19]; (ii) we then set up a rescattering model for the tunneled electrons, which well reproduced the plateau structure, cutoff energy and angular distribution of the photoelectrons as observed in experiments [20]; (iii) the above two models, one for bound electron and the other for tunneled electron, were combined to develop a semiclassical rescattering model for nonsequential double ionization [21–23]; (iv) the model was extended to the molecular case, to over-the-barrier regime [24], to below-the-recollision-threshold regime [25], and to relativistic regime [26].

In the following, we briefly present the theoretical methodology of our semiclassical model. We first consider a helium-like atom (two active electrons) interacting with an infrared linearly polarized laser field. One electron is released at the outer edge of the field-suppressed Coulomb potential through quantum tunneling with a rate given by the ADK formula [27]. The exit point is determined by the effective potential in parabolic coordinates [28]: $U(\eta) = -1/(4\eta) - 1/(8\eta^2) - \varepsilon(t)\eta/8 = -I_{p1}/4$, where I_{p1} is the first ionization energy of the atom, and η is the parabolic coordinate relating to the Cartesian coordinate through $\eta = -2z$, with z being the coordinate along the field direction. The tunneled electron has a Gaussian distribution on transverse (perpendicular to the field direction) velocity

$f(p_{x,y}) = \kappa/(\pi \varepsilon) \cdot \exp(-\kappa p_{x,y}^2/\varepsilon)$ and zero longitudinal velocity [22, 29], where $\kappa = \sqrt{|2I_{p1}|}$. For the bound electron, the initial position and momentum are sampled from microcanonical distribution [30]. The subsequent evolution of the two electrons with the above initial conditions are governed by Newton's equations of motion: $d^2\mathbf{r}_i/dt^2 = \varepsilon(t) - \nabla_{r_i}(V_{ne}^i + V_{ee})$. Here the index i denotes the two different electrons, $V_{ne}^i = -2/|\mathbf{r}_i|$ and $V_{ee} = 1/|\mathbf{r}_1 - \mathbf{r}_2|$, are Coulomb interactions between nucleus and electrons and between two electrons, respectively. The laser field $\varepsilon(t)$ has a constant amplitude for the first ten cycles and turns off with 3-cycles ramp. The above Newtonian equations are solved by employing the standard 4th–5th Runge-Kutta algorithm and DI events are identified by energy criterion. In our calculations, more than 10^7 weighted (i.e., by the tunneling rate) classical two-electron trajectories are traced until the end of the pulse. This results in more than 10^4 DI events for statistics.

This model has been proven to work well in the regime above the recollision threshold. To extend it to the BRT regime, we need to include the RIET effect in the second step. This is done by allowing the bound electron to tunnel through the potential barrier whenever it reaches the outer turning point, where $p_{i,z} = 0$ and $z_i\varepsilon(t) < 0$, with a tunneling probability P_i^{tul} given by the WKB approximation $P_i^{tul} = \exp[-2\sqrt{2}\int_{z_i^{in}}^{z_i^{out}} \sqrt{V(z_i) - V(z_i^{in})}dz_i]$. Here, z_i^{in} and z_i^{out} are the two roots of the equation for z_i: $V(z_i) = -2/r_i + z_i\varepsilon(t) = -2/r_i^{in} + z_i^{in}\varepsilon(t)$ ($|z_i^{out}| > |z_i^{in}|$) [25, 31].

3. The Surprising Fingerlike Structure in NSDI

In 2008, new high resolution and high statistics COLTRIMS experiments on double ionization of helium were performed independently by two groups, and a striking fingerlike (or V-shaped) structure was observed [12, 13] in the correlated electron momenta parallel to the laser polarization. The observation is in qualitative accordance with the prediction of S-matrix approach [32–34] and quantum mechanical calculation [35]. However, in the S-matrix model, a rather unphysical contact-type electron-electron interaction yields the agreement with the experimental data, while, the implementation of a realistic Coulomb interaction results in clear deviations from the data. Because the Volkov states are employed in the S-matrix theory, it is out of the question to consider the influence of the ion for the electrons propagating in the continuum. On the other aspect, the dynamical details of the electron recollision process are hardly extracted from the solution of the time-dependent Schrödinger equation, and for which a transparent physical interpretation of the fingerlike pattern is still lacking.

In this section, we address the influence of several physical mechanisms, such as the electron-electron and electron-ion interaction, on the fingerlike structure,

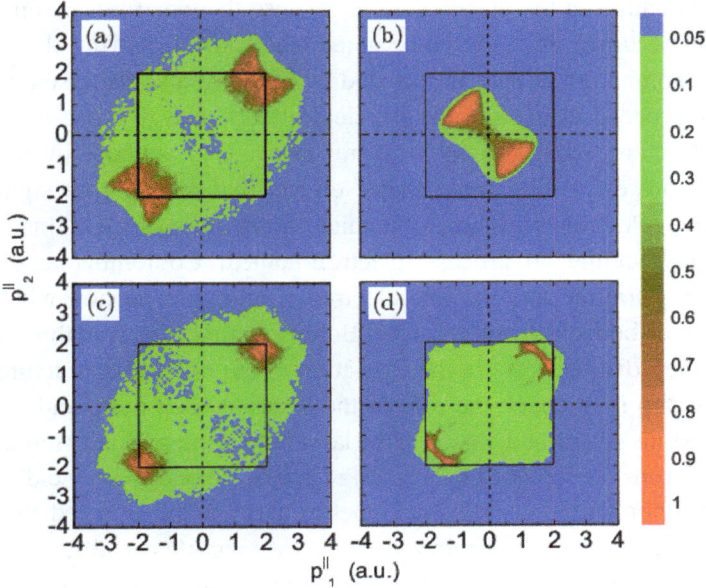

Fig. 1. (a) Distribution of correlated electron momenta along the laser polarization for Helium DI irradiated by 800 nm, 4.5×10^{14} W/cm^2 laser pulses. The black box indicates the $2\sqrt{U_p}$ boundary of electron momentum. The model calculations under various circumstances yield very different momentum distribution patterns (see text for details): (b) the laser field is removed and the tunneled electrons are replaced by a beam of projectile electrons; (c) the electron-electron Coulombic interaction is replaced with a Yukawa potential; (d) the nuclear Coulomb potential is softened.

by exploiting our 3D semiclassical model [36]. In contrast to previous theoretical studies, this model allows us to disentangle and thus thoroughly analyze the above-stated mechanisms. Without any controversy, our results unveil the significant role of the nuclear Coulomb attraction and the electron-electron Coulomb repulsion in forming the fingerlike pattern.

The resulting electron momentum distribution, calculated with our semiclassical model for the same parameters as in the experiment [12], is shown in Fig. 1(a). The calculation reproduces many key features observed in the experiment, including the emission of the two electrons primarily into the same hemisphere, the small circular accumulation around the zero momentum surrounded by four elliptical hard-to-reach regimes, and more importantly, the fingerlike structure beyond $2\sqrt{U_p}$. Here $U_p = \varepsilon_0^2/4\omega^2$ is the pondermotive energy with ε_0 and ω denoting the laser field amplitude and angular frequency, respectively. With assuming that the colliding electrons leave the atom with no significant energy and the motion of the individual electron is solely determined by the laser field, the parallel momentum $p_{1,2}^{\parallel}$ of each electron results exclusively from the acceleration in the optical field. Within this simple model $2\sqrt{U_p}$ should be the maximum momentum, and the momentum distribution favors accumulation in the diagonal zone because the electrons are emitted nearly simultaneously. However, it should be noted that $2\sqrt{U_p}$ is not the classical limit of the parallel momentum as clearly revealed by Fig. 1(a) and discussed in Ref. [37].

We now proceed to explore the physical effects that give rise to this peculiar fingerlike structure, using the classical trajectory (CT) method. In the context of NSDI, there are essentially three major effects that may play a significant role in the double electron emission dynamics: electron-laser field interaction that occurs throughout the DI process, electron-nuclear Coulomb interaction in the post-collision duration, and the interelectron Coulomb repulsion which becomes significant when both electrons get close. Below, we investigate all three interactions and clarify their distinct roles in the formation of the fingerlike structure.

The first step is to check the role of the external laser field, and an additional calculation is thus performed, in which the laser field is intentionally removed and the tunneled electrons are replaced by a beam of projectile ones with incident energy of $3.17U_p$, corresponding to the maximal kinetic energy of the tunneled electrons upon recollision. The result is shown in Fig. 1(b). Two significant differences from the complete model calculation in Fig. 1(a) are found: (i) the fingerlike structure beyond $2\sqrt{U_p}$ completely disappears; (ii) the two emitted electrons tend to distribute in the second and fourth quadrants, indicating that the incident electron transfers much of its momentum to the bound one while itself is backscattered into the opposite direction. The comparison between Figs. 1(a) and 1(b) shows the most important role of the laser field in turning the two back-to-back emitted electrons into the same direction and the post-collision acceleration of the electrons resulting in the fingerlike structure beyond $2\sqrt{U_p}$.

The next step concerns the question if this fingerlike structure is a fingerprint of a strong interelectron correlation between the ionizing electrons. We have performed another calculation in which the final-state electron Coulomb repulsion has been deliberately neglected by replacing the electron Coulombic interaction $V_{ee} = 1/|\mathbf{r}_1 - \mathbf{r}_2|$ with Yukawa repulsion potential of the form $V_{ee} = \exp[-\lambda r_b]/r_b$, where $r_b = \sqrt{|\mathbf{r}_{1,2}|^2 + b^2}$, $\lambda = 5.0$, and $b = 0.2$. The result of this calculation [Fig. 1(c)] shows that the prominent fingerlike structure is to a large extent reduced and thus provides clear evidence that the final-state electron correlation plays a significant role.

Last but not the least procedure is to justify the role of the electron-nuclear interaction, which is commonly believed to be the main reason for the recoil collision in field-free $(e, 2e)$ process. This interaction was also suggested to be the very ingredient for the field-assisted recoil collision in the context of intense field DI of atomic helium [12]. Accordingly, an additional calculation, in which we soften the nuclear Coulomb attraction by employing $V_{ne}^i = -2/\sqrt{|\mathbf{r}_i|^2 + a^2}$, where a is chosen as 1.0 to match the ground state energy of He^+, is performed.

Physically, the shielding of nuclear potential would to a great extent diminish the Coulomb focusing effect that has significant effects upon both electrons. Clearly, a Coulombic potential would attract the tunneled electron more dramatically when it moves near the atomic core. Such strong attraction may unambiguously bring the tunneled electron to share more kinetic energy with the bound one. For the bound electron, after achieving considerable transferred momentum upon collision, it may elastically backscatter from the Coulombic core on its way out of the atom. This double scattering process is coined as recoil collision [38] and was routinely found in traditional electron impact ionization experiments, especially when the projectile electron possesses the energy of only a few times of the binding energy of the inner one. The result shown in Fig. 1(d) indicates that the Coulomb focusing effect is decisive for the production of the electrons with high energy, and thus the fingerlike structure.

The dynamics behind the fingerlike structure observed in the experiments was supposed to be related with the recoil collision in the presence of the external laser field [12]. This conjecture has been justified by a 2-body 3D quantum mechanical simulation [39]. However, the dynamical details of such collision is hardly explored in the quantum mechanical treatment. With the CT approach, this becomes possible by back-analyzing the history of the DI events of interest.

It has been recognized that the characteristics of the DI trajectory can be well represented by the recollision and DI time [40–42]. We thus provide such an information for the trajectories that contribute to the fingerlike structure in Figs. 2(c) and 2(d). It is found that, in Fig. 2(c), the electron pairs contributing to the fingerlike structure tend to encounter right at zero field. Within rescattering picture, this can be understood as that these trajectories include the most energetic collisions for which the tunneled electrons are released at the laser phase of about 17°, and return around the zero crossing of the electric field at 270° with maximal energy of $3.17U_p$ [43, 44].

Upon recollision, the bound electron may be directly freed, a process termed as collision ionization (CI), or be excited and subsequently ionized by the next field maximum, known as collision-excitation ionization (CEI) [45]. The smaller and larger peaks in Fig. 2(d) correspond to these two mechanisms, respectively. The smaller ones around zero field represent the electron pairs emitted after a very short thermalization process (~attosecond) [46], and the larger ones correspond to CEI events with small time delay of 0.15T (i.e., 0.3π phase difference from the collision phase peaks) [24]. The results shown in Figs. 2(c) and 2(d) indicate that both CI and CEI with small time delay could contribute to the fingerlike structure.

To further unveil the microscopic mechanism underlying this unusual pattern, we collect all trajectories that constitute the fingerlike structure and trace back their dynamical evolution at collision and post-collision. With the CT diagnosis, we sketch two types of trajectory configurations that are responsible for the asymmetric

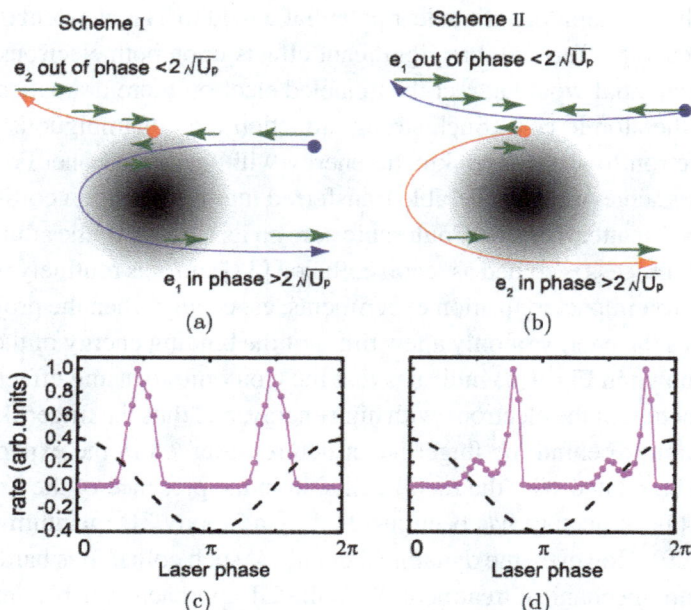

Fig. 2. (a), (b) Two trajectory configurations responsible for the fingerlike structure. DI yield versus laser phase at recollision (c) and at DI moment (d). Here, the statistics is only collected for the DI trajectories that contribute to the prominent fingerlike structure, i.e., the regimes of 1.5 a.u. $< |p_i^{||}| <$ 2.0 a.u. and 2.5 a.u. $< |p_j^{||}| <$ 3.0 a.u., where $i, j = 1, 2, i \neq j$. The dashed curves represent laser fields.

electron energy sharing after electron-electron collision and lead to the fingerlike structure beyond $2\sqrt{U_p}$ [see Figs. 2(a) and 2(b), correspond to type-I and II configurations, respectively]. While the tunneled electron is driven back to the nucleus by the laser field, the field strength reduces to zero and the collision is essentially a field-free three-body system under the pure Coulomb potential. For type-I configuration of Fig. 2(a), the electron-electron collision near nucleus could lead to the following consequences: the second electron acquires considerable momentum from the returned electron and emits in the forward direction, while the returned electron is slowed down. Under the influence of the nuclear attraction, the latter is transferred to a hyperbolic orbit around the nucleus with a large scattering angle. In this way, the returned electron reverses its direction in a time scale of attoseconds after the collision. Meanwhile, the laser field changes its direction. As the returned electron has nonzero residual momentum parallel to the instantaneous field direction and is further accelerated by the field, its final longitudinal momentum is expected to be above $2\sqrt{U_p}$. For the second electron after collision, its initial momentum is opposite to the instantaneous field direction; one expects that its final longitudinal momentum is below $2\sqrt{U_p}$. In Fig. 2(b), where a type-II trajectory is schematically shown, the situation is similar except that the roles of the two electrons have exchanged: Under assistance of nuclear Coulomb attraction, the second electron acquires a nonzero momentum parallel to the instantaneous field and therefore emits with a longitudinal momentum larger than $2\sqrt{U_p}$. Although the returned electron is slowed down by electron-electron repulsion, it still has a residual momentum opposite to the instantaneous field, resulting in a final longitudinal momentum below $2\sqrt{U_p}$. Our statistics reveals that the DI events in fingerlike structure consist of 70% type-I and 30% type-II configuration. It indicates that for the trajectories contributing to the fingerlike structure, the slower electron is usually the one ejected from the ion ground state, in agreement with the quantum calculation [35].

The above analysis reveals that the electron-electron collision assisted by the nuclear attraction is crucial for the emergence of fingerlike structure. In certain cases where the collision is strong, the fingerlike structure should be more prominent. To test this idea, we impose an additional confinement on our statistics to observe the variation of the correlated momentum patterns with respect to the relative perpendicular momentum between two electrons. The results are presented in Fig. 3.

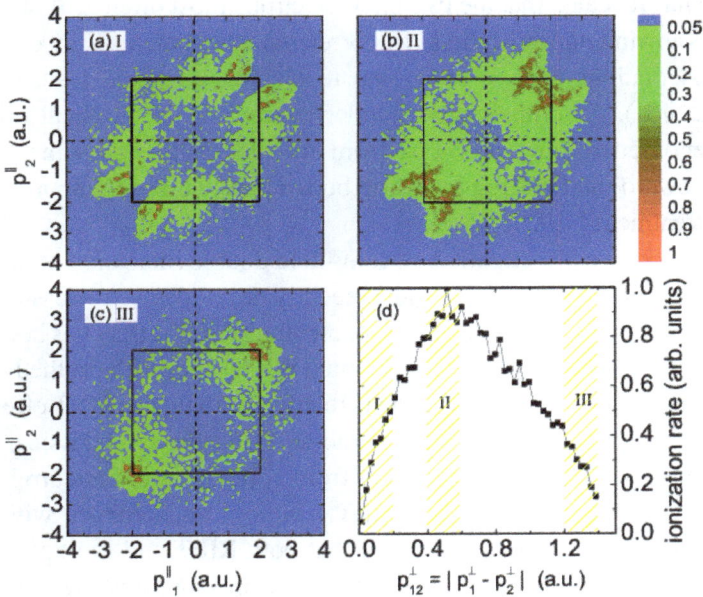

Fig. 3. Correlated parallel momentum distributions with additional conditions on the relative perpendicular momentum between two electrons, i.e., for (a) $0 \le p_{12}^\perp \le 0.2$, (b) $0.4 \le p_{12}^\perp \le 0.6$, (c) $1.2 \le p_{12}^\perp \le 1.4$. (d) The overall relative perpendicular momentum distribution.

It is shown in Fig. 3(d) that the relative perpendicular momentum $p_{12}^{\perp}$ for all DI events distributes over an interval of [0, 1.4] and exhibits a notable accumulation around 0.5 a.u. Moreover, the fingerlike pattern is more prominent for the case of small relative perpendicular momentum [see Fig. 3(a)]. As we increase the value of $p_{12}^{\perp}$, the two finger patterns start to merge [Fig. 3(b)] and finally totally disappear [Fig. 3(c)]. The above result is in agreement with the experimental observations [13], and provides a good explanation for the 1D quantum calculation in [47]. Because of dimensional restriction, the quantum calculation magnifies the electron-electron collision effects. As a result, a butterflylike structure similar to Fig. 3(a) emerges (see Fig. 1 of [47]).

4. NSDI Below the Recollision Threshold

When the idea of recollision was first introduced, it was anticipated that the double ionization yield would undergo a sudden drop supposing the maximum returned energy is smaller than the ionization potential of the inner electron, that is, in the regime below the recollision threshold (BRT). But, by now it is widely recognized that recollision excitation and field suppression effects might extend the process to lower intensities. More interestingly, a recent experiment in the deep BRT regime [14] has revealed that the two photoelectrons most often drift out in opposite directions, showing the so-called anticorrelated phenomenon, in contrast to all previous observations near or above the recollision threshold (see, for example, Refs. [6, 12, 13]). A comprehensive understanding of the physical origin of this striking phenomenon, however, is far from complete, although some progress has been made toward the BRT regime using both a purely classical approach [48] and a quantum treatment [49].

In this section, we investigate the strong-field double ionization of argon atom at the transition to the BRT regime using a recently developed semiclassical model, in which the recollision-induced excitation-tunneling (RIET) effect has been taken into account with the Wentzel-Kramers-Brillouin (WKB) approach. In the BRT regime, since the energy of the returned electron is relatively small, the RIET effect becomes important. Therefore, this extension is crucial. With this semiclassical model, we are capable of reproducing the transition from correlation to anticorrelation as the laser intensity decreases to the deep BRT regime, in accordance with the recent experiment on argon atoms. We identify that both RIET and multiple recollisions are responsible for the transition and these two mechanisms are found to leave distinct footprints on the correlated momentum spectra. As another signature of the transition, we find that the fingerlike structure in the correlated momentum spectra completely fades away.

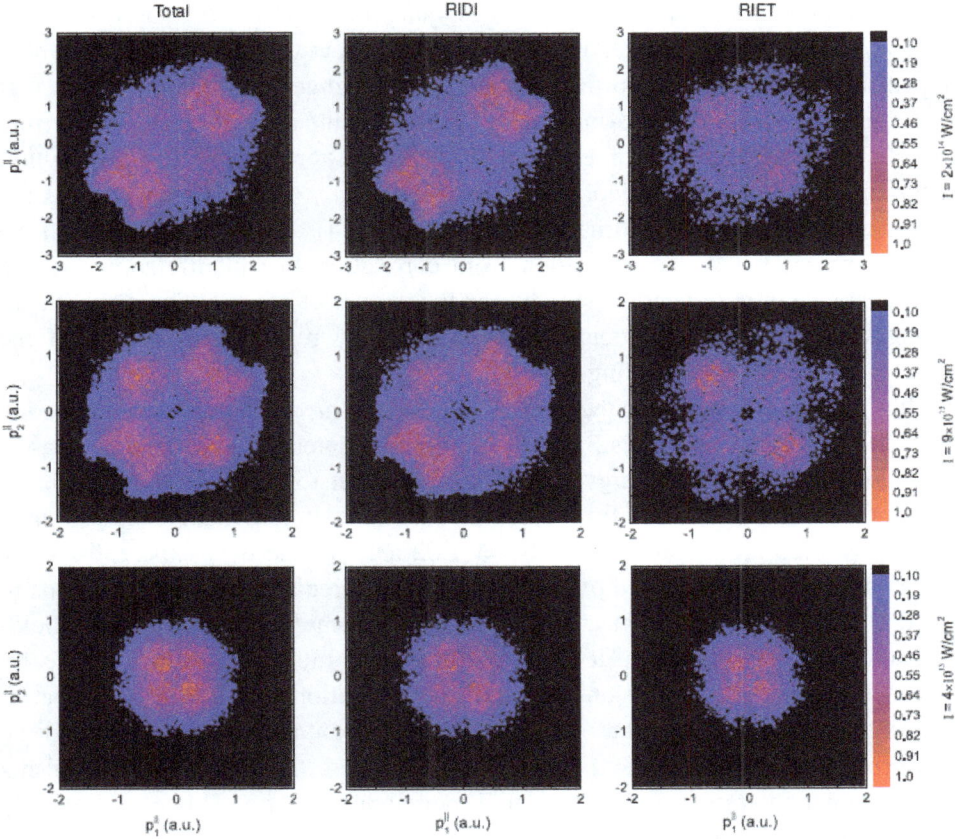

Fig. 4. Correlated momentum spectra for double ionization of argon at different laser intensities ranging from above to below the recollision threshold. Shown are the momentum components $p_{1,2}^{\parallel}$ along the laser polarization direction. The second and third columns are deduced from the first column by superimposing additional restrictions on our statistics according to different emission mechanisms.

The statistics on the momentum distribution at three different laser intensities $I = 2 \times 10^{14}$ W/cm^2 (high), 9×10^{13} W/cm^2 (moderate), and 4×10^{13} W/cm^2 (low), are presented in the first column of Fig. 4. A simple estimation of the intensity threshold is given by $3.17U_p = |I_{p2}|$, which gives 1.5×10^{14} W/cm^2. So, for the lowest laser intensity, the system has fallen deep into the BRT regime.

Our calculations have reproduced many key features observed in the experiments. At the highest laser intensity, the spectrum presents a correlated behavior, that is, the electron pairs have a higher probability of being emitted in the same direction parallel to the laser polarization. Moreover, the distribution exhibits an overall fingerlike structure, as discussed in the last section.

The momentum distribution becomes quite different as the laser intensity decreases. For moderate laser intensity, two red spots (high probability) start to appear in the second and fourth quadrants and the fingerlike structure in the first and third quadrants starts to fade away. When the intensity decreases deep into the BRT regime, the fingerlike structure totally disappears and the distribution tends to align along the anticorrelated diagonal of $p_1^{||} = -p_2^{||}$. These results are consistent with recent experimental observations [14]. With the simulations, we have demonstrated a clear transition from correlation to anticorrelation and the disappearance of the fingerlike structure in the deep BRT regime. These significant changes on the momentum distribution presumably reflect the variation of the underlying mechanisms leading to NSDI.

It is well known that when the tunneled electron is thrown back to the parent core, its maximal kinetic energy is $3.17U_p$. As the laser intensity gradually decreases, the maximal recolliding energy becomes insufficient to directly ionize the inner electron. This is the case for the lowest laser intensity in these calculations, where $3.17U_p = 0.28$ a.u. is far below $|I_{p2}| = 1.02$ a.u. As one may expect, the inner electron that has been excited by the first recollision requires other mechanisms to liberate it, for example, by means of a second (even multiple) recollision or field-assisted tunneling ionization. It is straightforward to model the former classically [48]. We term it as the recollision-induced direct ionization (RIDI). The latter, that is, the recollision-induced excitation tunneling, is of quantum nature and beyond the capability of previous classical treatments [48]. These two different mechanisms leave distinct footprints in the correlated momentum spectra. This can be clearly seen by comparing the second and third columns of Fig. 4, where we split the total correlated momentum spectra into two parts according to different emission mechanisms.

From the middle column of Fig. 4 for the RIDI mechanism, we see that the correlated momentum spectra exhibit a clear transition from the dominance of correlated emission to a situation where anticorrelated ejection has become much stronger and even the most important contribution, when the laser intensity decreases to the BRT regime. The patterns are to some extent analogous to that in the first column for the total distribution.

We plot two typical trajectories, corresponding to the single-recollision-induced correlated emission and double-recollision-induced anticorrelated emission, in Figs. 5(a) and 5(d) and 5(b) and 5(e), respectively.

In Fig. 5(a) and 5(d), the returned electron possesses high incident kinetic energy, thus a single recollision is enough to "kick out" the inner electron within half a cycle and push it to move along the recolliding direction. Then the laser field quickly reverses the struck electron, and finally both electrons drift out in

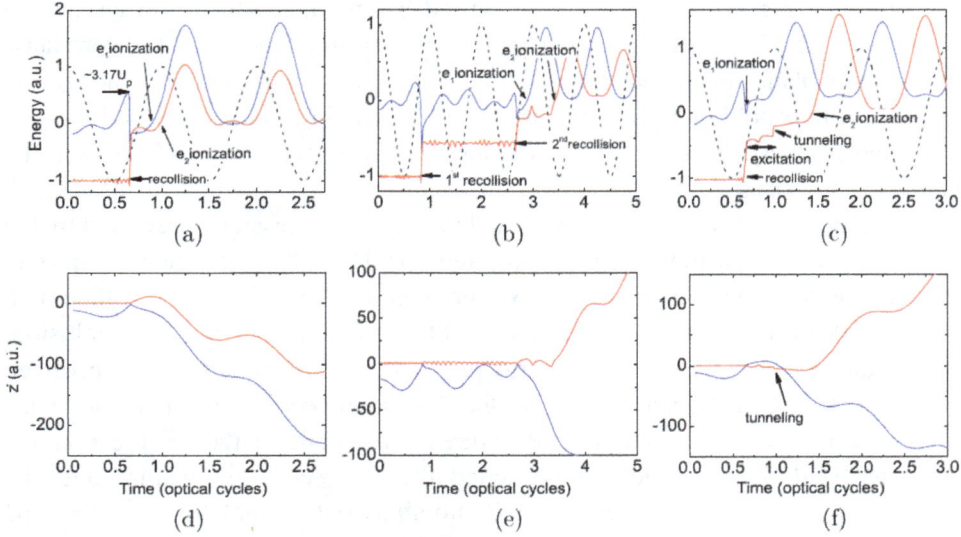

Fig. 5. Typical trajectories for NSDI: (a) Double ionization is induced directly by single recollision and by means of (b) two recollisions or (c) recollision-induced excitation-tunneling. (d)–(f) show the coordinate evolution of the two electrons along the laser field direction. The blue (dark gray) and red (light gray) curves, respectively, denote the tunneled electron and the inner one. The black dashed curves in (a)–(c) represent the laser electric field. Note that there is no collision at t = 2.0 o.c. in (e) and t = 1.15 o.c. in (f). In both cases, the distance in the lateral direction is very large. In (f) a small displacement along z toward negative values is just discernible as a result of the tunneling labeled by an arrow, placing the "classical" electron abruptly at a different position; see text for details.

the backward direction, with one following the other. This configuration usually emerges in the higher intensity regime. The "time delay" between the recollision and double ionization is less than 1/4 optical cycle [i.e., one optical cycle (o.c.) ~2.67 fs].

At low laser intensities, a second recollision is required to liberate the inner electron, as demonstrated explicitly by the classical trajectories in Figs. 5(b) and 5(e). Here, the opposite-hemisphere ionizations hinge on having a time delay of ~1/2 o.c. between the ionization moment of the first electron and the second one. During the 1/2 cycle, the field has reversed its direction, leading to back-to-back emission of the electrons. Physically, at low intensities the second recollision usually is weak; after the second recollision, both electrons still populate bound states. The strong *e-e* repulsion will push one electron to pass over the Coulomb barrier at the first coming field maximum, while leaving the other one being hindered until the second field maximum [see Figs. 5(b) and 5(e)]. A similar mechanism was also discussed in Ref. [48].

For the RIET mechanism, its percentage over the total double ionization events increases from 23.07% at 2×10^{14} W/cm^2 to 32.79% at 4×10^{13} W/cm^2.

We find amazingly that NSDI events caused by this mechanism always have a greater probability of occupying the second and fourth quadrants in the correlated momentum plane, no matter whether the laser intensity is below or above the recollision threshold. The percentage of doubly ionizing trajectories in these two quadrants is always beyond 50%, increasing slightly from 51.56% to 55.56% as the laser intensity rises.

The physical origin for the RIET-type back-to-back emission is revealed by the typical trajectory illustrated in Figs. 5(c) and 5(f). Here, the outer electron born in the $-z$ direction nearly at positive maximum field is thrown back to recollide with the inner electron at about 0.7 o.c., slightly before a field zero. After the recollision, the returned electron with positive energy runs to the $+z$ direction for a while, driven by the instantaneous negative field, but the field soon becomes positive and hence forces the returned electron to reverse its direction and emit in the $-z$ direction. On the other hand, the inner electron is excited after the recollision, waiting until the next peak field arrives. At that time, the Coulomb barrier is dramatically suppressed and the electron can be released through quantum tunneling, as evidenced by the energy jump at $t \simeq 1.0$ o.c. [50]. However, the excitation-tunneled electron could not be accelerated away from the nucleus, since it does not ionize until $t \simeq 1.45$ o.c., that is, after an additional half-cycle time delay. The reason is as follows. The electron tunneled slightly before the maximum of the laser field is only quasi-free (i.e., a long distance from the nucleus but with negative energy). As soon as the external field decreases, the electron feels the nuclear field, hindering the electron to be driven away. When the field reverses, however, it is effectively driven toward the nucleus and through a phase shift induced by electron-nucleus scattering it can now extract enough energy from the external field to escape. Here we notice the crucial role of the nuclear Coulomb attraction in the post-tunneling process. Without the nuclear attraction, according to the well-known simple-man model, we know that the electron tunneled before the field maximum would directly emit without returning to the core while the electron tunneled after the field maximum would return to the nucleus and then emit in the opposite direction in which it tunneled. The former gives the correlation while the latter leads to anticorrelation. These two processes are equally likely because the field in cosine form is symmetric with respect to its maximum. Nevertheless, the nuclear Coulomb attraction effect will break the balance by increasing the possibility of the latter process, which favors the anticorrelation. This has been evidenced by our calculation shown in Figs. 5(c) and 5(f), in which the electron tunneled slightly before the field maximum still returns to the nucleus and then emits in the opposite direction in which it tunneled.

The time over which the inner electron populates the excited state prolongs as the laser intensity decreases. For a very long time delay, both same-hemisphere and opposite-hemisphere emissions are possible and become equally important. Thus, there is no significant signal of anticorrelation at very low laser intensity, as shown in the bottom-right plot of Fig. 4.

According to this analysis, an interesting picture for the RIET-induced back-to-back emission emerges as follows: (i) The outer electron born in the $-z$ direction is thrown back by the field to recollide with the inner electron. After that, the returned electron is driven by the field to run to the $+z$ direction for a while, reverse its direction, and finally emit in the $-z$ direction. (ii) The recollision-excited electron tunnels out of the Coulomb barrier in the $-z$ direction. The electron feels the nuclear Coulomb attraction, which hinders the electron from being driven away. It waits for a half-cycle so that the laser field reverses its direction and drives it back to the nucleus. The electron finally emits in the $+z$ direction. This picture is different from the mechanism discussed in Ref. [48] where the escape of the excited electron at the first field maximum after recollision is hindered by the proximity of the other electron.

Based on these calculations and discussion, the physical picture for the double ionization at the transition to the BRT regime is summarized in the percentile map of Fig. 6. Our results unveil that (i) both the mechanisms of RIDI and RIET significantly contribute to the double ionization yield; (ii) the RIDI mechanism plays a decisive role in the transition from correlation to anticorrelation; and (iii) the RIET mechanism always prefers to cause back-to-back emission. Its percentage over the total NSDI events increases monotonically as the laser intensity decreases.

Another signature for the transition to the BRT regime, as we have pointed out, is the disappearance of the striking fingerlike structure in the correlated momentum spectra. Above the threshold, the fingerlike structure is shown to be closely related to the relative perpendicular momentum between two electrons (see Fig. 3). The

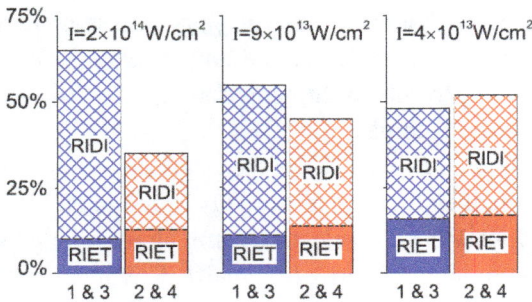

Fig. 6. Percentile map for doubly ionizing trajectories at three different laser intensities. All trajectories are classified into two categories based on whether the inner electron is freed through RIDI (hatched area) or RIET (color-filled area). The numbers at the bottom denote the quadrants of the correlated momentum plane.

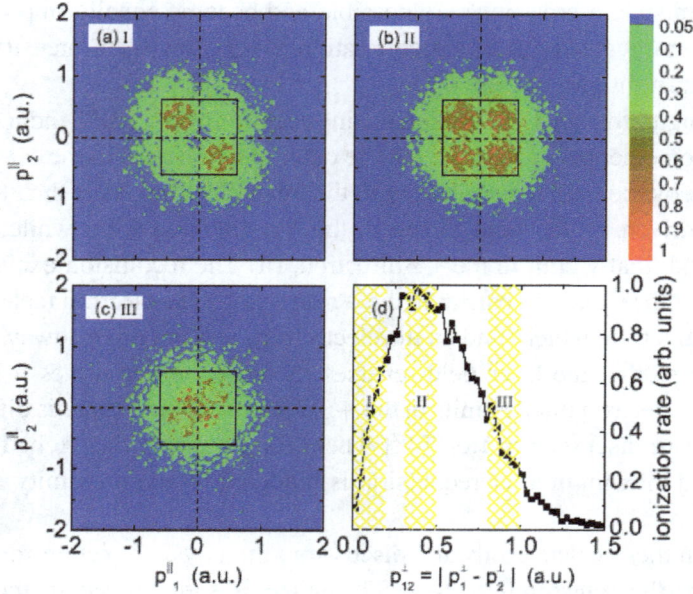

Fig. 7. Correlated parallel momentum distributions with additional conditions on the relative perpendicular momentum between two electrons: (a) $0 \leq p_{12}^{\perp} \leq 0.2$, (b) $0.3 \leq p_{12}^{\perp} \leq 0.5$, and (c) $0.8 \leq p_{12}^{\perp} \leq 1.0$. (d) The overall relative perpendicular momentum distribution. The laser intensity is 4×10^{13} W/cm^2.

question is how the anticorrelated pattern concerned in this paper depends on the relative perpendicular momentum. To answer it, we make some further calculations. We first show the distribution of the relative perpendicular momentum $p_{12}^{\perp}$ for all DI events in Fig. 7(d). It can be divided into three regimes: a rapid ascent, followed by a cambered top around $p_{12}^{\perp} = 0.4$ a.u., and finally a gentle slope. The correlated parallel momentum spectra for these three different regimes are, respectively, shown in Figs. 7(a)–7(c). We find that those electron pairs with small relative perpendicular momentum are mainly responsible for the anticorrelation. As $p_{12}^{\perp}$ increases, the distribution becomes uniformly occupying four quadrants, and finally the well-known correlated pattern emerges.

This calculation indicates that, if the electrons are emitted with small relative transverse momentum, then they are most likely found to be anticorrelated. In these DI events, we have found that both electrons emit almost along the polarization axis (i.e., their transverse momentum are also small). This indicates the important role of *e-e* Coulomb repulsion in forming the anticorrelation pattern.

As demonstrated in the previous section, our semiclassical model, which includes recollision-induced excitation tunneling but without considering the shielding effect and multiphoton processes, has captured the key feature of the double ionization at the transition to the BRT regime. Our model calculation clearly indicates that the threshold value is shifted downward in comparison with the simple criterion $3.17U_p = |I_{p2}|$. This point has been claimed theoretically in previous works [51, 52].

The underlying mechanism can be uncovered by the typical double ionization trajectories represented by Fig. 5(a) and 5(d). We see that, due to the field suppression of the Coulomb potential, electrons with negative energy can still escape from the atom as long as their energy is higher than the maximal barrier height. A quantitative estimation of the field suppression effect has been made using a simple 1D model [51]: Because the maximum of the barrier in the 1D potential $V(z) = -2/|z| + \varepsilon_0 z (z < 0)$ is given by $V_b = -2\sqrt{2\varepsilon_0}$, an incoming electron with kinetic energy E_{kin} has sufficient energy to eject a second electron when $-|I_{p2}| + E_{kin} > 2V_b = -4\sqrt{2\varepsilon_0}$. It is then expected that this formula should give the correct threshold intensity by replacing E_{kin} with the maximum kinetic energy of the incoming electron $3.17U_p$ [52].

However, the barrier height varies with the angle to the polarization axis (e.g., the barrier in other directions will be higher). To include this lateral effect, we need to consider the full 3D Schrödinger equation of an atom in a uniform field. Using atomic units and taking the nuclear charge to be two, we then have $(\nabla^2/2 + E + 2/r - \varepsilon_0 z)\psi = 0$ [28]. This allows separation of the variables in parabolic coordinates: $\xi = r + z$, $\eta = r - z$, $\phi = \arctan(y/x)$, and the eigen wave function is of the form $\psi = \frac{\chi_1(\xi)\chi_2(\eta)}{\sqrt{\xi\eta}}$. Variable separation leads to two equations: $\frac{1}{2}\frac{d^2\chi_1}{d\xi^2} + [\frac{E}{4} - U_1(\xi)]\chi_1 = 0$ and $\frac{1}{2}\frac{d^2\chi_2}{d\eta^2} + [\frac{E}{4} - U_2(\eta)]\chi_2 = 0$. Each of the equations takes the form of a 1D Schrödinger equation, where the effective energy is $E/4$ and the effective potentials take the form $U_1(\xi) = -\frac{\beta_1}{2\xi} - \frac{1}{8\xi^2} + \frac{\varepsilon_0\xi}{8}$ and $U_2(\eta) = -\frac{\beta_2}{2\eta} - \frac{1}{8\eta^2} - \frac{\varepsilon_0\eta}{8}$, respectively. Note that the separation constants fulfill $\beta_1 + \beta_2 = 2$, which equals the nuclear charge. Along the ξ coordinate the state is bounded while there exists a potential barrier along the η coordinate; the ionization of the electron from the atom in the direction $z \to -\infty$ corresponds to its passage into the region of large η. Approximately we take $\beta_1 = \beta_2 = 1$ [53]. The maximum of the barrier is estimated to be $V_b \simeq -\sqrt{\varepsilon_0}/2$.

On the other side, after the recollision, the kinetic energy of the returned electron will be shared by the struck one (i.e., $E_r + E_s \simeq 3.17U_p - |I_{p2}|$, where E_r and E_s denote the energy of returned electron and struck electron after recollision, respectively). Thus, single-recollision-induced DI becomes possible only when both electrons can pass over the suppressed barrier [i.e., $\min(E_r/4, E_s/4) \geq V_b$]. Obviously, the solution of this inequality depends on how the energy is shared by the electrons after recollision. We thus introduce the asymmetry parameter $\kappa = |(E_r - E_s)/(E_r + E_s)|$ to characterize the energy apportion. We consider the regime that $3.17U_p - |I_{p2}| < 0$ and both electrons populate bound states after

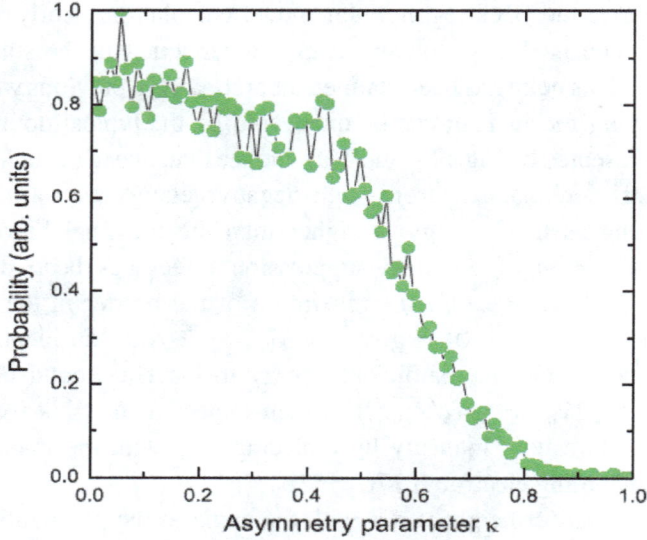

Fig. 8. Statistics on distribution probability with respect to the asymmetry parameter. The trajectories with the maximum returned energy $\simeq 3.17U_p$ are selected out to calculate the energy apportion after recollision. The laser parameters are 7×10^{13} W/cm^2 and 800 nm.

the recollision. $\kappa = 0$ corresponds to equal energy sharing while $\kappa = 1$ corresponds to the most uneven energy apportion. For a given κ the threshold intensity will be determined through following equation:

$$3.17U_p - |I_{p2}| \simeq -4/(1+\kappa)\sqrt{\varepsilon_0}, \quad 0 \leq \kappa \leq 1. \tag{1}$$

In practice, the energy is usually not equally allotted to the returned electron and struck electron after recollision. Numerically, we have compiled statistics on the distribution probability with respect to the asymmetry parameter κ in Fig. 8, where the parameters are chosen close to the threshold. It shows that the distribution function has a broad peak around zero with a long tail and decreases to zero at $\kappa = 1$. According to our calculation, the main profile of the distribution function is general while some detailed structures vary with the field parameters and the atomic structure.

In general, Eq. (1) defines a narrow transition zone whose upper boundary corresponds to $\kappa = 1$ and lower boundary corresponds to $\kappa = 0$ in the plot of intensity versus wavelength as shown in Fig. 9. Above the zone, the DI events are mainly induced by single recollision and therefore the side-by-side emission dominates, whereas, below the zone, single-recollision-induced DI is forbidden and the anticorrelation becomes prominent. The transition from correlation to anticorrelation is expected to emerge in the transition zone.

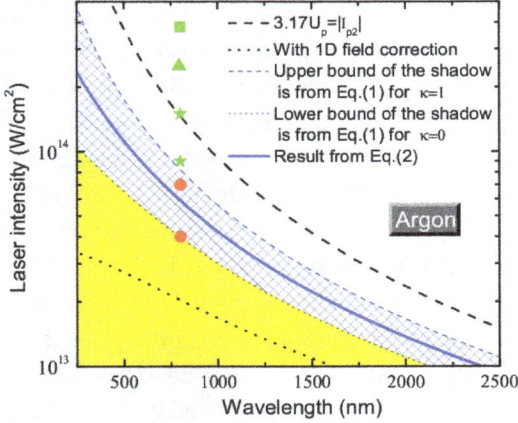

Fig. 9. Threshold intensity vs laser wavelength. Comparisons are made between various theories and experimental data taken from the literature: Ref. [6] (■), Ref. [45] (▲), Ref. [52] (⋆), and Ref. [14] (●). Green (light gray) and red (dark gray) indicate that the observed pattern is correlation and anticorrelation, respectively.

For practical application, we simplify our discussion by taking a simple algebraic average of the prefactor on the right-hand of Eq. (1) for the two limiting cases $\kappa = 0$ and 1; this gives an approximate threshold field through the following self-consistent equation:

$$3.17 \left(\frac{\varepsilon_{\text{th}}^2}{4\omega^2} \right) - |I_{\text{p2}}| \simeq -3\sqrt{\varepsilon_{\text{th}}}. \tag{2}$$

The threshold field ε_{th} could approximately demarcate the transition from correlation-dominated emission to a regime where anticorrelation plays an important and finally even dominant role. It is independent of the asymmetry parameter κ, and therefore it is general and feasible for the DI analysis of other helium-like atoms.

Our theory has been compared with existing data as shown in Fig. 9. Obviously, our theory is consistent with the experimental data to a large extent, while the simple criterion $3.17U_{\text{p}} = |I_{\text{p2}}|$ overestimates the threshold and the model with 1D field correction underestimates the threshold.

NSDI below the recollision threshold has attracted considerable attention during the past few years. In several experiments claims were made that this intriguing regime had been reached [35, 44, 52, 54]; however, the dominance of back-to-back emission was not observed. That is because, according to our present understanding, the laser intensities used did not penetrate into the BRT regime, whose boundary had shifted downward to a lower intensity regime due to the field suppression effect. This is also the reason why quantum calculations made by directly solving the Schrödinger equations also did not reveal any apparent anticorrelation [35, 49]. On the other hand, from a first intuitive glance, it is believed that BRT is a situation where classical considerations no longer hold [14]. However, a recent purely classical calculation did reproduce the observed feature of anticorrelation [48]. This raises a controversy as to whether the anticorrelation is caused by an effect that is inherent to quantum mechanics or can be captured by classical mechanics as well. The contradiction is eliminated by the quantitative calculations carried out by our semiclassical model that involves both the quantum effect and classical rescattering. We show that both the RIET (mainly quantum) and RIDI (mainly classical) mechanisms contribute to the NSDI at the transition to the BRT regime and that the latter is more important, contributing more than two-thirds to the total DI events. On the other hand, our calculations on the RIET mechanism contradict the simple picture proposed in Ref. [45], in which it is claimed that NSDI events caused by RIET should be averagely distributed among the four quadrants.

In conclusion, we have investigated the strong-field double ionization at the transition to below the recollision threshold using a semiclassical approach. It is found that the two-electron momentum distribution changes from correlation to anticorrelation and the fingerlike structure in the correlated momentum spectra fades away. Both RIET and multiple recollisions are responsible for the transition and are found to leave distinct footprints on the correlated momentum spectra. In addition, our model calculations reveal that the striking anticorrelation phenomenon is closely related to the relative perpendicular momentum distribution: Those electron pairs with small relative perpendicular momentum are mainly responsible for the anticorrelation. On the other hand, we claim that the appearance of the anticorrelated pattern signals the transition to the BRT regime. Discussion of the threshold intensity is made.

5. Atomic Species Dependence of NSDI

5.1. *Comparative Study on NSDI of Neon and Argon*

In this subsection, we investigate the target-species dependence of electron correlation in NSDI with comparative study on Ne and Ar. In kinematically complete measurements one find substantially different correlation behavior in the two-electron momentum distributions along the polarization direction [15]. For Ar emission into opposite hemispheres dominates ("anticorrelation," ACO) whereas for Ne the electrons predominantly emerge into the same hemisphere ("correlation," CO). Both can be partly explained within three-dimensional (3D) classical calculations including quantum effects, like tunneling of the second electron. Inspecting the transverse momenta we shed light on the dynamics, investigate Coulomb focusing effects, find nonclassical dynamics, and make predictions about threshold intensities, as well as CO- and ACO-dominated regimes.

In experiment, our collaborators from Max-Planck Institute for Nuclear Physics [15] recorded a total of 1780 double coincidence events (ion and one electron) for Ne^{2+} with a ratio of $Ne^{2+}/Ne^+ \sim 1.5 \times 10^{-4}$ at an intensity of 1.5×10^{14} W/cm^2 and 1160 counts for Ar at 3×10^{13} W/cm^2 with $Ar^{2+}/Ar^+ \sim 8 \times 10^{-4}$. To possibly even further penetrate into the multi-photon ionization (MPI) regime they continued to decrease the laser intensity to about 1.5×10^{13} W/cm^2 (Ar) and 8×10^{13} W/cm^2 (Ne). However, they could not find any indication of DI events while running the measurements for several weeks for each species. This means that DI rates decrease by at least 3 orders of magnitude between $1.5 \times 10^{14} (3 \times 10^{13})$ W/cm^2 and $8 \times 10^{13} (1.5 \times 10^{13})$ W/cm^2 for neon (argon).

In Fig. 10 we present $P_{||}$-correlation plots between both electrons (left column) and electron transverse momentum ($P_\perp$) distributions (right column) at intensities of 3×10^{13} W/cm^2 (Ar, upper row) and 1.5×10^{14} W/cm^2 (Ne, lower row), i.e., at the lowest intensities ever investigated. Even though the absolute single ionization count rate is rather comparable for those targets, both spectra, the correlation maps

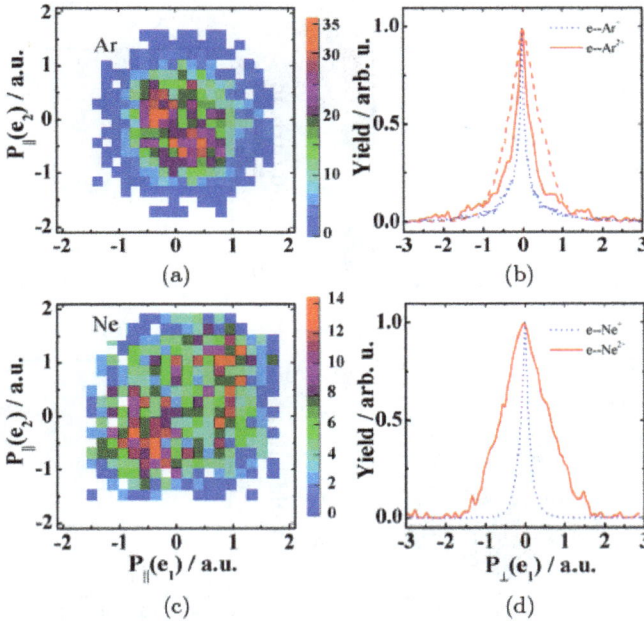

Fig. 10. Parallel momentum correlation of two electrons for argon (a) $(3 \times 10^{13}$ W/cm^2) and neon (c) $(1.5 \times 10^{14}$ W/cm^2). (b) Transverse momentum distributions of electrons in coincidence with Ar$^+$ (dotted line) and Ar^{2+} at 3×10^{13} W/cm^2 (solid line) and at 7×10^{13} W/cm^2 (dashed solid line) as well as (d) for Ne$^+$ and Ne^{2+} at 1.5×10^{14} W/cm^2.

as well as $P_\perp$ distributions show distinctly different shapes. Whereas for Ar a substantial amount of anticorrelated electrons are observed in line with previous measurements at slightly higher intensity [14] correlated emission is found to dominate in case of Ne. Differences in the dynamics become especially obvious if one inspects the transverse momenta of electrons coincident to double ionization (solid curve, right row) with a cusplike structure for Ar that is substantially broadened in the case of Ne. Even taking into account an at least 30% uncertainty in the absolute determination of the intensity and, thus, thinking we might be deeper in the MPI regime for Ar than for Ne, we still find similarly distinct differences if we compare the Ne spectrum to the one for Ar at 7×10^{13} W/cm^2 [dashed line in Fig. 10(b)].

The striking experimental structure dependence can be partially explained by our semiclassical model, in which the RIET effect has been taken into account within the Wentzel-Kramers-Brillouin (WKB) approach [25]. The energy level of the first excitation state of Ne$^+$ and the field-free DI threshold are both significantly higher than those of Ar$^+$. It is thus much more difficult to resonantly excite the second electron during recollision in a correct quantum-mechanical description. To take this effect approximately into account in our classical model, we have abandoned all DI trajectories for Ne if the inner electron is excited to an energy that is lower than the first excited state. Numerical results are shown in Figs. 11(a) and 11(c)

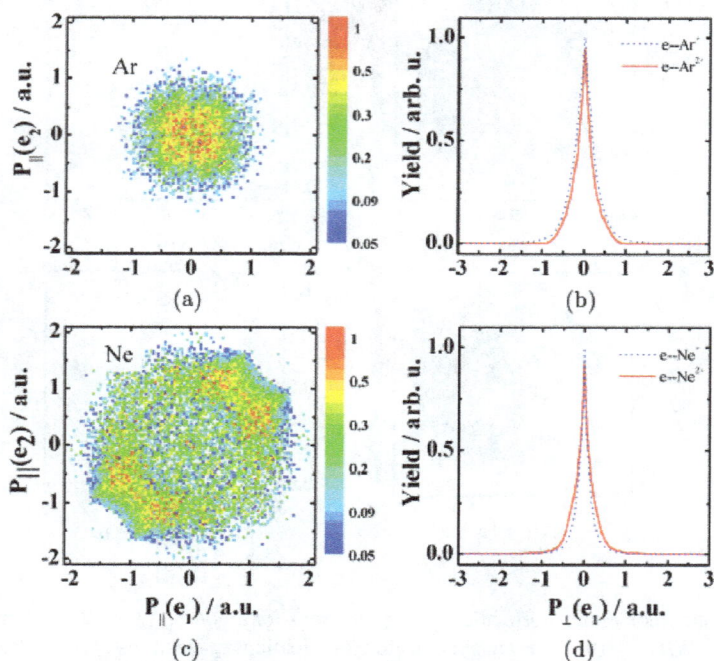

Fig. 11. Same as Fig. 10 but for theoretical calculations (see text).

and the correlated parallel momentum distributions for Ar and Ne are in quite good qualitative agreement with the observations, exhibiting ACO for Ar and CO for Ne, respectively.

Inspecting the electron transverse momentum distributions for single ionization and DI shown in Figs. 10(b) and 10(d) for the experiment as well as in Figs. 11(b) and 11(d) for our model calculations we find in general pronounced "cusplike" structures as observed before. It has been proposed that Coulomb focusing [55] or the Coulomb singularity [56] are responsible for their formation for single ionization in accordance with the present data [dotted curves in Figs. 10(b), 10(d), 11(b), and 11(d)]. Whereas the experimental momentum distribution of electrons coincident with Ar^{2+} ions (solid line) also shows a clear cusplike shape, only slightly broader than for single ionization and in good agreement with theory the respective distribution for Ne is significantly broader than predicted. In general, the cusplike shape is obtained as a result of the singularity in the electron continuum wave function at zero energy in the presence of a Coulomb potential [55]. Thus, we conclude that for Ar double ionization both electrons have very little energy in the final state. This is expected within the RIET picture, where the second electron tunnels, being essentially set into the continuum with zero momentum, independently from the first electron. Accordingly, our model calculations [Fig. 11(b)] largely reproduce the experimental data in Fig. 10(b).

On the other hand, electrons emerging coincident to Ne^{2+} production are observed to receive much larger transverse momenta and show a more Gaussian-like shape. This indicates that the recollision mechanism still dominates, where the recolliding electron could be scattered to a larger angle with its longitudinal momentum being transferred to the transverse direction. Surprisingly, such a feature is not found in our model calculations where the transverse momentum distribution of an electron coincident to the simulated Ne^{2+} ion, excluding trajectories that lead to unphysical low excitation energies, is shown in Fig. 11(d). Obviously such a model does not capture the full quantum dynamics of the system in all of its aspects correctly, underlining the quest for quantum calculations.

Finally, we investigate and predict the threshold behavior. In the same spirit of Sec. 4, we estimate a criterion $3.17U_p - |I_e(t)| = -2\sqrt{\varepsilon_0}$ where DI cutoff should occur, where I_e is the first excitation energy of the parent ion. Below that threshold, the maximum recollision energy is not enough to excite the bounded electron to at least the first excited state while the returning electron keeps just the minimum energy to escape. Therefore, DI is expected to decrease sharply. The results are plotted in Fig. 12 for both Ar and Ne together with the experimental data finding overall agreement. Especially, the ACO zone predicted by our theory for Ne is extremely narrow and, thus, might hardly be observable in experiments.

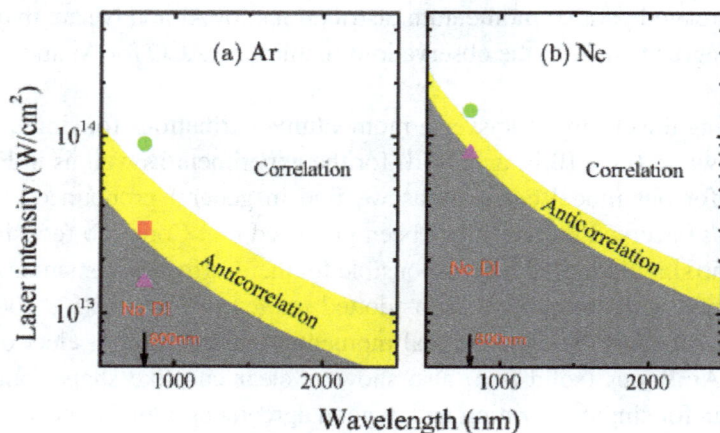

Fig. 12. Predicted areas where correlated or anticorrelated electron emission dominates as well as the boundary below which DI should be forbidden at all. Points indicate the present measurement where correlation (circle), anticorrelation (square), and no DI (triangle) have been observed, respectively.

In conclusion, we have presented kinematically complete measurements and model calculations for single and double ionization of Ar and Ne in infrared laser fields in the DI threshold regime, at the lowest intensities ever investigated. Whereas for Ar anticorrelation in the longitudinal two-electron momentum spectra is observed in agreement with previous measurements at slightly higher intensity, correlation dominates for the Ne target similarly close to the threshold. Further lowering the intensity led to the complete absence of any double ionization event over weeks of data taking time. We can reproduce that salient feature in the correlation plot with 3D classical calculations including tunneling for the second electron if, for the Ne target, all trajectories are rejected with recollision energies below the first excitation level of Ne^+. This provides strong evidence that the target structure, i.e., the low-lying excited levels in Ar^+ along with the large electron-impact excitation cross section are the reason for this observation. We have further made predictions on the regimes where one would expect correlation and anticorrelation behavior for both targets and provide a classical threshold below which DI should not occur in good agreement with the experimental results. From that it might be concluded (i) that we have really approached the DI threshold and (ii) that it might be impossible, at least very difficult to observe ACO for the Ne target since the allowed intensity range is very small.

Inspecting the transverse momentum distribution of electrons coincident with singly and doubly charged Ne and Ar ions, we shed light on the dynamics and provide clear evidence that our essentially classical calculations do not capture all of the dynamics: Whereas the cusplike behavior for single ionization is in good agreement with the experimental data, the experimental transverse electron momentum distribution for Ne^{2+} is significantly broader as predicted pointing to the existence of correlated quantum motion not taken into account by the classical model and underlining the quest for many-particle quantum approaches.

5.2. *New Mechanisms of Xenon NSDI*

Fully differential studies so far have been concentrated just on He, Ne, and Ar atoms. Until now, there has been no fully differential measurement on strong-field double ionization on high-Z atomic rare gases. If dating back the history of NSDI, the "knee" structure on the curve of the doubly charged ion yields vs laser intensity was first observed on Xe atoms [57]. Based on the measurements, several models, e.g., direct two-electron ejection [58], a high-order sequential mechanism [59], and a shake-off process during tunneling [60], were proposed. By measuring the photoelectron angular distribution of single ionization of Xe with the intensity dependence of the yields of double ionization, it was found that the nonresonant ionization process plays a role in the formation of doubly charged xenon ions [61]. By using the electron-ion coincident time-of-flight technique [62], the similarity of electron spectra from single ionization and double ionization of Xe was observed, which was attributed to the field-independent resonant excitation process [63]. Including the rescattering effect, the resonant model was subsequently invoked to explain the wavelength and intensity dependence of strong-field double ionization of Xe [64, 65]. However, a major question about electron correlation dynamics itself for Xe has not been explored. This is the most direct approach toward distinguishing the mechanisms of NSDI and thus learning about their characteristics and production process.

In this subsection, we perform semiclassical simulation to explore the physics behind the fully differential measurement on strong-field double ionization of Xe atoms. The experiment was recently performed at Peking university with near-infrared fields over a wide range of laser intensities ($4 \times 10^{13} \sim 3 \times 10^{14}$ W/cm^2) [16]. Several striking observations characterize this case when decreasing the laser intensity from sequential double ionization (SDI) to NSDI of Xe: (i) the momentum distributions of doubly charged ions always show a single peak distribution (Gaussian-like); (ii) the characteristic correlation momentum spectra of NSDI, i.e., "side-by-side emission" or "back-to-back emission", essentially disappear; (iii) the electron energy cutoffs of double ionization increase from 2.9Up to 7.8Up, decreasing the laser intensity from SDI to NSDI. The findings show that multiple rescatterings and electron mutual interaction in the final states contribute to the enhancement of double ionization of Xe atoms. Using the semiclassical two-electron

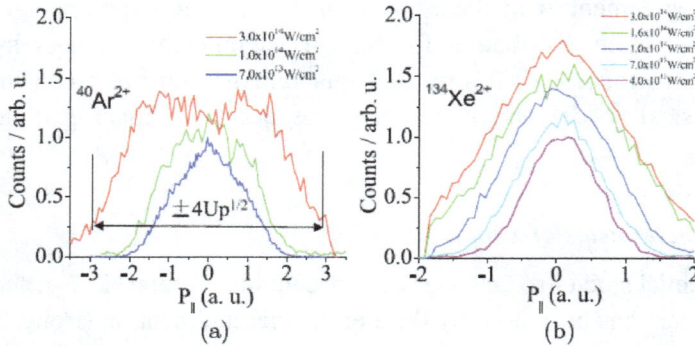

Fig. 13. The longitudinal momentum distributions of (a) ^{40}Ar^{2+} and (b) ^{134}Xe^{2+} at intensities of $0.4 \times 10^{14} - 3 \times 10^{14}$ W/cm^2 (25 fs, 790 nm).

atomic ensemble model, we have studied the electron shielding effect on strong-field ionization of high-Z atoms.

In Fig. 13(b), we present the longitudinal momentum distributions of ^{134}Xe^{2+} decreasing from 3×10^{14} W/cm^2 to 4×10^{13} W/cm^2. The width (FWHM) of the momentum distributions increases with the laser intensity, and it will be saturated above the intensity of 2×10^{14} W/cm^2. Interestingly, the longitudinal momentum distributions of ^{134}Xe^{2+} do not reveal the typical double-hump structure in the NSDI regime, and they always show the Gaussian-like distribution with the maximum at zero momentum over a wide range of laser intensities. This feature is similar to the longitudinal momenta of Ar^{2+} below the recollision threshold, i.e., 7×10^{13} W/cm^2. One may expect that the electron correlation behavior of strong-field double ionization of Xe could be similar with that of Ar below the recollision threshold, i.e., a dominant back-to-back emission in the laser polarization plane.

This straightforward picture, however, contradicts to further measurements on electron correlation spectra as shown in Fig. 14. Here, the momentum of electron "two" was calculated from those of electron "one" and of the ^{134}Xe^{2+} ion by the momentum conservation. The correlation momentum spectra reveal the typical sequential double ionization behavior at all laser intensities. The largest electron momentum probability of double ionization is populated around $[P_{||}(e_1), P_{||}(e_2)]$ $= (0, 0)$. One can easily understand the correlation spectrum in the SDI regime, i.e., at an intensity of 3×10^{14} W/cm^2 because the two electrons are released sequentially [6]. However, compared with other targets, i.e., He, Ne, and Ar, the electron correlation spectra are much different from other targets in the NSDI regime. Below the recollision threshold, the dominant back-to-back emission phenomenon was observed for Ar. However, the signature of both the side-by-side emission and back-to-back emission essentially disappears for strong-field double ionization of

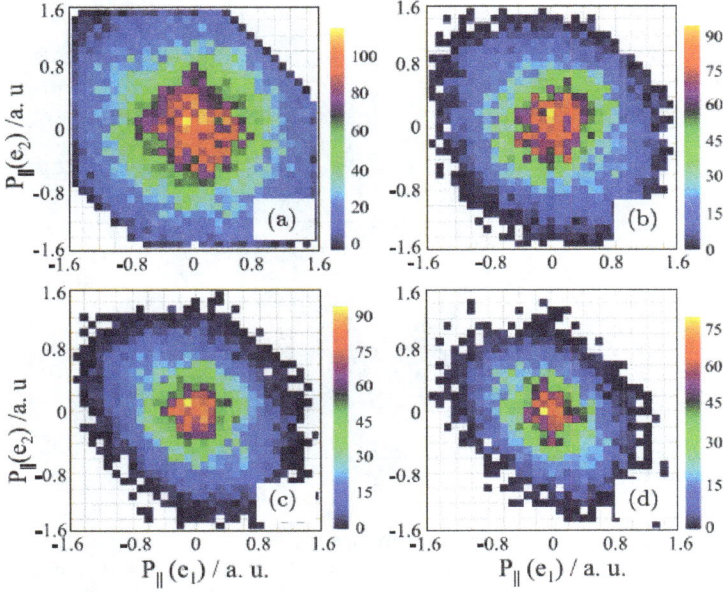

Fig. 14. The measured correlation momentum spectra of double ionization Xe at intensities of (a) 3×10^{14}, (b) 1.6×10^{14}, (c) 7×10^{13}, and (d) 4×10^{13} W/cm^2 (25 fs, 790 nm).

Xe. The correlation momentum spectra of Xe show a sequential feature over a wide of laser intensity, which suggests very different ionization dynamics from that of low-Z atoms.

In order to include all mechanisms naturally, we have performed a 3D semiclassical two-electron atomic ensemble calculation. This model has been rigorously tested and can work very well for strong-field double ionization of low-Z rare gas atoms. In the calculation, the first and second ionization potentials of the two-electron atom are chosen to match the Xe atom. As seen in Figs. 15(a) and 15(b), the simulations on Xe look more like the electron correlation dynamics of the low-Z rare gas atoms, i.e., He, Ne, and Ar [6, 12–15].

Strong-field double ionization has been usually studied by using the heliumlike model atom within the two-active-electron approximation. However, for those high-Z atoms, the inner electrons will leave a non-negligible influence on the strong-field double ionization. In atoms with many electrons, the outermost electrons are strongly shielded or screened from the nucleus by the inner electrons, known as the shielding or screening effect. The outermost electrons do not feel the complete charge of the nucleus. When the inner electrons are more, the screening effect will be larger. We further improved the two-electron semiclassical model with the screening potential, $V_{ne}^{i} = -[2 + (Z - 2)\Omega(r_i)]/r_i$, where $\Omega(r_i) = 1/[(\rho/\zeta)(e^{\zeta r_i} - 1) + 1]$ (ρ and ζ are the potential parameters) [66]. As seen in Figs. 15(c) and 15(d), by considering the

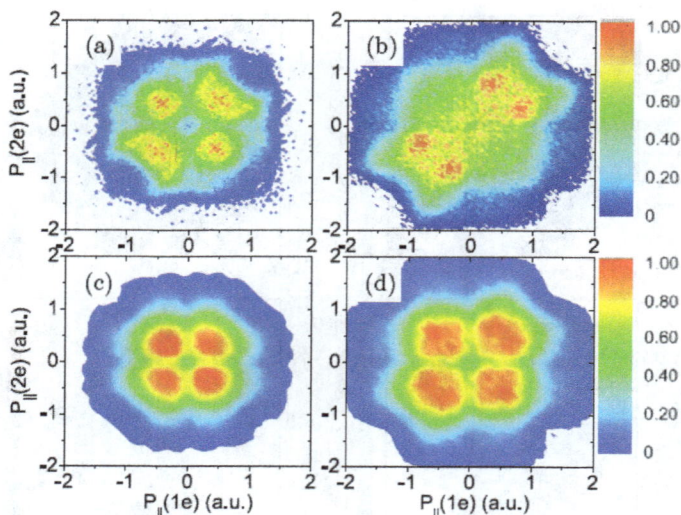

Fig. 15. The calculated correlation momentum spectra of double ionization Xe at intensities of 0.6×10^{14} (a) and 1.1×10^{14} W/cm^2 (b) with the Coulomb potential. The calculated correlation momentum spectra of double ionization Xe at intensities of 0.6×10^{14} (c) and 1.1×10^{14} W/cm^2 (d) using the shielding potential.

shielding effect, the electron correlation momenta are equally distributed in all four quadrants over a wide range of laser intensity and agree well with the measurement. The minimum structure along the coordinate axis is observed in the correlation momentum spectra of theoretical calculation. This is because, in the calculation, we have to stop the evolution of electrons when the laser field switches off to save computer time. The yields of the near-zero-momentum electron will decrease slightly.

In summary, recent experiment has comprehensively measured the strong-field double ionization of Xe atoms and revealed distinctively different behavior with respect to low-Z atoms. Although the curves of the integrated ion yields vs the laser intensity of all rare gas atoms show the general knee structure, the dramatically different electron correlation behavior of strong-field double ionization of Xe atoms was observed, as compared with other targets, i.e., He, Ne, and Ar. The pronounced double-hump structure that was observed in the longitudinal momentum distributions of doubly charged ions of low-Z atomic targets disappears for double ionization of Xe. Lacking the signature of the side-by-side and back-to-back emission, the electron momentum spectra reveal the remarkable non-correlation behavior. The fully differential data provide the most direct experimental evidence that the enhanced ionization rate of Xe^{2+} is due to the correlated high-order sequential process (sequential ionization from the excited states). We have shown that the recollision-induced excitation and the shielding effect are very important for strong-field double ionization of high-Z atoms.

Acknowledgments

This work is based on some of our published papers during the past 6 years. We acknowledge the contribution from all of our collaborators, with special emphasis on the invaluable discussions with Y. Liu, J. Chen, X. Liu, and J. Ullrich. We also thank the financial support by CAEP-FESTC under Grant No. J2014-0401-03.

References

1. I. S. Aleksakhin, I. P. Zapesochnyi and V. V. Suran, *JETP Lett.* **26**, 11 (1977).
2. P. B. Corkum, *Phys. Rev. Lett.* **71**, 1994 (1993).
3. B. Walker, B. Sheehy, L. F. DiMauro *et al.*, *Phys. Rev. Lett.* **73**, 1227 (1994).
4. Th. Weber, M. Weckenbrock, A. Staudte *et al.*, *Phys. Rev. Lett.* **84**, 443 (2000).
5. R. Moshammer, B. Feuerstein, W. Schmitt *et al.*, *Phys. Rev. Lett.* **84**, 447 (2000).
6. Th. Weber, H. Giessen, M. Weckenbrock *et al.*, *Nature* **405**, 658 (2000).
7. N. Camus, B. Fischer, M. Kremer *et al.*, *Phys. Rev. Lett.* **108**, 073003 (2012).
8. B. Bergues, M. Kübel, N. G. Johnson *et al.*, *Nat. Commun.* **3**, 813 (2012).
9. S. X. Hu, *Phys. Rev. Lett.* **111**, 123003 (2013).
10. Li Zhang, Xinhua Xie, Stefan Roither *et al.*, *Phys. Rev. Lett.* **112**, 193002 (2014).
11. J. Ullrich, R. Moshammer, A. Dorn *et al.*, *Rep. Prog. Phys.* **66**, 1463 (2003).
12. A. Staudte, C. Ruiz, M. Schöffler *et al.*, *Phys. Rev. Lett.* **99**, 263002 (2007).
13. A. Rudenko, V. L. B. de Jesus, Th. Ergler *et al.*, *Phys. Rev. Lett.* **99**, 263003 (2007).
14. Y. Q. Liu, S. Tschuch, A. Rudenko *et al.*, *Phys. Rev. Lett.* **101**, 053001 (2008).
15. Y. Q. Liu, Difa Ye, Jie Liu *et al.*, *Phys. Rev. Lett.* **104**, 173002 (2010).
16. X. F. Sun, Min Li, Difa Ye *et al.*, *Phys. Rev. Lett.* **113**, 103001 (2014).
17. W. Becker, X. Liu, P. Jo Ho and J. H. Eberly, *Rev. Mod. Phys.* **84**, 1011 (2012).
18. J. Liu, S. G. Chen and D. H. Bao, *Acta Phys. Sin.* (Overseas Eidtion) **4**, 881 (1995).
19. J. Liu and S. G. Chen, *Commun. Theor. Phys.* **28**, 475 (1997).
20. B. Hu, J. Liu and S. G. Chen, *Phys. Lett. A* **236**, 533 (1997).
21. J. Chen, J. Liu and S.G. Chen, *Phys. Rev. A* **61**, 033402(R) (2000).
22. L. B. Fu, J. Liu, J. Chen and S. G. Chen, *Phys. Rev. A* **63**, 043416 (2001).
23. L. B. Fu, J. Liu and S. G. Chen, *Phys. Rev. A* **65**, 021406(R) (2002).
24. J. Liu, D. F. Ye, J. Chen and X. Liu, *Phys. Rev. Lett.* **99**, 013003 (2007).
25. D. F. Ye and J. Liu, *Phys. Rev. A* **81**, 043402 (2010).
26. D. F. Ye, G. G. Xin, J. Liu and X. T. He, *J. Phys. B* **43**, 235601 (2010).
27. M. V. Ammosov, N. B. Delone and V. P. Krainov, *Zh. Eksp. Teor. Fiz.* **91**, 2008 (1986) [*Sov. Phys. JETP* **64**, 1191 (1986)].
28. L. D. Landau and E. M. Lifshitz, *Quantum Mechanics*, Pergamon, New York (1977).
29. N. B. Delone and V. P. Krainov, *J. Opt. Soc. Am. B* **8**, 1207 (1991).
30. J. G. Leopold and I. C. Percival, *J. Phys. B* **12**, 709 (1979).
31. J. S. Cohen, *Phys. Rev. A* **64**, 043412 (2001).
32. C. F. de Morisson Faria, H. Schomerus, X. Liu and W. Becker, *Phys. Rev. A* **69**, 043405 (2004).
33. C. F. de Morisson Faria, X. Liu, W. Becker and H. Schomerus, *Phys. Rev. A* **69**, 021402(R) (2004).
34. C. F. de Morisson Faria and M. Lewenstein, *J. Phys. B* **38**, 3251 (2005).
35. J. S. Parker, B. J. S. Doherty, K. T. Taylor *et al.*, *Phys. Rev. Lett.* **96**, 133001 (2006).

36. D. F. Ye, X. Liu and J. Liu, *Phys. Rev. Lett.* **101**, 233003 (2008).
37. D. Milosevic and W. Becker, *Phys. Rev. A* **68**, 065401 (2003).
38. J. Berakdar, J. Röder, J. S. Briggs and H. Ehrhardt, *J. Phys. B* **29**, 6203 (1996).
39. C. Ruiz, L. Plaja, L. Roso and A. Becker, *Phys. Rev. Lett.* **96**, 053001 (2006).
40. P. J. Ho, R. Panfili, S. L. Haan and J. H. Eberly, *Phys. Rev. Lett.* **94**, 093002 (2005).
41. S. L. Haan, L. Breen, A. Karim and J. H. Eberly, *Phys. Rev. Lett.* **97**, 103008 (2006).
42. The recollision time is defined as the instant when the two electrons get closest, while the DI time as the instant when both electron energies become greater than zero and never return to negative.
43. G. G. Paulus, W. Becker, W. Nicklich and H. Walther, *J. Phys. B* **27**, L703 (1994).
44. M. Weckenbrock, D. Zeidler, A. Staudte *et al.*, *Phys. Rev. Lett.* **92**, 213002 (2004).
45. B. Feuerstein, R. Moshammer, D. Fischer *et al.*, *Phys. Rev. Lett.* **87**, 043003 (2001).
46. X. Liu, C. Figueira de Morisson Faria, W. Becker and P. B. Corkum, *J. Phys. B* **39**, L305 (2006).
47. M. Lein, E. K. U. Gross and V. Engel, *Phys. Rev. Lett.* **85**, 4707 (2000).
48. S. L. Haan, Z. S. Smith, K. N. Shomsky and P. W. Plantinga, *J. Phys. B* **41**, 211002 (2008).
49. D. I. Bondar, W.-K. Liu and M. Yu. Ivanov, *Phys. Rev. A* **79**, 023417 (2009).
50. The energy shown in Fig. 5 is the sum of kinetic energy and electron-nuclear Coulomb energy. It does not include the electron-field interaction energy. Thus, there exists an energy jump $(\simeq |\varepsilon(t)[z_i^{out} - z_i^{in}]|)$ when the electron tunnels out of the Coulomb barrier. If we include the interaction energy, the total electron energy should be conserved during the tunneling process.
51. H. W. van der Hart and K. Burnett, *Phys. Rev. A* **62**, 013407 (2000).
52. E. Eremina, X. Liu, H. Rottke *et al.*, *J. Phys. B* **36**, 3269 (2003).
53. In analogy to the discussion in p. 293 of Ref. [28], the dependence of wave function ψ on variable ξ can be regarded as being of the same form as that when the external field is absent. It gives $\beta_1 = \beta_2 = 1$.
54. D. Zeidler, A. Staudte, A. B. Bardon *et al.*, *Phys. Rev. Lett.* **95**, 203003 (2005).
55. D. Comtois, D. Zeidler, H. Pépin *et al.*, *J. Phys. B* **38**, 1923 (2005).
56. A. Rudenko, K. Zrost, Th Ergler *et al.*, *J. Phys. B* **38**, L191 (2005).
57. A. L'Huillier, L. A. Lompre, G. Mainfray and C. Manus, *Phys. Rev. A* **27**, 2503 (1983).
58. D. N. Fittinghoff, B. Yang, L. F. DiMauro and K. C. Kulander, *Phys. Rev. Lett.* **69**, 2642 (1992).
59. D. Charalambidis, P. Lambropoulos, H. Schröder *et al.*, *Phys. Rev. A* **50**, 2822(R) (1994).
60. A. Talebpour, C.-Y. Chien, Y. Liang *et al.*, *J. Phys. B* **30**, 1721 (1997).
61. R. Wiehle and B. Witzel, *Phys. Rev. Lett.* **89**, 223002 (2002).
62. J. L. Chaloupka, J. Rudati, R. Lafon *et al.*, *Phys. Rev. Lett.* **90**, 033002 (2003).
63. J. Rudati, J. L. Chaloupka, P. Agostini *et al.*, *Phys. Rev. Lett.* **92**, 203001 (2004).
64. P. Kaminski, R. Wiehle, W. Kamke *et al.*, *Phys. Rev. A* **73**, 013413 (2006).
65. G. Gingras, A. Tripathi and B. Witzel, *Phys. Rev. Lett.* **103**, 173001 (2009).
66. R. H. Garvey, C. H. Jackman and A. E. S. Green, *Phys. Rev. A* **12**, 1144 (1975).

Chapter 5

Few-Photon Double Ionization of He and H_2

Wei-Chao Jiang*, Wei-Hao Xiong*, Ji-Wei Geng*, Qihuang Gong*,†,
and Liang-You Peng*,†,‡

*Department of Physics and State Key Laboratory for Mesoscopic Physics,
Peking University, Beijing 100871, China
†Collaborative Innovation Center of Quantum Matter,
Beijing 100871, China
‡liangyou.peng@pku.edu.cn

Studies on the few-photon double ionization of helium atom and hydrogen molecule have fundamental significances on the electronic correlation. In recent years, many theoretical efforts have been put on the numerical solution to the corresponding time-dependent Schrödinger equation (TDSE) in the real space. These *ab initio* calculations promise to produce accurate numerical results that can be directly compared with the experimental observations. With further developments of free electron lasers (FEL) and high harmonic generation (HHG) sources and relevant coincidence measurement techniques, these theoretical predictions have a greater chance to be experimentally checked. In this chapter, we will first introduce our numerical methodologies to solve the TDSE and then provide some numerical results for the few-photon double ionization of helium atom and hydrogen molecule. At the same time, existing experimental results will be reviewed briefly.

1. Introduction

Double ionization of atoms or molecules is a typical few-body coulomb problem in quantum mechanics, which has fundamental interests and has been widely studied for a long time. The basic approach to handle this problem is to solve the two-electron time-dependent Schrödinger equation (TDSE). In the past decades, a large number of accurate numerical methods have been developed to numerically solve the TDSE. These theoretical interests are also promoted by the experimental progresses in the various extreme ultraviolet (XUV) laser sources [1], such as free electron lasers (FEL) and high harmonic generation (HHG), which can make those theoretical predications experimentally observable. Great achievements have already been made under the cooperative efforts between the theoretical and experimental physicists. For example, one well-understood process is the one-photon double ionization of helium, for which theoretical predications and experimental measurements have achieved excellent agreements on the total and differential double ionization cross sections. More complex process, such as two-photon double ionization of helium, is continually attracting great interest.

In this chapter, we will firstly introduce our implementation to numerically solve TDSE of helium atom and H_2 molecule in Sec. 2, and then introduce several applications on particular physical processes in Sec. 3. In Sec. 3.1, the total and differential cross sections are discussed for the two-electron spectra in one-photon or two-photon double ionization. The method to extract recoil-ion momentum spectra in double ionization of helium is also introduced. In Sec. 3.2, the two-electron joint angular distributions are analyzed to look for the electron correlation effects in one- and two-photon double ionization of helium. In Sec. 3.3, the double ionization of helium by the time-delayed attesecond pulses is studied, where the two-electron interference phenomena and relative CEP effects are discussed.

2. Numerical Implementation to Solve Two-Electron TDSE

2.1. *General technologies in solving TDSE*

In this section, we introduce some technologies which are universal in solving TDSEs for different systems. Formally, the TDSE is generally expressed as

$$i\frac{\partial \Psi(\mathbf{r}, t)}{\partial t} = H\Psi(\mathbf{r}, t). \tag{1}$$

The wave function $\Psi(\mathbf{r}, t)$ of the system is the function of space coordinates $\mathbf{r}$ and time coordinate t. Note that, the vector $\mathbf{r}$ represents the space coordinates. To solve the TDSE shown in Eq. (1), one needs some special strategies to carefully treat the first-order partial differential term for time coordinate which appears in the left side of Eq. (1), and the second-order partial differential term for space coordinates which are implicated in the Hamiltonian H in the right side of Eq. (1).

2.1.1. *Time-propagation method*

Formally, the solution of the TDSE can be expressed as

$$\Psi(t) = \hat{T} \exp\left[-i \int_{t_0}^{t} H(t')dt'\right] \Psi(t_0), \qquad (2)$$

where $\hat{T}$ is the time-ordering operator, and $U(t, t_0) = \hat{T} \exp[-i \int_{t_0}^{t} H(t')dt']$ is the time evolution operator. Direct evaluation of the above equation is cumbersome, since the time evolution operator has to be expanded in a Dyson series to represent the time-ordering. However, for small time intervals Δt the Hamiltonian can be assumed to be time-independent, thus

$$U(t + \Delta t, t) \approx \exp[-i H(t) \Delta t], \qquad (3)$$

and

$$\Psi(t + \Delta t) \approx \exp[-i H(t) \Delta t] \Psi(t). \qquad (4)$$

The wave function at arbitrary time t can be obtained by iterating Eq. (4) from the initial wave function at time t_0. Many methods have been put forward for the time propagation of the wave function [2–10]. In the following, we introduce the Lanczos algorithm [7–10] to numerically evaluate Eq. (4).

Lanczos algorithm is related to the Krylov subspace, which is generated by the repeated action of H on a normalized initial state Ψ_0. The set of the basis functions in the Krylov subspace is

$$K_{N+1} = \{|\Psi_0\rangle, H|\Psi_0\rangle, H^2|\Psi_0\rangle, \ldots, H^N|\Psi_0\rangle\}. \qquad (5)$$

A new orthonormal set can be formed using the Gram-Schmidt orthogonalization procedure

$$Q_{N+1} = \{|q_0\rangle, |q_1\rangle, |q_2\rangle, \ldots, |q_N\rangle\}. \qquad (6)$$

Usually N is chosen much smaller than the dimension of the matrix representation of H. Generally, the matrix representation of a non-Hermitian operator in the subspace Q_{N+1} is an upper Hessenberg matrix. The Hamiltonian is a Hermitian operator, for which the matrix representation in the subspace Q_{N+1} is tridiagonal. Employing this character of the Hermitian operator, the procedure to construct the orthonormal subspace Q_{N+1} can be reduced with the following three-term recurrence relation

$$|q_0\rangle = \Psi_0\rangle, \qquad (7)$$

$$\beta_0|q_1\rangle = H|q_0\rangle - \alpha_0|q_0\rangle, \qquad (8)$$

$$\beta_j|q_{j+1}\rangle = H|q_j\rangle - \alpha_j|q_j\rangle - \beta_{j-1}|q_{j-1}\rangle, \qquad (9)$$

where

$$\alpha_j = \langle q_j | H | q_j \rangle, \tag{10}$$

$$\beta_j = \| H | q_j \rangle - \alpha_j | q_j \rangle - \beta_{j-1} q_{j-1} \rangle \|$$

$$= \langle q_{j-1} | H | q_j \rangle = \langle q_{j+1} | H | q_j \rangle. \tag{11}$$

The matrix of the Hamiltonian operator in the subspace Q_{N+1} is represented as

$$H_{ij}^{(Q)} = \langle q_i | H | q_j \rangle = \begin{bmatrix} \alpha_0 & \beta_0 & 0 & \cdots & & 0 \\ \beta_0 & \ddots & \ddots & \ddots & & \vdots \\ 0 & \ddots & \ddots & \ddots & & 0 \\ \vdots & \ddots & \ddots & \ddots & & \beta_{N-1} \\ 0 & \cdots & 0 & \beta_{N-1} & & \alpha_N \end{bmatrix}. \tag{12}$$

Now, the propagation of the wave function is performed in the subspace Q_{N+1}. The eigenvector $|Z_i\rangle$ and the eigenvalue h_i of $H_{ij}^{(Q)}$ can be obtained by direct diagonalization. The propagation operator in the subspace Q_{N+1} is then expressed as

$$U^{(Q)} = \exp(-iH^{(Q)}\Delta t) = \sum_l \exp(-ih_l\Delta t)|Z_l\rangle\langle Z_l|. \tag{13}$$

Applying propagation operator $U^{(Q)}$ to the initial wave function $|\Psi_0\rangle \equiv |\Psi(t)\rangle = |q_0\rangle$, one obtains the propagated wave function in the subspace Q_{N+1},

$$|\Psi(t + \Delta t)\rangle = \sum_l \langle Z_l | q_0 \rangle \exp(-ih_l\Delta t)|Z_l\rangle. \tag{14}$$

Expressed as the superpositions of the basis vector $|q_k\rangle$, the propagated wave function can be reformed as

$$|\Psi(t + \Delta t)\rangle = \sum_{k=0}^{N} a_k |q_k\rangle, \tag{15}$$

where

$$a_k = \sum_l \langle Z_l | q_0 \rangle \exp(-ih_l\Delta t)\langle q_k | Z_l \rangle. \tag{16}$$

The representation of the propagated wave function in the whole space can be obtained by actually calculating the summation in Eq. (15). Since $|q_k\rangle$ are linear combinations of $H^k|\Psi_0\rangle$, Eq. (15) is actually a N-th order polynomial expansion of the exponential in Eq. (4). In contrast to a standard N-th order Taylor or Chebyshev expansion, the unitarity of the propagation operator is conserved.

2.1.2. *Space discretization method*

We now turn to methods to handle the space coordinates. Here we will introduce the finite element discrete variable representation (FEM-DVR) method [11]. FEM-DVR is a basis-set method, in which the wave function is expanded by a set of orthonormal basis functions. The FEM-DVR method has the advantages that the Hamilton matrix in this representation is quite sparse and the number of grid points which are used to represent the wave function can be rather small.

In FEM-DVR methods, the space coordinates are firstly divided into a series of finite elements, the boundaries $x^{(i)}$ of which are marked as

$$x_{\min} \leq x^{(1)} < x^{(2)} < \cdots < x^{(\mathrm{Nelem})} \leq x_{\max}. \tag{17}$$

$x_{\min}$ and $x_{\max}$ are the boundaries of the space coordinate x, and they can be defined according to the particular problem. If x represents a rectangular coordinate, one can choose $x_{\min} = -x_{\max}$. If x represents a radial coordinate, naturally $x_{\min} = 0$, and $x_{\max}$ is usually chosen to be large enough to guarantee that the wave function beyond $x_{\max}$ can be safely treated to be zero, otherwise one needs to use absorbing boundary conditions to avoid nonphysical reflections. In each element, we further define a series of discrete variable representation (DVR) basis functions [12–15]. The DVR method provides a link between analytic basis set methods and the pure grid-based methods. The DVR basis functions in the element $x^{(i)} \leq x \leq x^{(i+1)}$ are defined as,

$$f^{(i,m)}(x) = \frac{L^{(i,m)}(x)}{\sqrt{w^{(i,m)}}}, \ 1 \leq m < N, \tag{18}$$

where the Lagrange interpolation polynomials are defined as

$$L^{(i,m)}(x) = \begin{cases} \displaystyle\prod_{k \neq m} \frac{x - x^{(i,k)}}{x^{(i,m)} - x^{(i,k)}}, & x^{(i)} \leq x^{(i,1)} \leq x \leq x^{(i,N)} \leq x^{(i+1)} \\ 0, & \text{otherwise} \end{cases}. \tag{19}$$

In the above, the index i represents the i-th element; $x^{(i,k)}$ and $w^{(i,k)}$ are the points and weights corresponding to the Gauss quadrature, respectively. Several types of Gauss quadratures are frequently employed. In Gauss–Legendre quadrature, no point coincides with the endpoints of the integral interval. In Gauss–Radau quadrature, one of the endpoints is included as a quadrature point and the other is excluded, i.e. $x^{(i,1)} > x^{(i)}$ and $x^{(i,N)} = x^{(i+1)}$. In Gauss–Lobatto quadrature, two of the points are constrained to coincide with the endpoints, i.e. $x^{(i,1)} = x^{(i)}$ and $x^{(i,N)} = x^{(i+1)}$. In present most cases, the Gauss–Lobatto quadrature is chosen to build FEM-DVR basis functions. In Gauss quadrature, the integral of an arbitrary function $g(x)$ in the interval $x^{(i)} \leq x \leq x^{(i+1)}$ is approximated to be

$$\int_{x^{(i)}}^{x^{(i+1)}} g(x)dx \approx \sum_{k=1}^{N} g(x^{(i,k)})w^{(i,k)}. \tag{20}$$

The weights $w^{(i,k)}$ and points $x^{(i,k)}$ of the Gauss quadrature in an arbitrary integral interval $x^{(i)} \leq x \leq x^{(i+1)}$ can be inferred by the weights $\omega^{(k)}$ and points $t^{(k)}$ of the Gauss quadrature in the integral interval $-1 \leq x \leq 1$,

$$w^{(i,k)} = \frac{x^{(i+1)} - x^{(i)}}{2}\omega^{(k)}, \tag{21}$$

$$x^{(i,k)} = \frac{x^{(i+1)} - x^{(i)}}{2}t^{(k)} + \frac{x^{(i+1)} + x^{(i)}}{2}. \tag{22}$$

To insure continuity of the wave function at the boundaries of two adjacent finite elements, we define the bridge functions as

$$f^{(i,N)}(x) = \frac{L^{(i,N)}(x) + L^{(i+1,1)}(x)}{\sqrt{w^{(i,N)} + w^{(i+1,1)}}}. \tag{23}$$

Those FEM-DVR basis functions defined above are orthonormalized in the sence of Gauss quadrature.

When any function $g(x)$ is expanded by the FEM-DVR basis functions

$$f^{(i,N)}(x) = \frac{L^{(i,N)}(x) + L^{(i+1,1)}(x)}{\sqrt{w^{(i,N)} + w^{(i+1,1)}}}, \tag{24}$$

the expansion coefficient for the basis function $f^{(i,m)}(x)$ is directly related to the value of the function $g(x)$ at the point $x^{(i,m)}$

$$g_{i,m} = \begin{cases} \dfrac{g(x^{(i,m)})}{\sqrt{w^{(i,m)}}}, & m \neq N \\[3mm] \dfrac{g(x^{(i,N)})}{\sqrt{w^{(i,N)} + w^{(i+1,1)}}}, & m = N \end{cases}. \tag{25}$$

Another attractive feature of DVR is that the potential matrix elements are diagonal and equal to the potential values at the grid points, i.e.,

$$V_{(i,m),(j,n)} = \int f^{(i,m)}(x) V(x) f^{(j,n)}(x) dx = V(x^{(i,m)}) \delta_{i,j} \delta_{m,n}. \qquad (26)$$

Similar conclusions hold for other local operators, such as the electron-laser interaction operator in the length gauge.

In the FEM-DVR basis, the kinetic energy matrix is not diagonal but rather sparse, and the matrix elements are analytically known. A wave function that is expanded in terms of the FEM-DVR basis functions will be continuous at the finite element boundaries, but will not have continuous derivatives. However, the derivative of the wave function need not be continuous to correctly define the kinetic energy [11]. The matrix element of the second derivative operator in the FEM-DVR basis can be evaluated as

$$\int f^{(i,m)}(x) \frac{d^2}{dx^2} f^{(j,n)}(x) dx = -(\delta_{i,j} + \delta_{i,j\pm1})$$

$$\times \int dx \frac{d}{dx} f^{(i,m)}(x) \frac{d}{dx} f^{(j,n)}(x). \qquad (27)$$

The integral in the right hand side of the above equation can be evaluated by the Gauss quadrature. To complete this, the following equations will be useful,

$$\frac{dL^{(i,m)}(x)}{dx} = \sum_{p(p\neq m)} \frac{1}{x^{(i,m)} - x^{(i,p)}} \prod_{k(k\neq m,p)} \frac{(x - x^{(i,k)})}{x^{(i,m)} - x^{(i,k)}}, \qquad (28)$$

$$\frac{dL^{(i,m)}(x^{(i'm')})}{dx} = \begin{cases} \delta_{ii'} \dfrac{1}{x^{(i,m)} - x^{(i,m')}} \displaystyle\prod_{k\neq m,m'} \dfrac{(x^{(i,m')} - x^{(i,k)})}{x^{(i,m)} - x^{(i,k)}}, & m' \neq m \\[6mm] \delta_{ii'} \displaystyle\sum_{k\neq m} \dfrac{1}{x^{(i,m)} - x^{(i,k)}}, & m' = m \end{cases}$$

$$(29)$$

In the case that $m' = m$, if the Gauss-Lobatto quadrature is used in the finite element, Eq. (15) can be further reduced

$$\frac{dL^{(i,m)}(x^{(i'm')})}{dx} = \begin{cases} \delta_{ii'} \dfrac{1}{x^{(i,m)} - x^{(i,m')}} \displaystyle\prod_{k\neq m,m'} \dfrac{(x^{(i,m')} - x^{(i,k)})}{x^{(i,m)} - x^{(i,k)}}, & m' \neq m \\[6mm] \delta_{ii'} \displaystyle\sum_{k\neq m} \dfrac{1}{x^{(i,m)} - x^{(i,k)}}, & m' = m \end{cases}$$

$$(31)$$

2.2. Methods to solve TDSE of helium

2.2.1. Time-dependent close coupling (TDCC) scheme

The TDSE of helium is given by

$$i\frac{\partial}{\partial t}\Psi(\mathbf{r}_1, \mathbf{r}_2, t) = H(t)\Psi(\mathbf{r}_1, \mathbf{r}_2, t), \tag{32}$$

where the Hamiltonian operator, in the dipole approximation (length gauge), can be written as

$$H(\mathbf{r}_1, \mathbf{r}_2, t) = \frac{\mathbf{p}_1^2}{2} + \frac{\mathbf{p}_2^2}{2} - \frac{2}{r_1} - \frac{2}{r_2} + \frac{1}{|\mathbf{r}_1 - \mathbf{r}_2|} + (\mathbf{r}_1 + \mathbf{r}_2) \cdot \mathbf{E}(t), \tag{33}$$

where $\mathbf{E}(t)$ is the electric field of the laser pulse.

In the close-coupling scheme [16], the two-electron wave function $\Psi(\mathbf{r}_1, \mathbf{r}_2, t)$ is expanded in the coupled spherical harmonics,

$$\Psi(\mathbf{r}_1, \mathbf{r}_2, t) = \sum_{L,M,l_1,l_2} \frac{R_{l_1,l_2}^{L,M}(r_1, r_2, t)}{r_1 r_2} Y_{l_1,l_2}^{L,M}(\hat{r}_1, \hat{r}_2), \tag{34}$$

in which

$$Y_{l_1,l_2}^{L,M}(\hat{r}_1, \hat{r}_2) = \sum_{m_1,m_2} \langle l_1 m_1 l_2 m_2 | l_1 l_2 L M \rangle Y_{l_1,m_1}(\hat{r}_1) Y_{l_2,m_2}(\hat{r}_2) \tag{35}$$

where $\langle l_1 m_1 l_2 m_2 | l_1 l_2 L M \rangle$ is the Clebsch-Gordan coefficient.

For a linearly polarized laser field, the quantum number M is conservative. If the initial state is chosen to be the ground state of helium, $M = 0$ always holds. In the following we only consider this case and neglect the index M for simplicity.

Substitution of Eq. (35) into Eq. (32) leads to a set of coupled equations for the radial wave function $R_{l_1,l_2}^{L}(r_1, r_2, t)$,

$$i\frac{\partial}{\partial t} R_{l_1 l_2}^{L}(r_1, r_2) = T_{l_1,l_2} R_{l_1 l_2}^{L}(r_1, r_2) + \sum_{l'_1,l'_2,L'} W_{l_1,l_2,l'_1,l'_2}^{L} R_{l'_1 l'_2}^{L}(r_1, r_2)$$

$$+ \sum_{l'_1,l'_2,L'} V_{l_1,l_2,l'_1,l'_2}^{LL'} R_{l'_1 l'_2}^{L'}(r_1, r_2) \tag{36}$$

In the above equation, T_{l_1,l_2} comes from the contributions of the kinetic energy operator and the Coulomb attractive potential operator, explicitly by

$$T_{l_1,l_2} = \sum_{i=1}^{2} \left(-\frac{1}{2} \frac{\partial^2}{\partial r_i^2} + \frac{l_i(l_i + 1)}{2r_i^2} - \frac{Z}{r_i} \right). \tag{37}$$

$W^L_{l_1,l_2,l'_1,l'_2}$ represents the electron correlation term. Neumman expansion is employed to calculate the matrix element of the electron-electron repulsive potential. The $W^L_{l_1,l_2,l'_1,l'_2}$ term is expressed as

$$W^L_{l_1,l_2,l'_1,l'_2} = \sqrt{(2l_1 + 1)(2l'_1 + 1)(2l_2 + 1)(2l'_2 + 1)}$$

$$\times \sum_{\lambda=0}^{\infty} (-1)^{L+\lambda} \frac{r^\lambda_<}{\lambda^{\lambda+1}_>} \begin{pmatrix} l_1 & \lambda & l'_1 \\ 0 & 0 & 0 \end{pmatrix} \begin{pmatrix} l_2 & \lambda & l'_2 \\ 0 & 0 & 0 \end{pmatrix}$$

$$\times \begin{Bmatrix} l'_1 & l'_2 & L \\ l_2 & l_1 & \lambda \end{Bmatrix}, \tag{38}$$

where $r_{>(<)} = \max(\min)(r_1, r_2)$. In the length gauge, the contribution of the electron-laser interaction term is

$$V^L_{l_1,l_2,l'_1,l'_2} = E(t)\sqrt{(2L + 1)(2L' + 1)} \begin{pmatrix} L & 1 & L' \\ 0 & 0 & 0 \end{pmatrix}$$

$$\times \left[r_1(-1)^{l_2}\sqrt{(2l_1 + 1)(2l'_1 + 1)} \begin{pmatrix} l_1 & 1 & l'_1 \\ 0 & 0 & 0 \end{pmatrix} \right.$$

$$\times \begin{Bmatrix} l_1 & l_2 & L \\ L' & 1 & l'_1 \end{Bmatrix} \delta_{l'_2 l_2} + r_2(-1)^{l_1}\sqrt{(2l_2 + 1)(2l'_2 + 1)}$$

$$\times \begin{pmatrix} l_2 & 1 & l'_2 \\ 0 & 0 & 0 \end{pmatrix} \begin{Bmatrix} l_2 & l_1 & L \\ L' & 1 & l'_2 \end{Bmatrix} \delta_{l'_1 l_1} \right]. \tag{39}$$

If the velocity gauge is used to describe the electron-laser interaction, one just needs to replace the $V^L_{l_1,l_2,l'_1,l'_2}$ term by

$$V^L_{l_1,l_2,l'_1,l'_2} = iA(t)\sqrt{(2L + 1)(2L' + 1)} \begin{pmatrix} L & 1 & L' \\ 0 & 0 & 0 \end{pmatrix}$$

$$\times \left[\left(\frac{\partial}{\partial r_1} + \frac{l'_1(l'_1 + 1) - l_1(l_1 + 1)}{2r_1} \right)(-1)^{l_2} \right.$$

$$\times \sqrt{(2l_1 + 1)(2l'_1 + 1)} \begin{pmatrix} l_1 & 1 & l'_1 \\ 0 & 0 & 0 \end{pmatrix} \begin{Bmatrix} l_1 & l_2 & L \\ L' & 1 & l'_1 \end{Bmatrix} \delta_{l'_2 l_2}$$

$$+ \left(\frac{\partial}{\partial r_2} + \frac{l'_2(l'_2+1) - l_2(l_2+1)}{2r_2} \right)(-1)^{l_1}$$

$$\times \sqrt{(2l_2+1)(2l'_2+1)} \times \begin{pmatrix} l_2 & 1 & l'_2 \\ 0 & 0 & 0 \end{pmatrix} \begin{Bmatrix} l_2 & l_1 & L \\ L' & 1 & l'_2 \end{Bmatrix} \delta_{l'_1 l_1} \Bigg].$$

$$\tag{40}$$

To solve the coupled equations for the radial wave function $R^L_{l_1,l_2}(r_1, r_2, t)$ [Eq. (36)], one needs to further discretize the radial coordinates r_1 and r_2, and employ some techniques to realize the time propagation. Those techniques have been introduced previously.

2.2.2. *Extraction of physical observables*

After one obtains the final wave function when the laser pulse ends, projection process is needed to extract the physical observables from the final wave function. In principle, one should project the final wave function $\Psi(\mathbf{r}_1, \mathbf{r}_2, t_f)$ to the double continuum $\Psi_{\mathbf{k}_1, \mathbf{k}_2}(\mathbf{r}_1, \mathbf{r}_2)$ to obtain the double ionization transition amplitude

$$f(\mathbf{k}_1, \mathbf{k}_2) = \langle \Psi_{\mathbf{k}_1, \mathbf{k}_2}(\mathbf{r}_1, \mathbf{r}_2) | \Psi(\mathbf{r}_1, \mathbf{r}_2, t_f) \rangle, \tag{41}$$

and the full differential double ionization probability is given by

$$P(\mathbf{k}_1, \mathbf{k}_2) = |f(\mathbf{k}_1, \mathbf{k}_2)|^2. \tag{42}$$

However, the accurate analytic expression of the double continuum $\Psi_{\mathbf{k}_1, \mathbf{k}_2}(\mathbf{r}_1, \mathbf{r}_2)$ is unknown. One method is to approximate the double continuum by the symmetric product of the one-electron scattering continuum state [18–21],

$$\Psi_{\mathbf{k}_1, \mathbf{k}_2}(\mathbf{r}_1, \mathbf{r}_2) = \frac{1}{\sqrt{2}} [\Psi_{\mathbf{k}_1}(\mathbf{r}_1)\Psi_{\mathbf{k}_2}(\mathbf{r}_2) + \Psi_{\mathbf{k}_1}(\mathbf{r}_2)\Psi_{\mathbf{k}_2}(\mathbf{r}_1)]. \tag{43}$$

The one-electron scattering continuum $\Psi_{\mathbf{k}}(\mathbf{r})$ is analytically given by

$$\Psi_{\mathbf{k}}(\mathbf{r}) = \frac{1}{k}\frac{1}{\sqrt{2\pi}} \sum_{l,m} i^l e^{-i(\sigma_l + \delta_l)} Y^*_{lm}(\hat{k}) R_{kl}(r) Y_{lm}(\hat{r}), \tag{44}$$

where $\sigma_l = \arg \Gamma\left(l + 1 - i\frac{Z}{k}\right)$ is the Coulomb phase shift, Z is the effective nuclear charge, and δ_l is the l-th partial wave phase shift (with respect to the Coulomb waves) due to any non-Coulomb short-range part of the potential. In the present situation, $Z = 2$ and $\delta_l = 0$. The radial wave function $R_{kl}(r)$ is analytically given by

$$R_{kl}(r) = \frac{2}{r} \frac{2^l e^{\pi Z/2k} |\Gamma(l+1-iZ/k)|}{(2l+1)!} e^{-ikr} (kr)^{(l+1)}$$

$$\times F_{11}(l+1+iZ/k, 2l+2, 2ikr). \tag{45}$$

In the above equation, F_{11} is the Kummer confluent hypergeometric function, and the radial wave function has been assumed to be normalized according to

$$\int_0^\infty r^2 R_{k'l}(r) R_{kl}(r) dr = 2\pi \delta(k - k').$$
(46)

If the final wave function obtained from solving TDSE is expressed as

$$\Psi(\mathbf{r}_1, \mathbf{r}_2, t_f) = \sum_{L,M,l_1,l_2} \frac{R^L_{l_1,l_2}(r_1, r_2, t_f)}{r_1 r_2} Y^L_{l_1,l_2}(\hat{r}_1, \hat{r}_2),$$
(47)

and the double continuum is approximated by Eq. (43), then the full differential double ionization probability will be given by

$$P(\mathbf{k}_1, \mathbf{k}_2) = \frac{2}{4\pi^2 k_1^2 k_2^2} \left| \sum_{L,l_1,l_2} (-i)^{l_1+l_2} e^{i(\sigma_{l_1}+\sigma_{l_2})} Y^L_{l_1,l_2}(\hat{k}_1, \hat{k}_2) M^L_{l_1,l_2}(k_1, k_2) \right|^2,$$
(48)

where,

$$M^L_{l_1,l_2}(k_1, k_2) = \iint dr_1 dr_2 r_1 r_2 R_{k_1 l_1}(r_1) R_{k_2 l_2}(r_2) R^L_{l_1,l_2}(r_1, r_2, t_f).$$
(49)

Equation (49) can be evaluated by the Gauss quadrature, and the quadrature points can be chosen to be the same set in obtaining $R^L_{l_1,l_2}(r_1, r_2, t_f)$ in the TDSE calculation. In the double continuum, due to the indistinguishability of the two electrons, exchanging the momentum of the two electrons will leave the two electrons in the same double continuum. Eq. (48) has taken this indistinguishability into account. However, when one calculates the total double ionization by integrating the full differential double ionization probability over the momentums $\mathbf{k}_1$ and $\mathbf{k}_2$, one should add a constraint such as $k_1 \leq k_2$ to avoid the double count of the same double continuum. One can also choose to avoid using the constraint by dividing Eq. (48) by 2,

$$P(\mathbf{k}_1, \mathbf{k}_2) = \frac{1}{4\pi^2 k_1^2 k_2^2} \left| \sum_{L,l_1,l_2} (-i)^{l_1+l_2} e^{i(\sigma_{l_1}+\sigma_{l_2})} Y^L_{l_1,l_2}(\hat{k}_1, \hat{k}_2) M^L_{l_1,l_2}(k_1, k_2) \right|^2.$$
(50)

Equation (50) can also be derived by projecting the final wave function $\Psi(\mathbf{r}_1, \mathbf{r}_2, t_f)$ to the simple product of one-electron scattering continuums $\Psi_{\mathbf{k}_1}(\mathbf{r}_1) \Psi_{\mathbf{k}_2}(\mathbf{r}_2)$.

Sometimes we are interested in the energy sharing between the two electrons regardless of the angular distributions. Integrating Eq. (50) over the ejection angles

$\hat{k}_1, \hat{k}_2$, we obtain the distributions on the magnitude of the momenta of the two electrons

$$P(k_1, k_2) = \iint P(\mathbf{k_1}, \mathbf{k_2})k_1^2 k_2^2 d\hat{k}_1 d\hat{k}_2$$

$$= \frac{1}{4\pi^2} \sum_{L,l_1,l_2} \left| M_{l_1,l_2}^L (k_1, k_2) \right|^2. \tag{51}$$

According to Eq. (51), one can further define the two-electron joint energy spectra

$$P(E_1, E_2) = \frac{1}{k_1 k_2} P(k_1, k_2)$$

$$= \frac{1}{4\pi^2 k_1 k_2} \sum_{L,l_1,l_2} \left| M_{l_1,l_2}^L (k_1, k_2) \right|^2, \tag{52}$$

where $k_1 = \sqrt{2E_1}$ and $k_2 = \sqrt{2E_2}$. The total double ionization probability can be obtained by integrating the joint energy spectra over the energies of the two electrons

$$P_{\text{total}} = \iint P(E_1, E_2) dE_1 dE_2. \tag{53}$$

2.3. *Methods to solve TDSE of H$_2$ molecule*

2.3.1. *Introduction of the prolate spheroidal coordinates*

The prolate spheroidal coordinates [22–26] are quite proper to solve the two-center problem of H$_2$ molecule. Before we list those equations for solving the TDSE of H$_2$ molecule, we firstly make a brief introduction for the prolate spheroidal coordinates. The position of a three-dimensional point can be marked by coordinates ξ, η and φ. φ is the azimuthal angle which is defined in the same way in the usual spherical coordinates frame. Both ξ and η are related to the distances between the space point and the two centers. In the TDSE calculation of H$_2$, the two centers are the nuclear positions. If we put the two centers on the z-axis symmetrically, and denote the vector which points from one center to the other as $\mathbf{R} = R\hat{\mathbf{z}}$, ξ and η are given by

$$\xi = \frac{|\mathbf{r} - \mathbf{R}/2| + |\mathbf{r} + \mathbf{R}/2|}{R}, \quad \eta = \frac{|\mathbf{r} - \mathbf{R}/2| - |\mathbf{r} + \mathbf{R}/2|}{R}, \tag{54}$$

where **r** is the position vector of any point. The transformation from the prolate spheroidal coordinates to the rectangular coordinates is given by

$$x = a\sqrt{(\xi^2 - 1)(1 - \eta^2)}\cos\varphi,$$
$$y = a\sqrt{(\xi^2 - 1)(1 - \eta^2)}\sin\varphi, \tag{55}$$
$$z = a\xi\eta,$$

where $a = R/2$.

In the following, we will introduce the expressions for several mathematic operators, which are involved in the TDSE calculation of H_2. The Laplacian operator, which appears in the kinetic energy operator, can be expressed as

$$\nabla^2 = \frac{1}{a^2(\xi^2 - \eta^2)}\left[\frac{\partial}{\partial\xi}(\xi^2 - 1)\frac{\partial}{\partial\xi} + \frac{\partial}{\partial\eta}(1 - \eta^2)\frac{\partial}{\partial\eta}\right.$$
$$\left. + \left(\frac{1}{\xi^2 - 1} + \frac{1}{1 - \eta^2}\right)\frac{\partial^2}{\partial\varphi^2}\right]. \tag{56}$$

The attractive potential between the electron and the two nuclei is given in the prolate spheroidal coordinates by

$$-\frac{1}{|\mathbf{r} - \mathbf{R}/2|} - \frac{1}{|\mathbf{r} + \mathbf{R}/2|} = -\frac{2\xi}{a(\xi^2 - \eta^2)}. \tag{57}$$

Finally, we give the expressions for the volume element

$$dV = a^3(\xi^2 - \eta^2)\,d\xi\,d\eta\,d\varphi, \tag{58}$$

2.3.2. *TDSE of H₂ molecule in the prolate spheroidal coordinates*

In the one-photon or two-photon double ionization of H_2 molecule, the removal of the electrons is much faster than the motion of the nuclei. Thus, it will be reasonable to treat the nuclei fixed when ionization takes place. The Hamiltonian of H_2 at a fixed internuclear distance in a linearly polarized laser field is given by

$$H(t) = \sum_{p=1}^{2}\left(-\frac{1}{2}\nabla_p^2 - \frac{1}{|\mathbf{r}_p - \mathbf{R}/2|} - \frac{1}{|\mathbf{r}_p + \mathbf{R}/2|} + \mathbf{E}(t)\cdot\mathbf{r}_p\right)$$
$$+ \frac{1}{|\mathbf{r}_1 - \mathbf{r}_2|}. \tag{59}$$

The two-electron wave function $\Psi(\mathbf{r}_1, \mathbf{r}_2, t)$ is expanded by a set of orthonormalized basis, which are defined by [22]

$$b_{i_1 i_2 j_1 j_2}^{m_1 m_2}(\mathbf{r}_1, \mathbf{r}_2) = \frac{1}{a^3} \frac{1}{\sqrt{(\xi_1^2 - \eta_1^2)(\xi_2^2 - \eta_2^2)}} f_{i_1}^{m_1}(\xi_1) f_{i_2}^{m_2}(\xi_2) g_{j_1}^{m_1}(\eta_1)$$

$$\times g_{j_2}^{m_2}(\eta_2) \frac{e^{i(m_1 \varphi_1 + m_2 \varphi_2)}}{2\pi}, \tag{59}$$

where $f_i^m(\xi)$ and $g_j^m(\eta)$ are the FEM-DVR functions. For even m, these FEM-DVR functions are defined as Eq. (18) and (23). Since the single electron wave function near the two centers behaves as $(\xi^2 - 1)^{|m|/2}(1 - \eta^2)^{|m|/2}$, the wave function will behave nonpolynomially for odd m. To explicitly describe the nonpolynomial behavior, a square-root factor is added to the usual FEM-DVR functions. If we note $f_i^m(\xi) = f_i^{\text{even}}(\xi)$ and $g_j^m(\eta) = g_j^{\text{even}}(\eta)$ for even m, and note $f_i^m(\xi) = f_i^{\text{odd}}(\xi)$ and $g_j^m(\eta) = g_j^{\text{odd}}(\eta)$ for odd m, then the basis functions $f_i^{\text{odd}}(\xi)$ and $g_j^{\text{odd}}(\eta)$ are defined by

$$f_i^{\text{odd}}(\xi) = \frac{\sqrt{\xi^2 - 1}}{\sqrt{\xi_i^2 - 1}} f_i^{\text{even}}(\xi),$$

$$\tag{59}$$

$$g_j^{\text{odd}}(\eta) = \frac{\sqrt{1 - \eta^2}}{\sqrt{1 - \eta_j^2}} g_j^{\text{even}}(\eta).$$

In above, ξ_i and η_j are the Gaussian quadrature points.

In constructing these FEM-DVR functions, we employ three kinds of Gaussian quadratures. Two kinds of Gaussian quadratures are employed in the ξ coordinate. The Gauss-Radau quadrature is employed in the first element to exclude the boundary point $\xi = 1$, and the Gauss-Lobatto quadrature is employed in the other elements to connect the adjacent elements. For the η coordinate, only one element is used, and the Gauss-Legendre quadrature is used in this element to exclude the boundary points $\eta = 1$ and $\eta = -1$.

In the following, we give the expressions for the Hamiltonian matrix in the representation of present basis functions. Firstly, we give the kinetic energy matrix of the first electron. For simplicity, we divide the kinetic operator to three parts $T_1 = T_{\xi_1} + T_{\eta_1} + T_{\varphi_1}$ by the differential terms of coordinates ξ_1, η_1 and φ_1,

$$T_{\varphi_1} = -\frac{1}{2} \frac{1}{a^2(\xi_1^2 - \eta_1^2)} \left(\frac{1}{\xi_1^2 - 1} \frac{\partial^2}{\partial \varphi_1^2} + \frac{1}{1 - \eta_1^2} \frac{\partial^2}{\partial \varphi_1^2} \right), \tag{60}$$

$$T_{\xi_1} = -\frac{1}{2}\frac{1}{a^2(\xi_1^2 - \eta_1^2)}\left(\frac{\partial}{\partial\xi_1}(\xi_1^2 - 1)\frac{\partial}{\partial\xi_1}\right), \tag{61}$$

$$T_{\eta_1} = -\frac{1}{2}\frac{1}{a^2(\xi_1^2 - \eta_1^2)}\left(\frac{\partial}{\partial\eta_1}(1 - \eta_1^2)\frac{\partial}{\partial\eta_1}\right). \tag{62}$$

The matrix of T_{φ_1} is diagonal,

$$\left\langle b_{i_1 i_2 j_1 j_2}^{m_1 m_2} \middle| T_{\varphi_1} \middle| b_{i_1' i_2' j_1' j_2'}^{m_1' m_2'} \right\rangle = \frac{\delta_{i_2 i_2'}\delta_{j_2 j_2'}\delta_{m_2 m_2'}\delta_{i_1 i_1'}\delta_{j_1 j_1'}\delta_{m_1 m_1'}}{2a^2(\xi_{i_1}^2 - \eta_{j_1}^2)}$$

$$\times \left(\frac{m_1^2}{\xi_{i_1}^2 - 1} + \frac{m_1^2}{1 - \eta_{j_1}^2}\right). \tag{63}$$

The matrixes of T_{ξ_1} and T_{η_1} are more complex and will depend on the parity of m_1. For even m_1,

$$\left\langle b_{i_1 i_2 j_1 j_2}^{m_1 m_2} \middle| T_{\xi_1} \middle| b_{i_1' i_2' j_1' j_2'}^{m_1' m_2'} \right\rangle$$

$$= \frac{\delta_{i_2 i_2'}\delta_{j_2 j_2'}\delta_{m_2 m_2'}\delta_{j_1 j_1'}\delta_{m_1 m_1'}}{2a^2\sqrt{\xi_{i_1}^2 - \eta_{j_1}^2}\sqrt{\xi_{i_1'}^2 - \eta_{j_1'}^2}} \int d\xi_1(\xi_1^2 - 1)$$

$$\times \frac{\partial}{\partial\xi_1}f_{i_1}^{\text{even}}(\xi_1)\frac{\partial}{\partial\xi_1}f_{i_1'}^{\text{even}}(\xi_1), \quad \left\langle b_{i_1 i_2 j_1 j_2}^{m_1 m_2} \middle| T_{\eta_1} \middle| b_{i_1' i_2' j_1' j_2'}^{m_1' m_2'} \right\rangle$$

$$= \frac{\delta_{i_2 i_2'}\delta_{j_2 j_2'}\delta_{m_2 m_2'}\delta_{j_1 j_1'}\delta_{m_1 m_1'}}{2a^2\sqrt{\xi_{i_1}^2 - \eta_{j_1}^2}\sqrt{\xi_{i_1'}^2 - \eta_{j_1'}^2}} \int d\eta_1(1 - \eta_1)^2$$

$$\times \frac{\partial}{\partial\eta_1}g_j^{\text{even}}(\eta_1)\frac{\partial}{\partial\eta_1}g_{j'}^{\text{even}}(\eta_1). \tag{64}$$

For odd m_1,

$$\left\langle b_{i_1 i_2 j_1 j_2}^{m_1 m_2} \middle| T_{\xi_1} \middle| b_{i_1' i_2' j_1' j_2'}^{m_1' m_2'} \right\rangle$$

$$= \frac{\delta_{i_2 i_2'}\delta_{j_2 j_2'}\delta_{m_2 m_2'}\delta_{j_1 j_1'}\delta_{m_1 m_1'}}{2a^2\sqrt{\xi_{i_1}^2 - \eta_{j_1}^2}\sqrt{\xi_{i_1'}^2 - \eta_{j_1'}^2}} \frac{1}{\sqrt{\xi_{i_1}^2 - 1}\sqrt{\xi_{i_1'}^2 - 1}}$$

$$\times \left[\int d\xi_1(\xi_1^2 - 1)^2\frac{\partial}{\partial\xi_1}f_{i_1}^{\text{even}}(\xi_1)\frac{\partial}{\partial\xi_1}f_{i'}^{\text{even}}(\xi_1)\right.$$

$$+ \xi_1 \left(f_{i_1}^{\text{even}}(\xi_1) \frac{\partial}{\partial \xi_1} f_{i_1'}^{\text{even}}(\xi_1) + f_{i_1'}^{\text{even}}(\xi_1) \frac{\partial}{\partial \xi_1} f_{i_1}^{\text{even}}(\xi_1) \right) \Bigg]$$

$$+ \frac{\delta_{i_2 i_2'} \delta_{j_2 j_2'} \delta_{m_2 m_2'} \delta_{i_1 i_1'} \delta_{j_1 j_1'} \delta_{m_1 m_1'}}{2a^2 \left(\xi_{i_1}^2 - \eta_{j_1}^2 \right)} \frac{\xi_{i_1}^2}{(\xi_{i_1}^2 - 1)}, \tag{65}$$

$$\left\langle b_{i_1 i_2 j_1 j_2}^{m_1 m_2} | T_{\eta_1} | b_{i_1' i_2' j_1' j_2'}^{m_1' m_2'} \right\rangle$$

$$= \frac{\delta_{i_2 i_2'} \delta_{j_2 j_2'} \delta_{m_2 m_2'} \delta_{j_1 j_1'} \delta_{m_1 m_1'}}{2a^2 \sqrt{\xi_{i_1}^2 - \eta_{j_1}^2} \sqrt{\xi_{i_1'}^2 - \eta_{j_1'}^2}} \frac{1}{\sqrt{1 - \eta_{i_1}^2} \sqrt{1 - \eta_{i_1'}^2}}$$

$$\times \left[\int d\eta_1 (1 - \eta_1)^2 \frac{\partial}{\partial \eta_1} g_{j_1}^{\text{even}}(\eta_1) \frac{\partial}{\partial \eta_1} g_{j_1'}^{\text{even}}(\eta_1) \right.$$

$$\left. - \eta_1 \left(g_{j_1}^{\text{even}}(\eta_1) \frac{\partial}{\partial \eta_1} g_{j_1'}^{\text{even}}(\eta_1) + g_{j_1'}^{\text{even}}(\eta_1) \frac{\partial}{\partial \eta_1} g_{j_1}^{\text{even}}(\eta_1) \right) \right]$$

$$+ \frac{\delta_{i_2 i_2'} \delta_{j_2 j_2'} \delta_{m_2 m_2'} \delta_{i_1 i_1'} \delta_{j_1 j_1'} \delta_{m_1 m_1'}}{2a^2 \left(\xi_{i_1}^2 - \eta_{j_1}^2 \right)} \frac{\eta_{j_1}^2}{(1 - \eta_{i_1}^2)}. \tag{66}$$

Those one-order differential terms for the FEM-DVR functions involved in Eqs. (65–67) can be calculated with Eqs. (28), (29) and (31), and the integrals involved in these equations are evaluated by Gaussian quadrature.

The Coulomb attractive potential between the first electron and the two nucleuses is $V_1 = -\frac{2\xi_1}{a(\xi_1^2 - \eta_1^2)}$, whose matrix is diagonal,

$$\left\langle b_{i_1 i_2 j_1 j_2}^{m_1 m_2} | V_1 | b_{i_1' i_2' j_1' j_2'}^{m_1' m_2'} \right\rangle = \frac{\delta_{i_2 i_2'} \delta_{j_2 j_2'} \delta_{m_2 m_2'} \delta_{i_1 i_1'} \delta_{j_1 j_1'} \delta_{m_1 m_1'}}{2a^2 (\xi_{i_1}^2 - \eta_{j_1}^2)} 4\xi_1 a. \tag{67}$$

If we describe the electron-laser interaction in the length gauge, the interaction operator will be given by $W_1 = W_{1x} + W_{1y} + W_{1z}$, where $W_{1x} = E_x(t)x$, $W_{1y} = E_y(t)y$, and $W_{1z} = E_z(t)z$. The expressions for x, y and z in the prolate spheroidal coordinates have been given by Eq. (55). The matrix for the electron-laser interaction operator is also diagonal,

$$\left\langle b_{i_1 i_2 j_1 j_2}^{m_1 m_2} | W_{1x} | b_{i_1' i_2' j_1' j_2'}^{m_1' m_2'} \right\rangle = \frac{1}{2} \delta_{i_2 i_2'} \delta_{j_2 j_2'} \delta_{m_2 m_2'} \delta_{i_1 i_1'} \delta_{j_1 j_1'} \delta_{m_1, m_1' \pm 1}$$

$$\times a E_x(t) \sqrt{(\xi_{i_1}^2 - 1)(1 - \eta_{j_1}^2)}, \tag{68}$$

$$\left\langle b^{m_1 m_2}_{i_1 i_2 j_1 j_2} \middle| W_{1y} \middle| b^{m'_1 m'_2}_{i'_1 i'_2 j'_1 j'_2} \right\rangle = \frac{1}{2i} \delta_{i_2 i'_2} \delta_{j_2 j'_2} \delta_{m_2 m'_2} \delta_{i_1 i'_1} \delta_{j_1 j'_1} (\delta_{m_1, m'_1 +} - \delta_{m_1, m'_1 - 1})$$

$$\times a E_y(t) \sqrt{(\xi^2_{i_1} - 1)(1 - \eta^2_{j_1})}, \tag{69}$$

$$\left\langle b^{m_1 m_2}_{i_1 i_2 j_1 j_2} \middle| W_{1z} \middle| b^{m'_1 m'_2}_{i'_1 i'_2 j'_1 j'_2} \right\rangle = \frac{1}{2} \delta_{i_2 i'_2} \delta_{j_2 j'_2} \delta_{m_2 m'_2} \delta_{i_1 i'_1} \delta_{j_1 j'_1} \delta_{m_1, m'_1} a E_z(t) \xi_{i_1} \eta_{j_1}, \tag{70}$$

The kinetic matrix, electron-nucleus potential matrix and the electron-laser interaction matrix for the second electron can be written in a similar way to those matrixes for the first electron.

Now we turn to handle the electron correlation term. The electron-electron Coulomb repulsive potential can be expressed as a Neumann expansion in the prolate spheroidal coordinates [22, 25, 26]

$$\frac{1}{|\mathbf{r}_1 - \mathbf{r}_2|} = \frac{1}{a} \sum_{l=0}^{\infty} \sum_{m=-l}^{l} (-1)^{|m|} (2l + 1) \left[\frac{(l - |m|)!}{(l + |m|)!} \right]^2$$

$$\times P^{|m|}_l(\xi_<) Q^{|m|}_l(\xi_>) P^{|m|}_l(\eta_1) P^{|m|}_l(\eta_2) e^{im(\varphi_1 - \varphi_2)}, \tag{71}$$

where $\xi_{>(<)} = \max(\min)(\xi_1, \xi_2)$, and $P^{|m|}_l(\eta)$ and $Q^{|m|}_l(\xi)$ are regular and irregular Legendre functions respectively. The matrix element of the electron-electron repulsive potential can be given by [22]

$$\left\langle b^{m_1 m_2}_{i_1 i_2 j_1 j_2} \middle| \frac{1}{|\mathbf{r}_1 - \mathbf{r}_2|} \middle| b^{m'_1 m'_2}_{i'_1 i'_2 j'_1 j'_2} \right\rangle$$

$$= \delta_{i_1 i'_1} \delta_{j_1 j'_1} \delta_{i_2 i'_2} \delta_{j_2 j'_2} \delta_{m_1 - m'_1, m'_2 - m_2}$$

$$\times \frac{1}{a} \sum_{l=|m|}^{\infty} (2l + 1) \frac{(l - |m|)!}{(l + |m|)!} P^{|m|}_l(\eta_{j_1}) P^{|m|}_l(\eta_{j_2})$$

$$\times \left[\frac{[T^{|m|}]^{-1}_{i_2 i_1}}{\sqrt{\omega^{i_1}_\xi \omega^{i_2}_\xi}} + (-1)^{|m|} \frac{(l - |m|)!}{(l + |m|)!} P^{|m|}_l(\xi_{i_1}) P^{|m|}_l(\xi_{i_2}) \frac{Q^{|m|}_l(\xi_{max})}{P^{|m|}_l(\xi_{max})} \right], \tag{72}$$

where $m = m_1 - m'_1 = m'_2 - m_2$, and ω^i_ξ is the i-th Gaussian quadrature weight for ξ coordinate. For the boundary points between two adjacent finite elements, $\omega^i_\xi = \omega^{(n,N)}_\xi + \omega^{(n+1,1)}_\xi$, where $\omega^{(n,N)}_\xi$ is the last Gaussian quadrature weight in the n-th finite element, and $\omega^{(n+1,1)}_\xi$ is the first Gaussian quadrature weight in the $(n+1)$-th finite element. ξ_{max} is the truncated maxima value of the ξ coordinate. The matrix element of $T^{|m|}$ is given by

$$T^{|m|}_{\mu\mu'} = -\int_1^{\xi_{max}} d\xi \, f^m_\mu(\xi) \left[(\xi^2 - 1)\frac{d^2}{d\xi^2} + 2\xi\frac{d}{d\xi} \right.$$

$$\left. - l(l+1) - \frac{m^2}{\xi^2 - 1} \right] f^m_{\mu'}(\xi), \tag{73}$$

and $\left[T^{|m|} \right]^{-1}_{i_2 i_1}$ denotes the element of the inverse matrix of $T^{|m|}$.

2.3.3. *Extraction of physical observable*

The process to extract differential double ionization probability of H_2 is quite similar to that of helium. The double continuum of H_2 is also approximated by the product of the one-electron scattering state, which is expressed as [25]

$$\psi^{(-)}_{\mathbf{k}}(\xi, \eta, \varphi) = \frac{1}{\sqrt{2\pi k}} \sum_{l=0}^{\infty} \sum_{m=-l}^{l} i^l e^{-i\delta_{lm}(k)} S^*_{lm}(\cos\theta_k, k)$$

$$\times e^{-im\varphi_k} R_{lm}(\xi, k) S_{lm}(\eta, k) \frac{e^{im\varphi}}{\sqrt{2\pi}}, \tag{74}$$

where $R_{lm}(\xi, k)$ and $S_{lm}(\eta, k)$ are the radial and angular parts of the generalized spheroidal wave functions [24], and $\delta_{lm}(k)$ is the two-center Coulomb phase shift.

If we denote the final wave function in TDSE calculation as

$$\Psi_f = \sum_{m_1 m_2} \Psi_f^{m_1 m_2}(\xi_1, \eta_1, \xi_2, \eta_2) \frac{e^{i(m_1\varphi_1 + m_2\varphi_2)}}{2\pi}, \tag{75}$$

then the fully differential probability density is given by

$$P(\mathbf{k}_1, \mathbf{k}_2) = \left| \frac{1}{2\pi k_1 k_2} \sum_{l_1=0}^{+\infty} \sum_{m_1=-l_1}^{l_1} \sum_{l_2=0}^{+\infty} \sum_{m_2=-l_2}^{l_2} (-i)^{l_1+l_2} e^{i(\delta_{l_1 m_1} + \delta_{l_1 m_1})} M_{l_1 l_2 m_1 m_2} \right.$$

$$\left. \times S_{l_1 m_1}(\cos\theta_{k_1}, k_1) S_{l_2 m_2}(\cos\theta_{k_2}, k_2) e^{i(m_1\varphi_{k_1} + m_2\varphi_{k_2})} \right|^2, \tag{76}$$

where

$$M_{l_1 l_2 m_1 m_2} = \int_1^\infty d\xi_1 \int_{-1}^1 d\eta_1 \int_1^\infty d\xi_2 \int_{-1}^1 d\eta_2 a^6 (\xi_1^2 - \eta_1^2)(\xi_2^2 - \eta_2^2)$$

$$\times \Psi_f^{m_1 m_2}(\xi_1, \eta_1, \xi_2, \eta_2) R_{l_1 m_1}(\xi_1, k_1) R_{l_2 m_2}(\xi_2, k_2)$$

$$\times S_{l_1 m_1}(\eta_1, k_1) S_{l_2 m_2}(\eta_2, k_2). \tag{77}$$

In Eq. (77), the indistinguishability of the two electrons in the double continuum is not taken into account, see Sec. 2.3.2 for similar discussion. Similar to Eq. (51) and (52) for helium, for H_2 molecule we also have

$$P(k_1, k_2) = \sum_{l_1=0}^{+\infty} \sum_{m_1=-l_1}^{l_1} \sum_{l_2=0}^{+\infty} \sum_{m_2=-l_2}^{l_2} |M_{l_1 l_2 m_1 m_2}|^2, \tag{78}$$

and

$$P(E_1, E_2) = \frac{1}{k_1 k_2} \sum_{l_1=0}^{+\infty} \sum_{m_1=-l_1}^{l_1} \sum_{l_2=0}^{+\infty} \sum_{m_2=-l_2}^{l_2} |M_{l_1 l_2 m_1 m_2}|^2, \tag{79}$$

3. Applications of TDSE Methods on Double Ionization Processes

3.1. *One-photon and two-photon double ionization cross sections*

3.1.1. *Definitions for photon-double ionization cross sections*

One-photon or two-photon double ionization can be analyzed in the framework of perturbation theory. In the long pulse limit, the dependence of the double ionization probability on the laser parameters will show several simple rules, which can be known by perturbation analysis. For example, for n-photon process, the ionization probability is proportional to I^n, where I is the laser intensity. The photon ionization cross section, which is proportional to the ionization probability, can explicitly take these rules into account.

More specifically, the n-photon double ionization cross section is defined by the ratio between the ionization rate and the n-th power of the photon flux. The total double ionization cross section $\sigma^{(n)}$ is related to the total double ionization probability P_{total} by

$$\sigma^{(n)} = \left(\frac{\omega}{I_0}\right)^n \frac{P_{\text{total}}}{T_{\text{eff}}^{(n)}}, \tag{80}$$

where ω is the photon energy, I_0 is the peak intensity of the laser pulse, and $T_{\text{eff}}^{(n)}$ is the effective pulse duration, which is defined by

$$T_{\text{eff}}^{(n)} = \int_{-\infty}^{\infty} f^{2n}(t)dt. \tag{81}$$

$f(t)$ is the envelop shape function of the laser pulse. For a $\sin^2$ shape pulse with duration of T

$$
f(t) = \begin{cases} \sin^2\left(\dfrac{\pi t}{T}\right), & 0 \leq t \leq T \\[2mm] 0, & t > T \text{ or } t < 0 \end{cases}, \tag{82}
$$

$T_{\text{eff}}^{(1)} = 3T/8$ and $T_{\text{eff}}^{(2)} = 35T/128$.

The differential cross section is related to the differential double ionization probability. However it will be a little more complex to obtain the differential cross section from the differential double ionization probability. For infinitely long pulse, the energies of the ionized two electrons are not independent due to the energy conservation law $E_1 + E_2 = n\omega - E_B$, where E_B is the double ionization potential. In the TDSE calculation, one can only use a finitely long pulse, for which a broad range of photon energy will be covered. Some strategies are needed to account for this photon-energy broadening effect. The physical quantity which gives the most detailed information on the photon-double ionization process will be the triply differential cross section (TDCS), which reflects the energy sharing between the two electrons and their angular distributions.

For one-photon double ionization, we give three methods [22] to extract the triply differential cross section (TDCS). In the first method (denoted as Method 1), the TDCS is given by

$$
\frac{d\sigma^{(1)}}{dE_1 d\Omega_1 d\Omega_2} = \frac{2\pi k_1 k_2 \omega P(E_1, E_2, \Omega_1, \Omega_2)}{I_0 |F(\Delta\omega)|^2}, \tag{83}
$$

where $P(E_1, E_2, \Omega_1, \Omega_2) = P(\mathbf{k}_1, \mathbf{k}_2)$ has been given in Eqs. (50) and (76), $\Delta\omega = E_1 + E_2 + E_B - \omega$, and the function $F(\omega')$ is defined by

$$
F(\omega') = \int_{-\infty}^{\infty} f(t) e^{i\omega' t} dt. \tag{84}
$$

If we choose to calculate the TDCS at the central photon energy of the laser pulse, we will have $\Delta\omega = 0$. For a $\sin^2$ shape pulse with duration of T, we can obtain that $F(0) = \int_0^T \sin^2(\frac{\pi t}{T}) dt = T/2$.

The other two methods will involve some kind of integral operation to account for the photon-energy broadening effect. Before giving these equations, we introduce some definitions to characterize the energies of the two electrons. The excess energy refers to the energy that can be shared by the two electrons after absorbing the photon energy, $E_{\text{exc}} = E_1 + E_2 = \omega - E_B$. To account for the energy-sharing between the two electrons, one can define a hyperangle α by the implicit functions

$E_1 = E_{\text{exc}} \cos^2 \alpha$ and $E_2 = E_{\text{exc}} \sin^2 \alpha$. In the second method to extract the one-photon TDCS (denoted as Method 2), the TDCS is expressed as

$$\frac{d\sigma^{(1)}}{dE_1 d\Omega_1 d\Omega_2} = 2 \frac{1}{k_1 k_2} \frac{\omega}{I_0} \frac{1}{T_{\text{eff}}} \int_0^\infty dE' k_1'^2 k_2'^2 P(E'_1, E'_2, \Omega_1, \Omega_2), \qquad (85)$$

where $E'_1 = E' \cos^2 \alpha$, $E'_2 = E' \sin^2 \alpha$, $k'_1 = \sqrt{2E'_1}$, and $k'_2 = \sqrt{2E'_2}$.

The third method (denoted as Method 3) is realized by another kind of integral operation,

$$\frac{d\sigma^{(1)}}{dE_1 d\Omega_1 d\Omega_2} = \frac{\omega}{I_0} \frac{1}{T_{\text{eff}}} \int_0^\infty dE_2 k_1 k_2 P(E_1, E_2, \Omega_1, \Omega_2). \qquad (86)$$

The last two methods can be easily generalized to the two-photon double ionization process, and we will have

$$\frac{d\sigma^{(2)}}{dE_1 d\Omega_1 d\Omega_2} = 2 \frac{1}{k_1 k_2} \left(\frac{\omega}{I_0}\right)^2 \frac{1}{T_{\text{eff}}} \int_0^\infty dE' k_1'^2 k_2'^2 P(E'_1, E'_2, \Omega_1, \Omega_2), \qquad (87)$$

or

$$\frac{d\sigma^{(2)}}{dE_1 d\Omega_1 d\Omega_2} = \left(\frac{\omega}{I_0}\right)^2 \frac{1}{T_{\text{eff}}} \int_0^\infty dE_2 k_1 k_2 P(E_1, E_2, \Omega_1, \Omega_2). \qquad (88)$$

One may also be interested in the so called single differential cross section, which is defined by integrating the TDCS over the ejection angles of the two electrons.

3.1.2. *Numerical results for electron spectrums of helium atom and H$_2$ molecule*

The one-photon double ionization (PDI) of helium has now been treated as a well understood process after decades of investigations [27,28]. Both the total and differential double ionization cross sections have been measured for a large range of photon energies. Various *ab initio* calculations have also achieved excellent agreements with those measurements. For an example, in Fig. 1 we compare present TDSE calculation with the experimental measurements for the total double ionization cross section.

In spite of the success of the studies on the one-photon double ionization of helium, the two-photon double ionization (TPDI) of helium is still not well understood. On one side, experimental measurements on TDPI are more difficult due to the much smaller signal than PDI. Though the total cross sections of TPDI have been measured with the help of high harmonic generation (HHG) [29, 30] or free electron lasers (FEL) [31], these measurements have been pointed out that there are large uncertainties [32]. Thus the total cross sections of TPDI still calls for more accurate measurements. As to the triply differential cross section

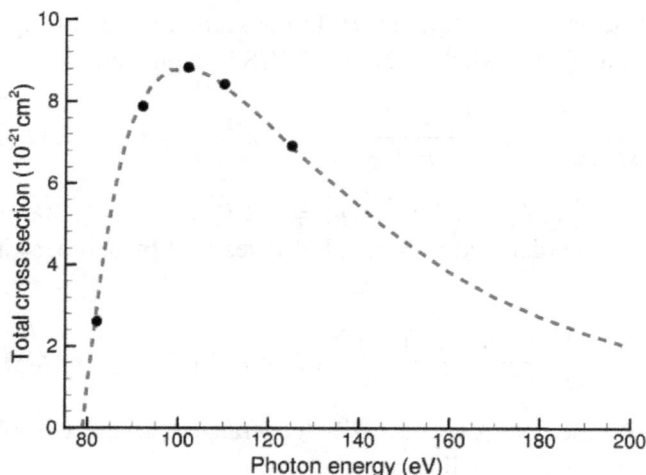

Fig. 1. Comparison of the total one-photon double-ionization cross section of He, calculated using the present approach for solving the two-electron TDSE (filled circles), with the experimental measurements (dashed line). (from Ref. [18])

(TDCS), no experimental measurement has been performed until now. Instead, several measurements on the recoil ion spectrums have been performed [33, 34]. On the other hand, large distinctions exist in the theoretical predictions even on the total cross section. In the past decade or so, a large number of theoretical efforts have been put on the topic of TPDI of helium [35–58]. Those authors have reached a consensus that the final wave functions, which are obtained by solving TDSE, should be consistent among different methods. The main debate is on the accuracy of extracting the double ionization information from the final wave function. To complete the information-extracting process, those researchers have developed various strategies, in which the main task is to include electron correlation reasonably.

Two-photon double ionization of helium is used to be discussed in two different regions, sequential double ionization region (SDI) and nonsequential double ionization (NSDI) region. Since the ground state energy of He and He^+ are -79.0 eV and -54.4 eV respectively, the first and second ionization potentials will be respectively 24.6 eV and 54.4 eV. If the photon energy is larger than 54.4 eV, TPDI of helium can be treated as a sequential process, in which the first electron can be ionized by absorbing one photon and the second electron can be ionized by absorbing another photon. If the photon energy is smaller than 54.4 eV but larger than 39.5 eV, the energy of one photon is not large enough to induce the ionization of He^+, which means the sequential picture is inapplicable; but the sum energy of two photons is large enough to induce the double ionization of helium. So here

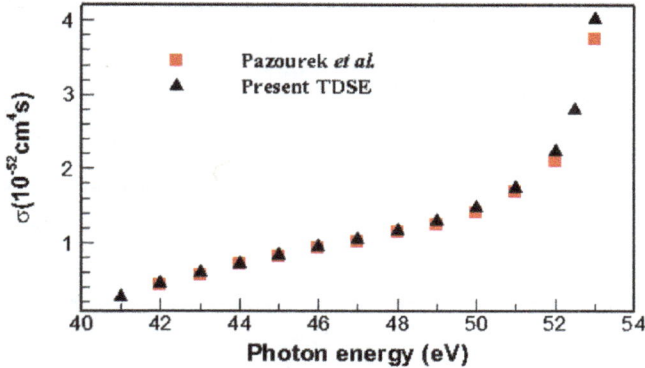

Fig. 2. The total two-photon double ionization cross section of helium. The black deltas are present TDSE calculations, and the red squares are the TDSE calculations by Pazourek *et al.* [36] Both calculations use the sin^2 shape pulse with a duration of 4 fs.

the term nonsequential double ionization (NSDI) refers to the TPDI in the photon-energy range of 39.5 eV to 54.4 eV. Cross section of TPDI is defined in the NSDI region. Present calculations for the total TPDI cross section are shown in Fig. 2, compared with the calculations by Pazourek *et al.* [36] The excellent agreement between these two calculations is unsurprising, since quite similar strategies are used to solve the two-electron TDSE. Various calculations for the total TPDI cross section can be divided into two groups. These results shown in Fig. 2 belong to the group which has more supporters [35–53]. The results of this group are also much closer to the only two experimental measurements, though the experimental results are not decisive due to the uncertainties in experimental conditions. The other group predicts about ten times high values [54–56].

The calculations for the triply differential cross section (TDCS) of TPDI of helium at the photon of 42 eV are shown in Fig. 3. The TDCS is calculated in the coplane that the vectors of two electron momentums and laser polarization lie in. By fixing the electron-energy sharing ($E_1 = E_2 = 2.5$ eV) and the ejection direction of the first electron, the distribution of TDCS as the function of the ejection direction of the second electron is studied. Different pulse durations have been used to extract the TDCS, and we can see that the extracted TDCS show some degree of pulse-duration dependence for not long enough pulses. Several characters of the TDCS for TPDI of helium can be identified in Fig. 3. For example, the two electrons tend to be ejected back to back, which is opposite to the PDI of helium. In PDI of helium, the equal-energy sharing and back-to-back ejection mode is forbidden. In addition, in Fig. 3 we can see another common character that the two electrons cannot be ejected in the same direction. This observation is consistent with our intuitive prediction,

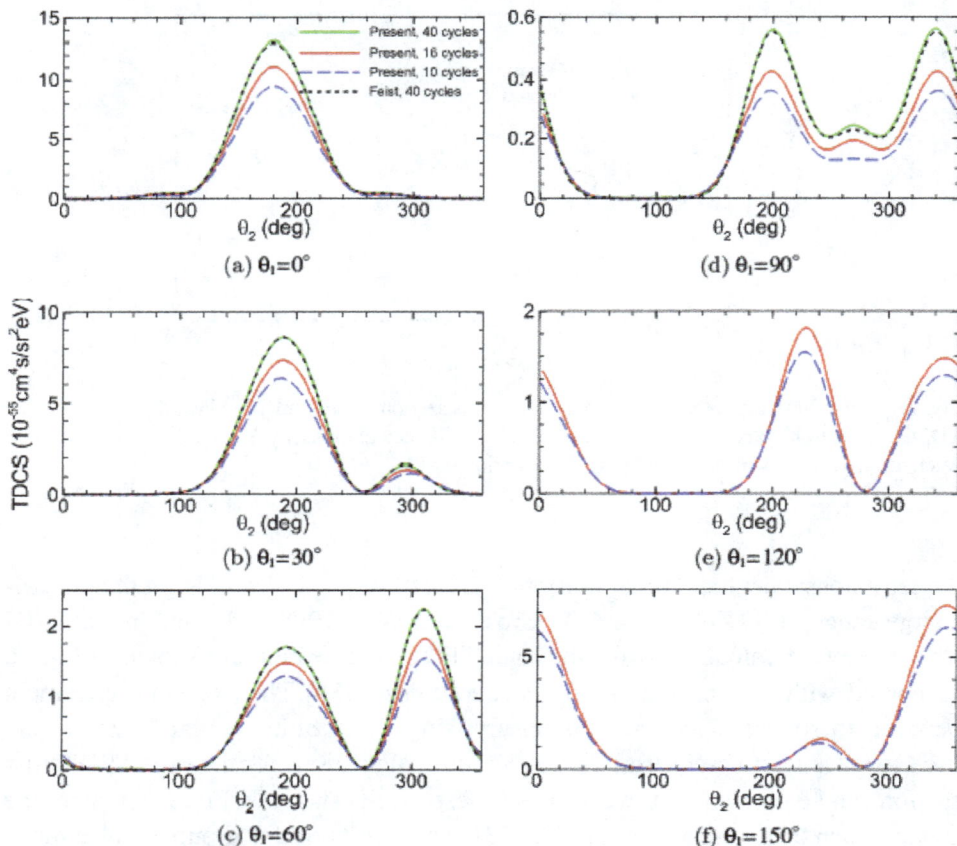

Fig. 3. Triply differential cross section (TDCS) for two-photon double ionization of He plotted as a function of the ejection angle θ_2 of the second electron (with respect to the laser polarization direction) for the photon energy $\omega = 42$ eV. The kinetic energies of the two electrons are $E_1 = E_2 = 2.5$ eV. The ejection angle of the first electron is fixed respectively at (a) $\theta_1 = 0°$, (b) $\theta_1 = 30°$, (c) $\theta_1 = 60°$, (d) $\theta_1 = 90°$, (e) $\theta_1 = 120°$, and (f) $\theta_1 = 150°$. The blue long-dashed lines are for the present results for a 10-cycle pulse, the red solid lines are for the present results for a 16-cycle pulse, the green solid lines are for the present results for a 40-cycle pulse, and the black short-dashed lines are for Feist *et al.* [40] for a 40-cycle pulse.

since the Coulomb repulsive correlation between the two electrons will always force the two electrons to depart from each other.

Compared with the photon double ionization of helium, the study for photon double ionization of H_2 is more difficult both theoretically and experimentally. The completely break up of H_2 produces four particles, i.e., two electrons and two protons. Early experimental measurement for one-photon double ionization of H_2 (D_2) only detects parts of the four particles. Researchers found that in the condition that only the two electrons are detected (i.e., the molecular orientation relative to the laser polarization is random), the angular distributions of the two electrons can be quite similar to those distributions in PDI of helium [59, 60]. Until 2004, Weber *et al.* [61–63] realized the simultaneous measurements for all the four particles, which made it possible to obtain the two-electron angular distributions for a particular molecular orientation and a particular nuclear distance. They found interesting molecular effects that the two-electron angular distributions can be quite sensitive to the molecular orientation and the nuclear distance [61–65].

Reliable theoretical treatments for PDI of H_2 were also developed in recent years [22, 25, 26, 66–72]. These theories are based on the *ab initio* calculations by solving the two-electron TDSE by fixing the nuclear motions. The molecular effects observed in experimental measurements can be addressed by these *ab initio* calculations. In Fig. 4, present implementation of the TDSE in the prolate spheroidal coordinates is compared with a recent implementation [72] in the spherical coordinates for the TDCS of double ionization of H_2 at equilibrium nuclear distance $R = 1.4$ a.u. We choose the photon energy as 75 eV, and study the case that the two electrons share the excess energy equally. Since the double ionization potential of H_2 is 51.4 eV, this will lead to $E_1 = E_2 = 11.8$ eV. The ejection direction of the

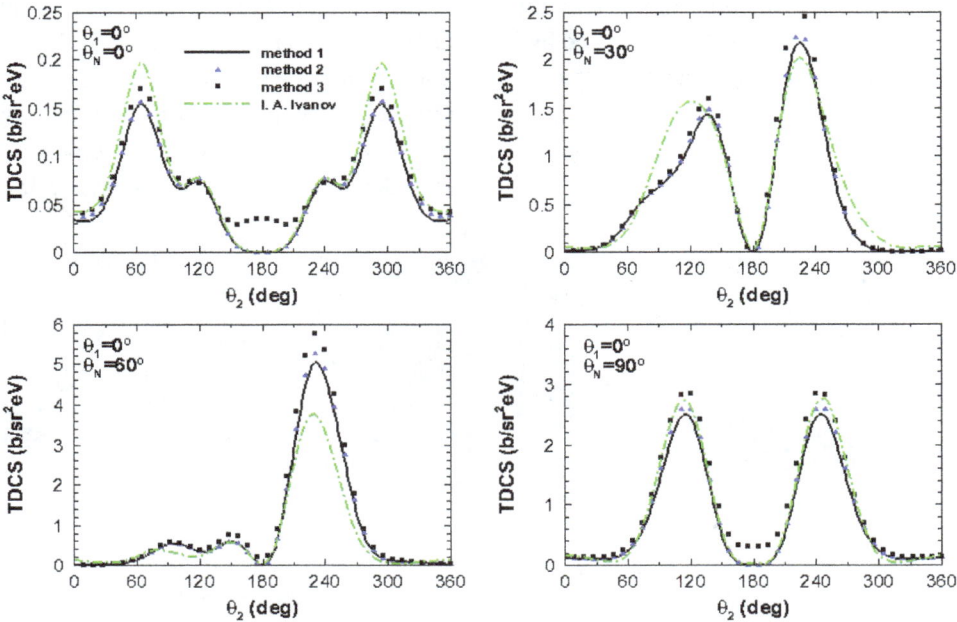

Fig. 4. Coplanar TDCSs of H_2 in the equal-energy sharing case at photon energy of 75 eV. One electron is detected at fixed direction of $\theta_1 = 0°$ with respect to the laser polarization direction. The TDCSs extracted by three methods are compared with the calculations of Ivanov and Kheifets [72]. (from Ref. [22])

first electron is fixed along the laser-polarization direction. In each panel of Fig. 4, the molecular orientation is fixed to a particular value. θ_N is the angle between the molecular orientation and laser polarization. Four kinds of molecular orientations are shown. The TDCS is shown as a function of the ejection direction of the second electrons. Three methods, which have been introduced in Sec. 3.1.1, have been used to extract the TDCS, and we see general consistent results for these three methods and the calculations of Ivanov and Kheifets [72]. Comparing the four panels shown in Fig. 4, we see obvious dependence of the TDCS on the molecular orientation, and the TDCS of the perpendicular configuration ($\theta_N = 90°$) is larger than the parallel configuration ($\theta_N = 0°$) by more than a magnitude. Except for the one-photon double ionization of H_2 discussed here, two-photon double ionization of H_2 has also been calculated by theoretical physicists [73–76].

3.1.3. *Recoil ion spectrum of photon double ionization of helium*

In the study of double ionization of helium, except for the two-electron spectrum, the recoil ion spectrum will be another tool to understand that process. Experimental measurement for the momentum spectrum of the recoil ion is easier than the measurement of the two-electron momentum spectrum. In Fig. 5(b) and (c), we show the experimental measurement [33] for the momentum spectrum in TPDI by 44 eV soft-x-ray which is produced by free-electron laser in Hamburg (FLASH). The recoil-ion momentum distributions for single ionization are also shown in Fig. 5(a). The recoil ion spectrum is connected to the two-electron spectrum by momentum conservation. Since the photon momentum is usually small enough to be neglected, the sum momentum of the two electrons and the ion is zero. So the recoil ion

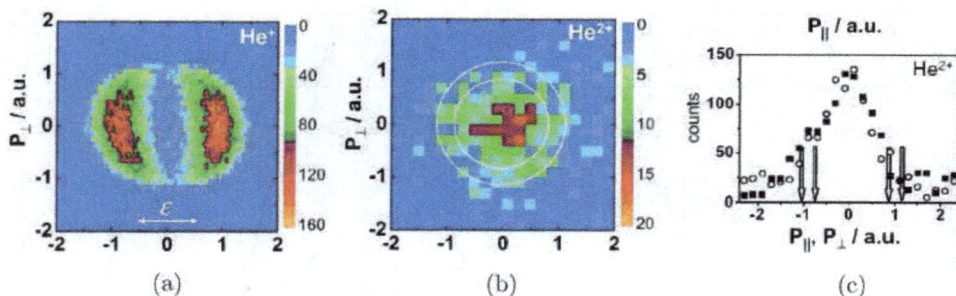

Fig. 5. Density plot of the measured recoil-ion momentum distributions for (a) single and (b) double ionization of He by 44 eV FLASH photons. The arrow in (a) indicates the direction of the FLASH polarization. Inner and outer circles in (b) mark the maximum He^{2+} momentum for the cases where one electron would have taken all the excess energy and for the equal energy sharing with the emission of both electrons in the same direction, respectively. (c) Projections of the 2D distribution of panel (b) onto the axis parallel (solid squares) and perpendicular (open circles) to the polarization direction. Arrows indicate the positions of the circles shown in panel (b). (taken from the Ref. [33])

spectrum is equal to the spectrum of sum momentum of the two electrons. For the single ionization process showed in Fig. 5(a), the recoil ion spectrum simply mimics the electron spectrum. For the nonsequential two-photon double ionization process, we can clear see that the dominative recoil ion signal is distributed at the zero momentum point, as we can see clearly in Fig. 5(b) and (c). This observation indicates that two electrons favor to be ejected back-to-back with equal energies in the nonsequential two-photon double ionization process.

Several *ab initio* calculations have already been performed to predict the recoil ion spectrum [21, 34, 42, 77–80]. Due to the momentum conservation, the three vectors are dependent by $\mathbf{k}_1 + \mathbf{k}_2 + \mathbf{Q} = 0$, where $\mathbf{k}_1$ and $\mathbf{k}_2$ represent the momentum vectors of the two electrons and $\mathbf{Q}$ represents the momentum of the recoil ion. Any two of the three vectors can be chosen to represent the final double ionization continuum. Instead of using $\mathbf{k}_1$ and $\mathbf{k}_2$, now we use $\mathbf{k}_1$ and $\mathbf{Q}$ to represent the double ionization. The fully differential double ionization probability $P(\mathbf{k}_1, \mathbf{Q}) = P(\mathbf{k}_1, \mathbf{k}_2)$ has already been given in Eq. (50). The recoil-ion momentum spectrum is then given by

$$P(\mathbf{Q}) = \int P(\mathbf{k}_1, \mathbf{Q}) d\mathbf{k}_1. \tag{89}$$

The integral in Eq. (89) is not easy to be numerically realized. In the following, we introduce a method to evaluate the integral.

We notice that the most time-consuming part in computing Eq. (50) is the term $M_{l_1,l_2}^L(k_1, k_2)$ expressed in Eq. (49). The term $M_{l_1,l_2}^L(k_1, k_2)$ is dependent on the magnitudes of the vectors $\mathbf{k}_1$ and $\mathbf{k}_2$ but not on their angles. Thus if we can choose k_1 and k_2 as two of the three integral coordinates in Eq. (89), it will be possible to reduce the computational burden. When the third integral coordinate is changed, the term $M_{l_1,l_2}^L(k_1, k_2)$ will not need to be repeatedly calculated.

We now need to choose another integral coordinate, which should be convenient to determine the angels of the vectors $\mathbf{k}_1$ and $\mathbf{k}_2$ in the condition that k_1, k_2 and Q are known. Due to the relation that $\mathbf{k}_1 + \mathbf{k}_2 + \mathbf{Q} = 0$, the vectors $\mathbf{k}_1$, $\mathbf{k}_2$ and $\mathbf{Q}$ can determine a triangle in the momentum space by locating the start point of one vector on the end point of another. The magnitudes k_1, k_2 and Q can determine the shape of the triangle, leaving the position of triangle undetermined. Though the vector $\mathbf{Q}$ is determined, the triangle can still be rotated around the vector $\mathbf{Q}$. To determine the vectors $\mathbf{k}_1$ and $\mathbf{k}_2$, we just need to fix the triangle by determining the rotation angle of the triangle. So we choose the rotation angle as the third integral coordinate. To define the rotation angel, it will be convenient to rotate these vectors to a new frame in which the direction of the recoil ion momentum is along the z axis. We denote the azimuthal and polar angle of $\mathbf{Q}$ as φ_q and θ_q respectively. The coordinate transformation of vectors is performed by two steps of rotations, around z axis by

φ_q and around y axis by θ_q. By the rotations, a vector $\mathbf{k}_i = (k_{xi}, k_{yi}, k_{zi})^T$ will be transformed to

$$\mathbf{k}'_i = \begin{bmatrix} \cos\theta_q & 0 & -\sin\theta_q \\ 0 & 1 & 0 \\ \sin\theta_q & 0 & \cos\theta_q \end{bmatrix} \begin{bmatrix} \cos\varphi_q & \sin\varphi_q & 0 \\ -\sin\varphi_q & \cos\varphi_q & 0 \\ 0 & 0 & 1 \end{bmatrix} \begin{pmatrix} k_{xi} \\ k_{yi} \\ k_{zi} \end{pmatrix}. \tag{90}$$

The new vector $\mathbf{k}'_i = (k'_{xi}, k'_{yi}, k'_{zi})^T$ can be transformed backward to the original vector by

$$\mathbf{k_i} = \begin{bmatrix} \cos\varphi_q & -\sin\varphi_q & 0 \\ \sin\varphi_q & \cos\varphi_q & 0 \\ 0 & 0 & 1 \end{bmatrix} \begin{bmatrix} \cos\theta_q & 0 & \sin\theta_q \\ 0 & 1 & 0 \\ -\sin\theta_q & 0 & \cos\theta_q \end{bmatrix} \begin{pmatrix} k'_{xi} \\ k'_{yi} \\ k'_{zi} \end{pmatrix}. \tag{91}$$

The transformation defined in Eq. (90) can redirect the vector $\mathbf{Q}$ to the z axis. The polar angles of rotated momentum vectors $\mathbf{k}'_1$ and $\mathbf{k}'_2$ will be simply given by the law of cosines

$$\theta'_1 = \arccos\frac{Q^2 + k_1^2 - k_2^2}{2Qk_1}, \tag{92}$$

$$\theta'_2 = \arccos\frac{Q^2 + k_2^2 - k_1^2}{2Qk_2}. \tag{93}$$

The azimuthal angles of $\mathbf{k}'_1$ and $\mathbf{k}'_2$ are correlated by $\varphi'_1 + \varphi'_2 = 0$. We choose φ'_1 as the third integral coordinate, and the vectors $\mathbf{k}_1$ and $\mathbf{k}_2$ can be determined by Eqs. (91)–(93). The volume element $d\mathbf{k}'_1$ can be expressed as,

$$d\mathbf{k}'_1 = dk_1 dk_2 d\varphi'_1 k_1 k_2 / Q. \tag{94}$$

Finally, the recoil-ion momentum spectrum is reformulated as

$$P(\mathbf{Q}) = \int\int\int_{|k_1 - k_2| \leq Q \leq k_1 + k_2} \frac{P(\mathbf{k}_1, \mathbf{Q})k_1 k_2}{Q} dk_1 dk_2 d\varphi'_1. \tag{95}$$

It is not difficult to see that $P(Q, \theta_q, \varphi_q) = P(Q, \theta_q, 0)$, by a simple symmetry analysis.

For n-photon double ionization process, we can also define the triply differential cross section for the recoil ion,

$$\frac{d\sigma^{(n)}}{dQ_x dQ_y dQ_x} = \left(\frac{\omega}{I_0}\right)^n \frac{1}{T_{\text{eff}}^{(n)}} P(\mathbf{Q}), \tag{96}$$

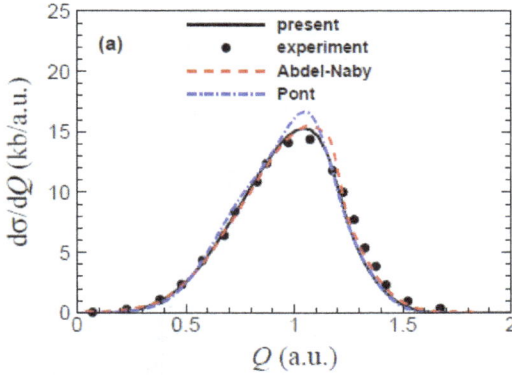

Fig. 6. Recoil-ion singly differential cross section for the one-photon double ionization at a photon energy of 99 eV. The present numerical calculation is compared with the experimental measurement [81] and other two theoretical calculations (Pont *et al.* [82] and Abdel-Naby *et al.* [79]). (from Ref. [21])

where $T_{\text{eff}}^{(n)}$ has been defined in Eq. (82). The singly differential cross section for recoil ion can be defined by

$$\frac{d\sigma^{(n)}}{dQ_x dQ_y dQ_x} = \left(\frac{\omega}{I_0}\right)^n \frac{1}{T_{\text{eff}}^{(n)}} P(\mathbf{Q}). \tag{97}$$

In the following, we give some numerical results for the recoil-ion momentum spectrum. In Fig. 6, we show the recoil-ion singlely differential cross section for PDI at photon energy of 99 eV. Present result is compared with the experimental measurement [81] and other two theoretical calculations [79, 82], and we see these results are consistent with each other rather well.

The recoil-ion triply differential cross sections are shown in Fig. 7 for PDI at the photon energies of 99 eV, 125 eV and 225 eV. Present results generally coincide with the calculations of Abdel-Naby *et al.* [79] except for some discrepancies in the low momentum region. It is widely known that the equal-energy sharing back-to-back ejection mode for the two electrons is forbidden in one-photon double ionization. According to this selection rule, the recoil-ion signal at zero momentum will be zero, which can be clearly seen in our calculations shown in Fig. 6. The maxima of recoil ion can be predicted by $Q_{\max} = 2\sqrt{E_{\text{exc}}}$ (shown as thick black vertical lines in Fig. 6), which corresponds to the two electrons are ejected with the same energy sharing and in the same direction. In addition, we see that double-peak structure is a general character for all the photon energies in our calculations. The positions of the peaks can be estimated by studying the equal-energy coplanar distribution of the two electrons. The two electrons favor to be ejected with a special inter-angle, and the recoil ion momentums in such ejection modes are marked as thin blue vertical lines in Fig. 7.

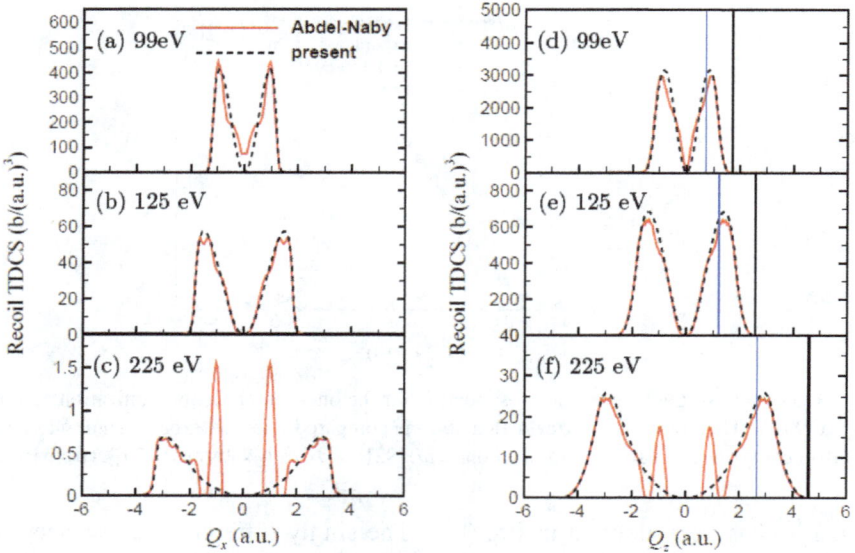

Fig. 7. The recoil ion triply differential cross section for one-photon double ionization at the photon energies of 99 eV, 125 eV, and 225 eV. In the left (right) column, we show the signals along the axis which is perpendicular to (parallel with) the laser polarization, where $Q_y = Q_z = 0$ ($Q_x = Q_y = 0$). The present numerical calculations are compared with those by Abdel-Naby *et al.* [79] (from Ref. [21])

The recoil-ion triply differential cross sections are shown in Fig. 8 for TPDI at the photon energies of 44 eV. Consistent with the experimental measurement shown in Fig. 5, the most intense recoil-ion signal is around the zero momentum. Similar to Fig. 7, the predicted maxima of recoil ion is marked as thick black vertical lines, and the thin blue vertical lines represent the ejection modes that the two electrons are ejected with equal-energy sharing and a favorite inter-angel.

In Fig. 9, we show the recoil ion momentum distribution in the x-z plane for one-photon [Fig. 9(a)], nonsequential two-photon [Fig. 9(b)], and sequential two-photon [Fig. 9(c)] double ionization. We see that different distribution characters can be identified for these three different double ionization processes. In above, we have analyzed the characters for one-phonon and nonsequential two-photon double ionization. For the sequential two-photon [Fig. 9(c)] double ionization, we see four-peak structures in the recoil-ion momentum spectrum. In the sequential two-photon double ionization, the two electrons favor to be ejected with two particular energies, i.e. $E_1 = \omega - 24.6\,\text{eV}$ and $E_2 = \omega - 54.4\,\text{eV}$. When the two electrons with energies $E_1 = \omega - 24.6\,\text{eV}$ and $E_2 = \omega - 54.4$ are ejected in the same (opposite) direction, the recoil ion will be ejected with the momentum $Q = \sqrt{2E_1} + \sqrt{2E_2}$ ($Q = \sqrt{2E_1} - \sqrt{2E_2}$), which is marked as the bigger (smaller) black dashed circle. The black dashed circles roughly coincide with the four-peak positions, and so the

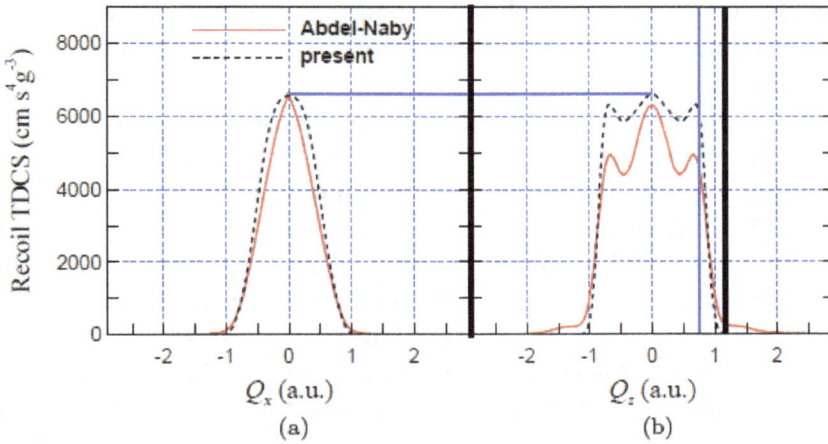

Fig. 8. Same as Fig. 7, but for two-photon double ionization at the photon energy of 44 eV. The present numerical calculations are compared with the recent numerical calculations by Abdel-Naby *et al.* [80] (from Ref. [21])

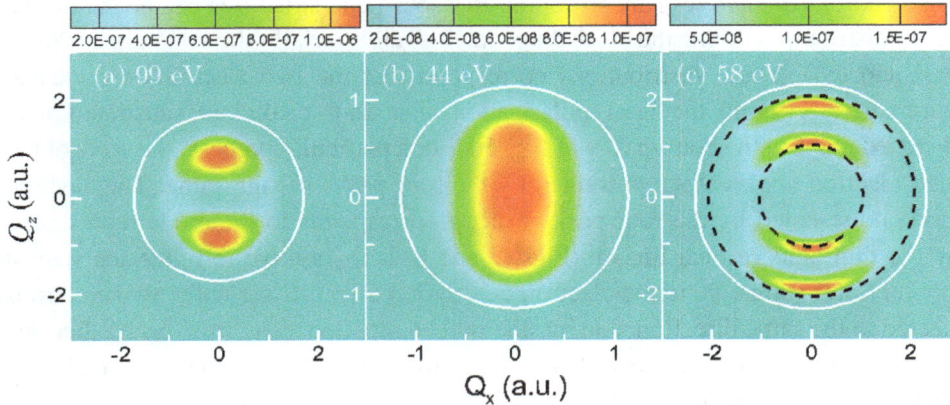

Fig. 9. Recoil ion momentum distribution in the x-z plane for (a) one-photon, (b) two-photon nonsequential, and (c) two-photon sequential double ionization. (from Ref. [21])

origin of the four-peak structure is addressed. In Fig. 9, those white circles mark the predicted maxima, similar to the thick black vertical lines in Figs. 7 and 8.

3.2. *Electron correlation in the joint angular distributions*

The most detailed information on the double ionization process is the two-electron angular distributions. Electron correlation effect is more important in the calculations of two-electron angular distributions than in the calculation of total double ionization or recoil ion spectrum. In the following, we will analyze the two-electron angular distribution in the co-plane which is determined by the momentum vectors of the two electrons and the laser polarization direction. The special equal-energy sharing case is considered.

In Fig. 10, the two-electron angular distributions are shown for three different double ionization processes, i.e. one-photon double ionization at 85 eV [Figs. 10(a1), (a2) and (a3)], nonseqential two-photon double ionization at 42.5 eV [Figs. 10(b1), (b2) and (b3)], and sequential two-color two-photon double ionization at photon energies of 30 eV and 55 eV [Figs. 10(c1), (c2) and (c3)]. We see that the two-electron distributions are quite different in different ionization processes, even though the excess energies are the same in the processes. The nonsequential and sequential two-photon ionization processes produce quite similar two-electron angular distributions, which, however, are obvious different from the distributions in one-photon double ionization. The back-to-back ejection is the most favorable ejection mode in the two-photon double ionization, but this mode is forbidden in one-photon double ionization. Common characters can also been identified in the three ionization processes. The distributions along those black diagonal lines in Figs. 10(a1), (b1) and (c1) are so similar that four-peak structures can be identified. These distributions along the black diagonal lines are specially shown in Figs. 10(a3), (b3) and (c3). These distributions correspond that the two electrons are ejected symmetrically along the laser polarization. The four peaks are illustrated with sketched maps in the bottom row of Fig. 10. For a particular process, the sizes of the inter-electron angles $\Delta\theta$ represented by the four peaks are the same. The ejection mode represented by the four peaks exactly corresponds to the mode represented by the thin blue vertical lines in Figs. 7 and 8. It is intuitive to relate the size of the inter-electron angle to the electron correlation effect. Classically, two electrons ejected in the same direction side by side will be repelled away from each other due to the Coulomb force between them. In the following we will make comparisons for the sizes of inter-electron angles in different double ionizations.

Before the comparisons for the sizes of inter-electron angles, it is worthy to mention a widely used fitting formula for the two-electron angular distributions in the one-photon double ionization process. This formula is based on the Wannier theory. In the equal energy sharing case, the dependence of the differential probability density on the ejected angles of the two electrons on the co-plane satisfies [83–85]

$$P(\theta_1, \theta_2) \propto (\cos\theta_1 + \cos\theta_2)^2 e^{-4\ln 2\left(\frac{\theta_{12}-\pi}{\theta_{1/2}}\right)^2}, \tag{98}$$

where θ_{12} ($0 \le \theta_{12} \le \pi$) is the angle between the two electrons, and $\theta_{1/2}$ is a fitting parameter, which stands for the full width at half maxima (FWHM) of the Gaussian-shape envelope centered at π. The parameter $\theta_{1/2}$ changes with the variations of the photon energy, and it was often used to measure the electron correlation. We

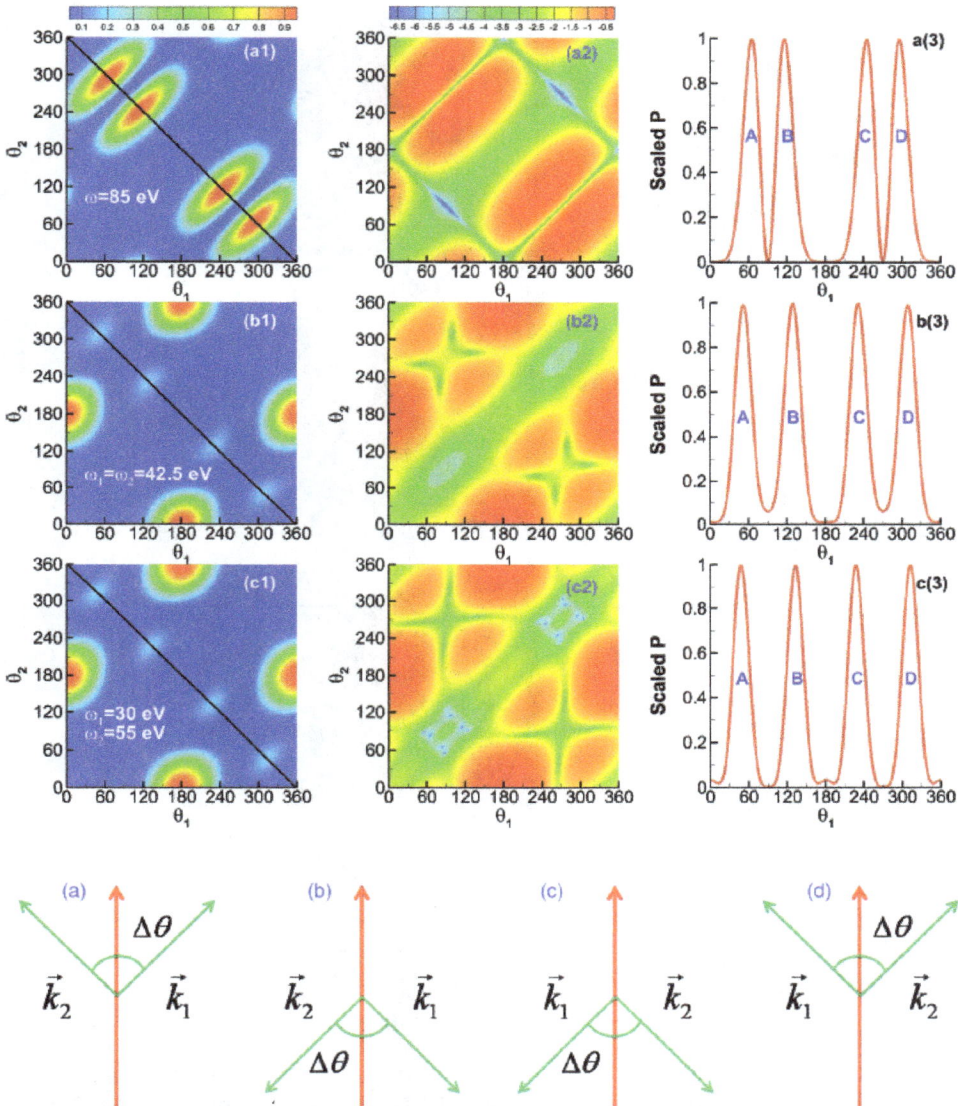

Fig. 10. The coplanar angular distributions of the two equal energy sharing electrons for the one-photon double-ionization process (the first row), the two-photon nonsequential double-ionization process (the second row), and the two-photon sequential double ionization process (the third row). The photon energies are marked in each panel of the first column, in which the differential probability density is shown in the same normalized linear scale. The second column shows the corresponding differential probability density on the same normalized log scale. The third column shows the scaled differential probability along the line $\theta_1 + \theta_2 = 360°$, which is marked as a black line in each panel of the first column. The data in the third column have been scaled to unit at the maxima value for clarity. The four peaks (marked as A, B, C, and D) are illustrated with sketched maps in the bottom row. (from Ref. [20])

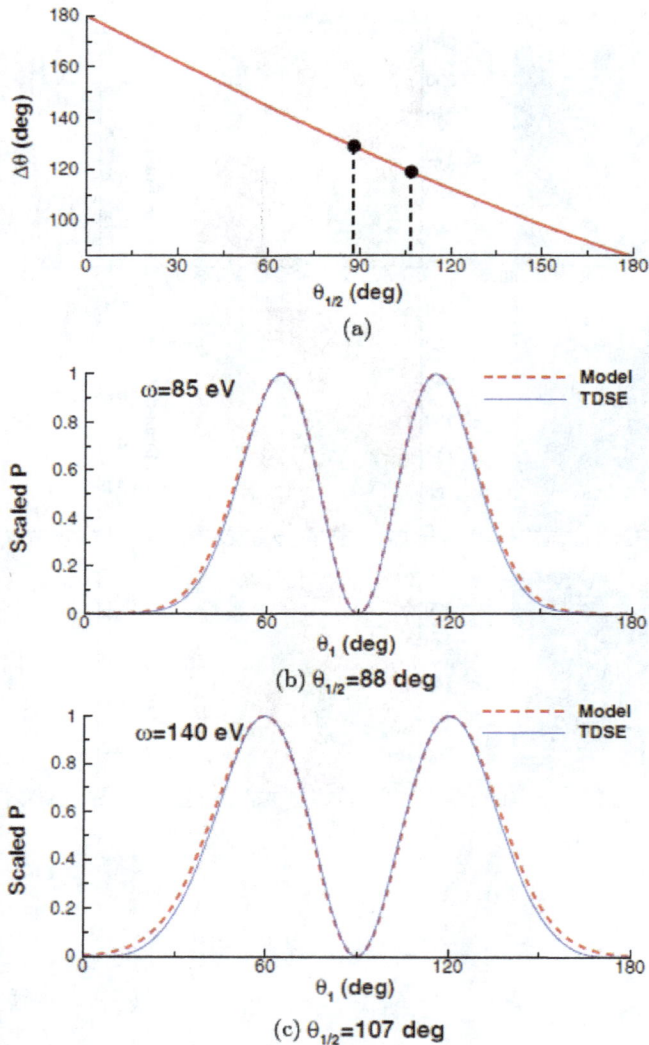

Fig. 11. (Color online) (a) The relation of the interelectron angle $\Delta\theta$ with the fitting parameter $\theta_{1/2}$ built from Eq. (98). (b) and (c) Comparisons of the results according to Eq. (98) and from the TDSE calculations for the one-photon double-ionization process. The photon energy ω and the fitting parameter $\theta_{1/2}$ are marked in the corresponding panel. (from Ref. [20])

can predict the same two-electron angular distributions as shown in Fig. 10(a3) by Eq. (98). So it is not difficult to built a one-to-one correspondence between the inter-electronic angle $\Delta\theta$ and the fitting parameter $\theta_{1/2}$, see Fig. 11(a). This means that the present inter-electronic angle can be a substitute of the more popular parameter $\theta_{1/2}$. In Figs. 11(b) and (c), we can see that the fitting formula in Eq. (98) can reproduce the TDSE results rather well.

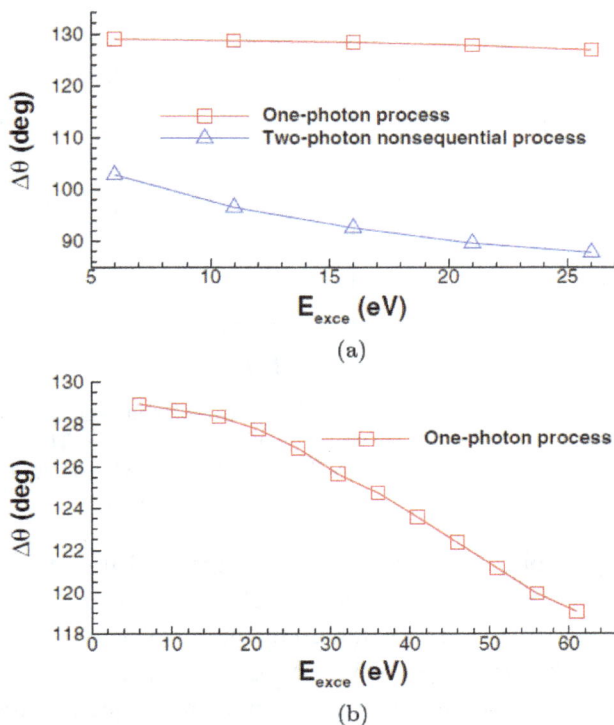

Fig. 12. (a) The comparison of the interelectron angles $\Delta\theta$ in the one-photon double-ionization process (squares) and the two-photon nonsequential double-ionization process (triangles) for a series of excess energies. (b) Interelectron angles for a wide range of excess energy are shown for the one-photon process. (from Ref. [20])

In Fig. 12, the inter-electron angle $\Delta\theta$ as a function of different excess energy E_{exc} for the one-photon and two-photon nonsequential double ionization. For the one-photon double ionization process, the series of pulses have center frequencies ranging from 85 eV to 105 eV, which means that the excess energies range from 6 eV to 26 eV. For the two-photon nonsequential double ionization, the series of pulses have center frequencies ranging from 42.5 eV to 52.5 eV, which produce the same excess energies as those in the one-photon process. From the comparison in Fig. 3(a), for the same excess energy, it is clear that the inter-electron angle $\Delta\theta$ in the one-photon process is much larger ($25°$ or more) than those in the two-photon nonsequential process.

To compare the nonsequential and sequential two-photon double ionization, it will be instructive to look at the inter-electron angle for the same excess energy. To achieve this goal, we employ two-color pulses. The sum of the two-photon energies is fixed to keep the excess energies the same in two-photon double ionization. In Fig. 13, the sum of energies is taken to be 85 eV. We allow the relative

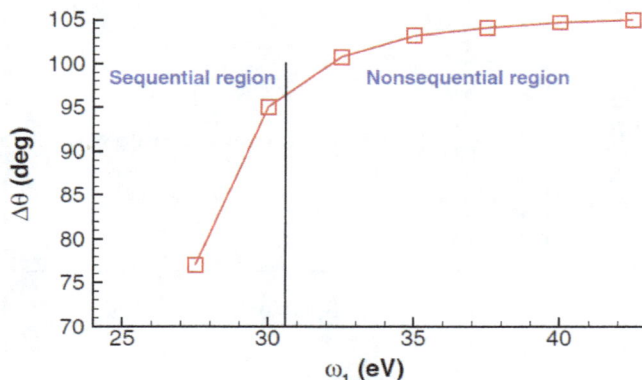

Fig. 13. Comparison of the interelectron angle in the two-photon sequential and nonsequential double-ionization processes. The pulse durations are all taken to be 3 fs. The sum of the energies of the two photons are fixed as $\omega_1 + \omega_2 = 85$ eV. The boundary ($\omega_1 = 30.6$ eV) of the sequential and nonsequential regions is marked as a vertical black line. (from Ref. [20])

size of the energies of the two photons to change, and then two-photon double ionization will change from sequential (24.6 eV $< \omega_1 < 30.6$ eV) to nonsequential (30.6 eV $< \omega_1 < 42.5$ eV). Here ω_1 represent the smaller one of the two photon energies. The boundary ($\omega_1 = 30.6$ eV) of sequential and nonsequential processes is marked as a vertical black line in Fig. 13. In the nonsequential region, the interelectron angle $\Delta\theta$ is not sensitive to the ratio of the energies of two photons. But when it enters into the sequential region as ω_1 becomes smaller, inter-electron angle $\Delta\theta$ decreases drastically.

3.3. *Double ionization of helium by double attosecond pulses*

The applications of attosecond pulses are attracting more and more attentions many of the applications will involve double attosecond pulses [88–91]. In the following, we study a two-electron interference phenomenon in the double ionization of helium by time-delayed double attosecond pulses, and the effects of carrier envelop phases (CEP) in the interference process. These predications from TDSE calculations have a great chance to be demonstrated by the future experimental measurements.

Our proposal is instructed in Fig. 14. Mathematically, the attosecond pulses are described by giving the expressions for the electric fileds,

$$\mathbf{F}(t) = [F_1(t) + F_2(t)]\hat{\mathbf{z}} = -\frac{\partial}{\partial t}[A_1(t) + A_2(t)]\hat{\mathbf{z}} \qquad (99)$$

where,

$$A_i(t) = A_0 \cos^2\left[\frac{(t - T_{0i})\pi}{T}\right] \sin\left[\omega_0(t - T_{0i}) + \phi_i\right], \qquad (100)$$

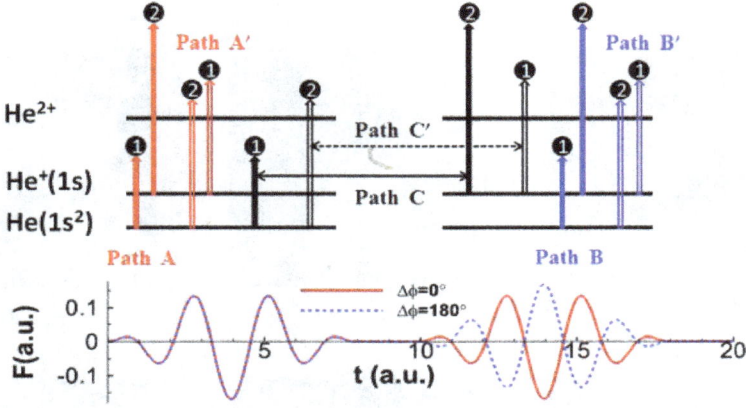

Fig. 14. Six pathways of the sequential double ionization of He [upper row] by the two time-delayed pulses [lower row, cf. Eq. (1)]. Path A (B): double ionization by sequentially absorbing two photons from the first (second) pulse. Path C: double ionization by absorbing one photon from each pulse. Paths A′, B′ and C′ are respectively the exchange symmetry pathways of A, B and C. Note that () represents the electron with the same energy E_1 (E_2). (from Ref. [19])

for $-T/2 + T_{0i} \leq t \leq T/2 + T_{0i}$, and $A_i(t) = 0$ for other t. In the present calculations, we take the two pulses to have the same center frequency $\omega_0 = 65\,\text{eV}$, the same pulse duration $T = 3$ cycles, and the same peak amplitude A_0. The time delay and relative CEP between the two pulses are respectively defined as $\tau = T_{02} - T_{01}$ and $\Delta\phi = \phi_2 - \phi_1$.

As can be seen in Figs. 15(a) and (b), interesting interference patterns can be observed in the two-electron energy spectrum from TDSE calculations. These interference patterns can be addressed by a model calculation base on the second-order time-dependent perturbation theory, in which the two-electron energy spectrum is given by [19, 39, 86]

$$P(E_1, E_2) = \frac{1}{2}\left(\frac{c}{4\pi^2}\right)^2 \left| \sqrt{\frac{\sigma^{\text{He}}(E_1)}{\omega_{ai}}} \sqrt{\frac{\sigma^{\text{He}^+}(E_2)}{\omega_{fa}}} K(E_a) \right.$$

$$\left. + \sqrt{\frac{\sigma^{\text{He}}(E_2)}{\omega_{bi}}} \sqrt{\frac{\sigma^{\text{He}^+}(E_1)}{\omega_{fb}}} K(E_b) \right|^2 , \tag{101}$$

where $\omega_{ai} = E_a - E_i, \omega_{fa} = E_f - E_a, \omega_{bi} = E_b - E_i, \omega_{fb} = E_f - E_b, E_i = -79\,\text{eV}, E_a = E_1 - 54.4\,\text{eV}, E_b = E_1 - 54.4\,\text{eV} , E_f = E_1 + E_2, \sigma^{\text{He}}$ and σ^{He^+} are the one-photon single ionization cross section of He and He$^+$ respectively [87],

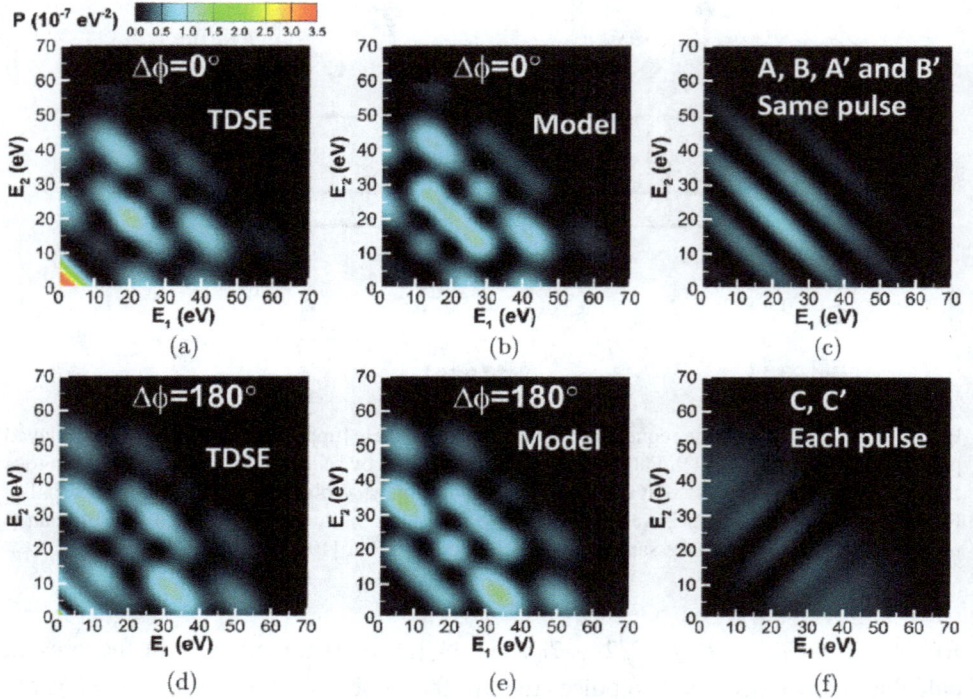

Fig. 15. Angle-integrated joint energy spectra $P(E_1, E_2)$ of the two electrons. (a) and (b): results of the TDSE calculations for $\Delta\phi = 0°$ and $180°$ respectively. (c) and (d): corresponding results of the perturbation model by Eq. (4), including all pathwayssketched in Fig. 14. (e): results of the perturbation model for $\Delta\phi = 0°$, only including paths A, B, A′ and B′. (f): results of the perturbation model for $\Delta\phi = 0°$, only including paths C and C′. All the panels share the same color bar. (from Ref. [19])

and the field-integral function $K(E_a)$ is given by

$$K(E_a) = \int_{-\infty}^{\infty} d\tau_1 F(\tau_1) e^{i\omega_{fa}\tau_1} \int_{-\infty}^{\tau_1} d\tau_2 F(\tau_2) e^{i\omega_{ai}\tau_2}. \tag{102}$$

The results from calculating Eq. (102) is given in Figs. 15(c) and (d), which reproduce the corresponding TDSE results rather well in the high energy region $E_1 + E_2 > 15\,\text{eV}$ where two-photon process dominates. To make the physics behind the interference patterns more clear, we make more detailed analysis with the help of present model.

For cases considered in the present work, the overlap between the two pulses is zero or negligible so that field-integral function defined in Eq. (102) can be broken up into three terms:

$$K(E_a) = K_1(E_a) + e^{i[(\omega_{fa}+\omega_{ai})\tau - 2\Delta\phi]} K_1(E_a)$$

$$+ e^{i(\omega_{fa}\tau - \Delta\phi)} \mathcal{F}_1(\omega_{ai}) \mathcal{F}_1(\omega_{fa}), \tag{103}$$

where $K_1(E_a)$ is defined by replacing $F(t)$ with $F_1(t)$ in Eq. (102), and $\mathcal{F}_1(\omega)$ is the Fourier transform of $F_1(t)$. Each term in Eq. (103) has an apparent physical meaning and corresponds to the ionization paths A, B and C shown in Fig. 14 respectively. Similarly the $K(E_b)$ can be broken up into three terms, which corresponds to the ionization paths A', B' and C' respectively. The phase difference between paths A (A') and B (B') is $\Delta\Phi_{AB} = (\omega_{fa} + \omega_{ai})\tau - 2\Delta\phi = (E_1 + E_2 + I_p)\tau - 2\Delta\phi$, which leads to strip-like interference structures parallel to the line $E_1 + E_2 = 0$, as shown in Fig. 15(e). For path C and C', the phase difference is $\Delta\Phi_{CC'} = (\omega_{fb} - \omega_{fa})\tau = (E_2 - E_1)\tau$, which leads to strip-like interference structures parallel to the line $E_1 - E_2 = 0$, as shown in Fig. 15(f).

Compared (a) and (b) in Fig. 15, we can identify the relative CEP effects in the interference patterns. Next we explore the relative CEP effects more deeply. In Fig. 16, the energy spectra for the sum energy of the two electrons, which is defined by $P(E) = \int_0^E P(E_1, E - E_1) dE_1$, are shown for a series of relative CEPs. We can clearly see from Fig. 16(b) that the changing of the relative CEP leads the shift of those interference peaks by $\Delta\phi = \Delta E' \pi / \Delta E$, where $\Delta E'$ is peak shift and ΔE is separation distance of two adjacent interference peaks. The ΔE is read off to be 15.8 eV, which approximately corresponds to the roughly predicted value $2\pi/\tau = 17.1$ eV.

In Fig. 17 we show that not only the interference patterns on the joint energy spectra, but also the total double ionization probability is dependent on the relative CEP between the two pulses. The total double ionization probability is calculated by $P_{\text{total}} = \int_0^\infty P(E) dE$. To compare the TDSE results and the model results, we also calculated the total double ionization in the high-energy region by $P_{\text{high}} = \int_{E_b}^\infty P(E) dE$, where $E_b = 15$ eV. The magnitude of oscillation for the total double ionization as the relative CEP is related to the time delay between the two attodecond pulses, as can be seen in Figs. 17(c) and (d). This observation can be easily understood by making comparison with the following hypothetical double-slit interference experiment.

In our hypothetical double-slit interference experiment, the optical wave passing through the double slits interfere, and interference fringes appear on a screen. Now we open a slit on the screen, and the width of this silt is a little smaller than the fringe spacing. There are two sources of photons passing through the slit on the screen, which are ejected from the different slit of the double slits. The number of the photons passing through the silt per unit time depends on the relative phase

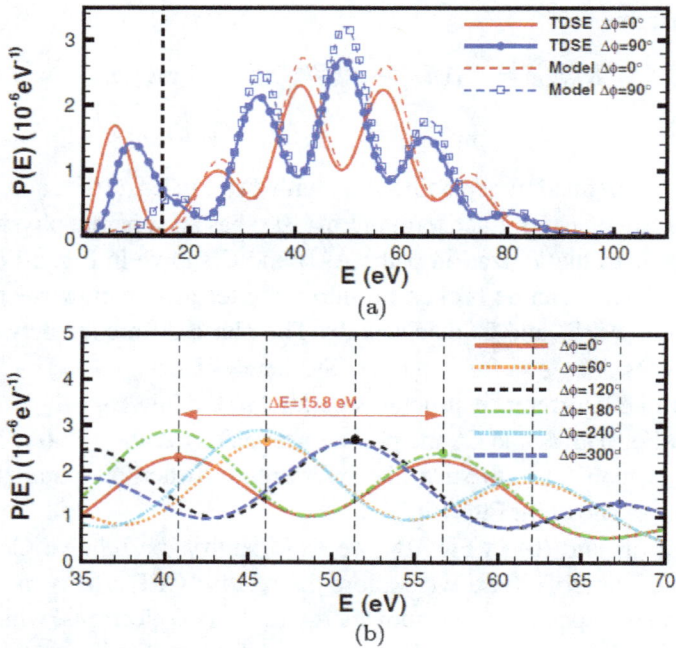

Fig. 16. Total energy spectra $P(E)$ of the two electrons for $I_0 = 1\,\mathrm{PW/cm^2}$ and $\tau = 10$ a.u. (a): comparison between the TDSE method and the perturbation model for $\Delta\phi = 0°$ and $90°$. (b): TDSE results for a series of relative CEP $\Delta\phi$. The horizontal red arrow indicates the fringe space $\Delta E = 15.8\,\mathrm{eV}$. The vertical dashed lines are even spaced by $\Delta E/3$, and the leftmost line is at $E = 40.8\,\mathrm{eV}$, which is a peak position for $\Delta\phi = 0°$. The circles indicate the cross points between the vertical dashed lines and the curves of the total energy spectra. (from Ref. [19])

of the two-photon wave packets. In this case, by adjusting the relative phase of the two-photon beams ejected from the double slits, the interference fringe on the screen will be moved. When a bright (dark) fringe happens to move to the slit on the screen, there are more (less) photons passing through it. The present double ionization of the electrons is similar to this double-slit interference experiment. The common point is that interference patterns appear in a limited region. The finite width of the joint energy spectrum of the two electrons has a similar effect to the third slit in the above hypothetical double-slit experiments. Generally speaking, when interference fringes form in a limited region, the relative phases of different wave packets can affect the interference patterns, usually in the form of fringe shifts. As a result, the total number of the particles (represented by the wave packets) that pass through the limited region can be modulated by the relative phases between the different wave packets.

Let us go back to the problem on the magnitude of oscillation for the total double ionization. In the above double-slit interference experiment, the modulation degree

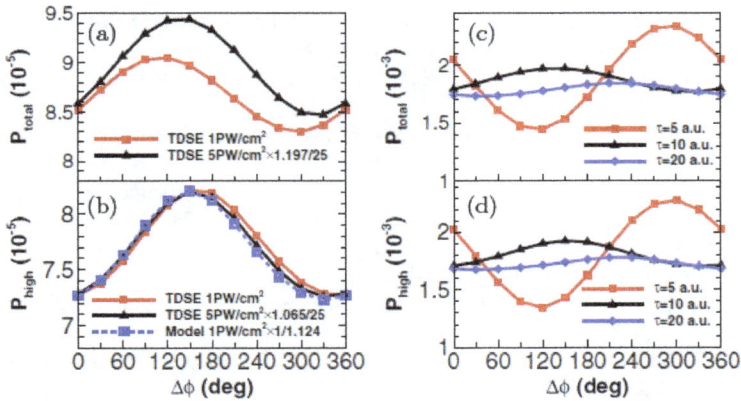

Fig. 17. Total double ionization probability P_{total} and partial double ionization probability P_{high} in the high region as a function of the relative CEP. (a) and (b): results for the time delay $\tau = 10$ a.u. and peak intensity $I_0 = 1\,PW/cm^2$ or $5\,PW/cm^2$ from the TDSE calculations. Also shown in (b) is the result from the perturbation model for $\tau = 10$ a.u. and peak intensity $I0 = 1\,PW/cm^2$. (c) and (d): results for $I0 = 5\,PW/cm^2$ at time delay $\tau = 5$, 10, and 20 a.u respectively. (from Ref. [19])

for the number of photons passing through the slit on the screen is clearly decided by the relative sizes between the fringe spacing and the width of the slit on the screen. In the present case, the width of the slit on the screen corresponds to the width of the energy spectra of ionized electrons or the width of the frequency spectra of the attosecond pulses, which is fixed. Then the modulation degree (or magnitude of oscillation) is expected to be proportional to interference fringe spacing in the joint energy spectra of the two electrons, which is in turn decided by the time delay between the two attosecond pulses.

References

1. M. Couprie, New generation of light sources: Present and future, *J. Electron. Spectros. Relat. Phenom.* **196**, (2014) 3.
2. C. Leforestier, R. Bisseling, C. Cerjan, M. Feit, R. Friesner, A. Guldberg, A. Hammerich, G. Jolicard, W. Karrlein, H. Meyer, N. Lipkin, O. Roncero, R. Kosloff, A comparison of different propagation schemes for the time-dependent schrodinger-equation, *J. Comput. 3220 Phys.* **94**, (1991) 59.
3. S. Billeter, W. VanGunsteren, A comparison of different numerical propagation schemes for solving the time-dependent Schrodinger equation in the position representation in one dimension for mixed quantum- and molecular dynamics simulations, *Mol. Simulat.* **15**, (1995) 301.
4. H. G. Muller, An efficient propagation scheme for the time-dependent schrödinger equation in the velocity gauge, *Laser Phys.* **9**, (1999) 138.
5. D. Bauer, P. Koval, QPROP: A Schrodinger-solver for intense laser-atom interaction, Comput. *Phys. Commun.* **174**, (2006) 396.
6. J. Parker, E. Smyth, K. Taylor, Intense-field multiphoton ionization of helium, *J. Phys. 3195 B* **31**, (1998) L571.

7. E. Smyth, J. Parker, K. Taylor, Numerical integration of the time-dependent Schrodinger equation for laser-driven helium, *Comput. Phys. Commun.* **114**, (1998) 1.
8. K. Taylor, J. Parker, D. Dundas, E. Smyth, S. Vivirito, Laser-driven helium in fulldimensionality, *Laser Phys.* **9**, (1999) 98.
9. J. Parker, L. Moore, E. Smyth, K. Taylor, One- and two-electron numerical models of multiphoton ionization of helium, *J. Phys. B* **33**, (2000) 1057.117.
10. J. Parker, D. Glass, L. Moore, E. Smyth, K. Taylor, P. Burke, Time-dependent and time-independent methods applied to multiphoton ionization of helium, *J. Phys. B* **33**, (2000) L239.
11. T. N. Rescigno, C. W. McCurdy., Numerical grid methods for quantum-mechanical scattering problems, *Phys. Rev. A* **62**, (2000) 032706.
12. J. Lill, G. Parker, J. Light, Discrete variable representations and sudden models in 3210 quantum scattering-theory, *Chem. Phys. Lett.* **89**, (1982) 483.
13. J. Lill, G. Parker, J. Light, The discrete variable finite basis approach to quantum scattering, *J. Chem. Phys.* **85**, (1986) 900.
14. J. Light, I. Hamilton, J. Lill, Generalized discrete variable approximation in quantum mechanics, *J. Chem. Phys.* **82**, (1985) 1400.
15. G. Corey, J. Tromp, Variational discrete variable representation, *J. Chem. Phys.* **103**, (1995) 1812.
16. M. S. Pindzola, F. Robicheaux, S. D. Loch, J. C. Berengut, T. Topcu, J. Colgan, M. Foster, D. C. Griffin, C. P. Ballance, D. R. Schultz, T. Minami, N. R. Badnell, M. C. Witthoeft, D. R. Plante, D. M. Mitnik, J. A. Ludlow, U. Kleiman, The time-dependent close-coupling method for atomic and molecular collision processes, *J. Phys. B* **40**, (2007) R39.
17. L.-Y. Peng, Z. Zhang, W.-C. Jiang, G.-Q. Zhang, Q. Gong, Probe of the electron correlation in sequential double ionization of helium by two-color attosecond pulses, *Phys. Rev. A* **86**, (2012) 063401.
18. Z. Zhang, L.-Y. Peng, M.-H. Xu, A.F. Starace, T. Morishita, and Q. Gong, Two-photon double ionization of helium: Evolution of the joint angular distribution with photon energy and two-electron energy sharing. *Phys. Rev. A* **84**, (2011) 043409.
19. W.-C. Jiang, W.-H. Xiong, T.-S. Zhu, L.-Y. Peng, Q. Gong, Double ionization of He by time-delayed attosecond pulses, *J. Phys. B* **47**, (2014) 091001.
20. W.-C. Jiang, L.-Y. Peng, W.-H. Xiong, Q. Gong, Comparison study of electron correlation in one-photon and two-photon double ionization of helium, *Phys. Rev. A* **88**, (2013) 023410.
21. W.-C. Jiang, Y. Tong, Q. Gong, L.-Y. Peng, Recoil-ion-momentum spectrum for fewphoton double ionization of helium, *Phys. Rev. A* **89**, (2014) 043422.
22. W.-C. Jiang, L.-Y. Peng, J.-W. Geng, Q. Gong, One-photon double ionization of H2 with arbitrary orientation, *Phys. Rev. A* **88**, (2013) 063408.
23. X.-F. Hou, L.-Y. Peng, Q.-C. Ning, Q. Gong, Attosecond streaking of molecules in the low-energy region studied by a wavefunction splitting scheme, *J. Phys. B* **45**, (2012) 074019.
24. D. Yan, L.-Y. Peng, Q. Gong., Grid method for computation of generalized spheroidal wave functions based on discrete variable representation, *Phys. Rev. E* **79**, (2009) 036710.
25. X. Guan, K. Bartschat, B. I. Schneider, Breakup of the aligned H2 molecule by xuv laser pulses: A time-dependent treatment in prolate spheroidal coordinates, *Phys. Rev. A* **83**, (2011) 043403.
26. L. Tao, C. W. McCurdy, T. N. Rescigno, Grid-based methods for diatomic quantum scattering problems. III. Double photoionization of molecular hydrogen in prolate spheroidal coordinates, *Phys. Rev. A* **82**, (2010) 023423.
27. L. Avaldi, A. Huetz, Photodouble ionization and the dynamics of electron pairs in the continuum, *J. Phys. B* **38**, (2005) S861.
28. J. Briggs, V. Schmidt, Differential cross sections for photo-double-ionization of the helium atom, *J. Phys. B* **33**, (2000) R1.
29. H. Hasegawa, E. Takahashi, Y. Nabekawa, K. Ishikawa, K. Midorikawa, Multiphoton ionization of He by using intense high-order harmonics in the soft-x-ray region, *Phys. Rev. A* **71**, (2005) 023407.

30. Y. Nabekawa, H. Hasegawa, E. Takahashi, K. Midorikawa, Production of doubly charged helium ions by two-photon absorption of an intense sub-10-fs soft x-ray pulse at 42 eV photon energy, *Phys. Rev. Lett.* **94**, (2005) 043001.

31. A. A. Sorokin, M. Wellhoefer, S. V. Bobashev, K. Tiedtke, M. Richter, X-ray-laser interaction with matter and the role of multiphoton ionization: Free-electron-laser studies on neon and helium, *Phys. Rev. A* **75**, (2007) 051402.

32. L. A. A. Nikolopoulos, P. Lambropoulos, Comment on "Production of doubly charged helium ions by two-photon absorption of an intense sub-10-fs soft X-ray pulse at 42 eV photon energy", *Phys. Rev. Lett.* **97**, (2006) 169301.

33. A. Rudenko, L. Foucar, M. Kurka, T. Ergler, K. U. Kuehnel, Y. H. Jiang, A. Voitkiv, B. Najjari, A. Kheifets, S. Luedemann, T. Havermeier, M. Smolarski, S. Schoessler, K. Cole, M. Schoeffler, R. Doerner, S. Duesterer, W. Li, B. Keitel, R. Treusch, M. Gensch, C. D. Schroeter, R. Moshammer, J. Ullrich, Recoil-ion momentum distributions for two-photon double ionization of he and ne by 44 eV free-electron laser radiation, *Phys. Rev. Lett.* **101**, (2008) 073003.

34. M. Kurka, J. Feist, D. A. Horner, A. Rudenko, Y. H. Jiang, K. U. Kuehnel, L. Foucar, T. N. Rescigno, C. W. McCurdy, R. Pazourek, S. Nagele, M. Schulz, O. Herrwerth, M. Lezius, M. F. Kling, M. Schoeffler, A. Belkacem, S. Duesterer, R. Treusch, B. I. Schneider, L. A. Collins, J. Burgdoerfer, C. D. Schroeter, R. Moshammer, J. Ullrich, Differential cross sections for non-sequential double ionization of He by 52 eV photons from the Free Electron Laser in Hamburg, FLASH, *New J. Phys.* **12**, (2010) 073035.

35. L. Malegat, H. Bachau, B. Piraux, F. Reynal, A novel estimate of the two-photon double-ionization cross section of helium, *J. Phys. B* **45**, (2012) 175601.

36. R. Pazourek, J. Feist, S. Nagele, E. Persson, B. I. Schneider, L. A. Collins, J. Burgdörfer, Universal features in sequential and nonsequential two-photon double ionization of helium, *Phys. Rev. A* **83**, (2011) 053418.

37. R. Nepstad, T. Birkeland, M. Forre, Numerical study of two-photon ionization of helium using an ab initio numerical framework, *Phys. Rev. A* **81**, (2010) 063402.

38. A. Palacios, D. A. Horner, T. N. Rescigno, C. W. McCurdy, Two-photon double ionization of the helium atom by ultrashort pulses, *J. Phys. B* **43**, (2010) 194003.

39. A. Palacios, T. N. Rescigno, C. W. McCurdy, Time-dependent treatment of two-photon resonant single and double ionization of helium by ultrashort laser pulses, *Phys. Rev. A* 4080 **79**, (2009) 033402.

40. J. Feist, S. Nagele, R. Pazourek, E. Persson, B. I. Schneider, L. A. Collins, J. Burgdörfer, Nonsequential two-photon double ionization of helium, *Phys. Rev. A* **77**, (2008) 043420.

41. X. Guan, K. Bartschat, B. I. Schneider, Dynamics of two-photon double ionization of helium in short intense xuv laser pulses, *Phys. Rev. A* **77**, (2008) 043421.

42. D. A. Horner, C. W. McCurdy, T. N. Rescigno, Triple differential cross sections and nuclear recoil in two-photon double ionization of helium, *Phys. Rev. A* **78**, (2008) 043416.

43. D. A. Horner, F. Morales, T. N. Rescigno, F. Martin, C. W. McCurdy, Two-photon double ionization of helium above and below the threshold for sequential ionization, *Phys. Rev. A* **76**, (2007) 030701.

44. A. Ivanov, A. S. Kheifets, Two-photon double ionization of helium in the region of photon energies 42–50 eV, *Phys. Rev. A* **75**, (2007) 033411.

45. A. S. Kheifets, I. A. Ivanov, Convergent close-coupling calculations of two-photon double ionization of helium, *J. Phys. B* **39**, (2006) 1731.

46. S. Hu, J. Colgan, L. Collins, Triple-differential cross-sections for two-photon doubleionization of He near threshold, *J. Phys. B* **38**, (2005) L35.

47. H. Bachau, Theory of two-photon double ionization of helium at the sequential threshold, *Phys. Rev. A* **83**, (2011) 033403.

48. R. Shakeshaft, Two-photon single and double ionization of helium, *Phys. Rev. A* **76**, (2007) 063405.

49. J. Colgan, M. Pindzola, Core-excited resonance enhancement in the two-photon complete fragmentation of helium, *Phys. Rev. Lett.* **88**, (2002) 173002.
50. L. Feng, H. van der Hart, Two-photon double ionization of He, *J. Phys. B* **36**, (2003) L1.
51. H. W. van der Hart, Time-dependent R-matrix theory applied to two-photon double ionization of He, *Phys. Rev. A* **89**, (2014) 053407.
52. S. Laulan, H. Bachau, Correlation effects in two-photon single and double ionization of helium, *Phys. Rev. A* **68**, (2003) 013409.
53. B. Piraux, J. Bauer, S. Laulan, H. Bachau, Probing electron-electron correlation with attosecond pulses, *EPJ D* **26**, (2003) 7.
54. E. Foumouo, P. Antoine, B. Piraux, L. Malegat, H. Bachau, R. Shakeshaft, Evidence for highly correlated electron dynamics in two-photon double ionization of helium, *J. Phys. B* **41**, (2008) 051001.
55. L. A. A. Nikolopoulos, P. Lambropoulos, Time-dependent theory of double ionization of helium under XUV radiation, *J. Phys. B* **40**, (2007) 1347.
56. L. Nikolopoulos, P. Lambropoulos, Multichannel theory of two-photon single and double ionization of helium, *J. Phys. B* **34**, (2001) 545.
57. I. A. Ivanov, A. S. Kheifets, J. Dubau, On the account of final state correlation in double ionization processes, *EPJ D* **61**, (2011) 563.
58. E. Foumouo, H. Bachau, B. Piraux, (2 gamma, 2e) total and differential cross-section calculations for helium with (h)over-bar omega = 40–50 eV, *EPJ ST* **175**, (2009) 175.
59. T. Reddish, J. Wightman, M. MacDonald, S. Cvejanovic, Triple differential cross section measurements for double photoionization of D$_2$, *Phys. Rev. Lett.* **79**, (1997) 2438.
60. J. Wightman, S. Cvejanovic, T. Reddish, (gamma, 2e) cross section measurements of D2 and He, *J. Phys. B* **31**, (1998) 1753.
61. T. Weber, A. O. Czasch, O. Jagutzki, A. K. Mueller, V. Mergel, A. Kheifets, E. Rotenberg, G. Meigs, M. H. Prior, S. Daveau, A. Landers, C. L. Cocke, T. Osipov, R. Diez Muino, H. Schmidt-Boecking, R. Doerner, Complete photo-fragmentation of 4235 the deuterium molecule, *Nature* **443**, (2006) 1014.
62. T. Weber, A. Czasch, O. Jagutzki, A. Muller, V. Mergel, A. Kheifets, E. Rotenberg, G. Meigs, M. Prior, S. Daveau, A. Landers, C. Cocke, T. Osipov, R. Muino, H. Schmidt-Bocking, R. Dorner, Complete photo-fragmentation of the deuterium molecule, *Nature* **431**, (2004) 437.
63. T. Weber, A. Czasch, O. Jagutzki, A. Muller, V.Mergel, A. Kheifets, J. Feagin, E. Rotenberg, G. Meigs, M. Prior, S. Daveau, A. Landers, C. Cocke, T. Osipov, H. Schmidt-Bocking, R. Dorner, Fully differential cross sections for photo-double-ionization of D$_2$, *Phys. Rev. Lett.* **92**, (2004) 163001.
64. T. J. Reddish, J. Colgan, P. Bolognesi, L. Avaldi, M. Gisselbrecht, M. Lavollee, M. S. 4245 Pindzola, A. Huetz, Physical interpretation of the "Kinetic Energy Release" effect in the double photoionization of H2, *Phys. Rev. Lett.* **100**, (2008) 193001.
65. M. Gisselbrecht, M. Lavollee, A. Huetz, P. Bolognesi, L. Avaldi, D. Seccombe, T. Reddish, Photodouble ionization dynamics for fixed-in-space H$_2$, *Phys. Rev. Lett.* **96**, (2006) 153002.
66. D. A. Horner, S. Miyabe, T. N. Rescigno, C. W. McCurdy, F. Morales, F. Martin, Classical Two-Slit Interference Effects in Double Photoionization of Molecular Hydrogen at High Energies, *Phys. Rev. Lett.* **101**, (2008) 183002.
67. D. A. Horner, W. Vanroose, T. N. Rescigno, F. Martin, C. W. McCurdy, Role of nuclear motion in double ionization of molecular hydrogen by a single photon, *Phys. Rev. Lett.* 4255 **98**, (2007) 073001.
68. W. Vanroose, D. A. Horner, F. Martin, T. N. Rescigno, C. W. McCurdy, Double photoionization of aligned molecular hydrogen, *Phys. Rev. A* **74**, (2006) 052702.
69. W. Vanroose, F. Martin, T. Rescigno, C. McCurdy, Complete photo-induced breakup of the H2 molecule as a probe of molecular electron correlation, *Science* **310**, (2005) 1787.

70. T. J. Reddish, J. Colgan, P. Bolognesi, L. Avaldi, M. Gisselbrecht, M. Lavollee, M. S. 4245 Pindzola, A. Huetz, Physical interpretation of the "Kinetic Energy Release" effect in the double photoionization of H_2, *Phys. Rev. Lett.* **100**, (2008) 193001.

71. J. Colgan, M. S. Pindzola, F. Robicheaux, Triple differential cross sections for the double photoionization of H2, *Phys. Rev. Lett.* **98**, (2007) 153001.

72. I. A. Ivanov, A. S. Kheifets, Time-dependent calculations of double photoionization of the aligned H2 molecule, *Phys. Rev. A* **85**, (2012) 013406.

73. J. Colgan, M. S. Pindzola, F. Robicheaux, Two-photon double ionization of the hydrogen molecule, *J. Phys. B* **41**, (2008) 121002.

74. X. Guan, K. Bartschat, B. I. Schneider, Two-photon double ionization of H2 in intense femtosecond laser pulses, *Phys. Rev. A* **82**, (2010) 041404.

75. I. A. Ivanov, A. S. Kheifets, Two-photon double ionization of the H2 molecule: Cross sections and amplitude analysis, *Phys. Rev. A* **87**, (2013) 023414.

76. F. Morales, F. Martin, D. A. Horner, T. N. Rescigno, C. W. McCurdy, Two-photon double ionization of H2 at 30 eV using exterior complex scaling, *J. Phys. B* **42**, (2009) 134013.

77. D. A. Horner, T. N. Rescigno, C. W. McCurdy, Nuclear recoil cross sections from time- dependent studies of two-photon double ionization of helium, *Phys. Rev. A* **81**, (2010) 023410.

78. D. A. Horner, T. N. Rescigno, C. W. McCurdy, Decoding sequential versus nonsequential two-photon double ionization of helium using nuclear recoil, *Phys. Rev. A* **77**, (2008) 030703.

79. S. A. Abdel-Naby, M. S. Pindzola, J. Colgan, Nuclear-recoil differential cross sections for the double photoionization of helium, *Phys. Rev. A* **86**, (2012) 013424.

80. S. A. Abdel-Naby, M. F. Ciappina, M. S. Pindzola, J. Colgan, Nuclear-recoil differential cross sections for the two-photon double ionization of helium, *Phys. Rev. A* **87**, (2013) 063425.

81. R. Dörner *et al.*, *Phys. Rev. Lett.* **77**, (1996) 1024.

82. M. Pont and R. Shakeshaft, *Phys. Rev. A* **54**, (1996) 1448.

83. L. Avaldi and A. Huetz, *J. Phys. B: At., Mol. Opt. Phys.* **38**, (2005) S861.

84. A. Huetz, P. Selles, D. Waymel, and J. Mazeau, *J. Phys. B: At., Mol. Opt. Phys.* **24**, (1991) 1917.

85. J. M. Feagin, *J. Phys. B: At., Mol. Opt. Phys.* **17**, **2433**, (1984).

86. A. Palacios, T. N. Rescigno and C. W. McCurdy, *Phys. Rev. Lett.* **103**, (2009) 253001.

87. D. A. Verner, G. J. Ferland, K. T. Korista and D. G. Yakovlev, *Astrophys. J.* **465**, (1996) 487.

88. A. Palacios, T. N. Rescigno, C. W. McCurdy, Two-Electron Time-Delay Interference in Atomic Double Ionization by Attosecond Pulses, *Phys. Rev. Lett.* **103**, (2009) 253001.

89. J. Feist, S. Nagele, C. Ticknor, B. I. Schneider, L. A. Collins, J. Burgdörfer, Attosecond Two-Photon Interferometry for Doubly Excited States of Helium, *Phys. Rev. Lett.* **107**, (2011) 093005.

90. S. Hu, L. Collins, Attosecond pump probe: Exploring ultrafast electron motion inside an atom, *Phys. Rev. Lett.* **96**, (2006) 073004.

91. T. Morishita, S. Watanabe, C. D. Lin, Attosecond light pulses for probing two-electron dynamics of helium in the time domain, *Phys. Rev. Lett.* **98**, (2007) 083003.

Chapter 6

Probing Orbital Symmetry of Molecules Via Alignment-Dependent Ionization Probability and High-Order Harmonic Generation by Intense Lasers

Song-Feng Zhao*,†, Xiao-Xin Zhou* and C. D. Lin‡

*College of Physics and Electronic Engineering,
Northwest Normal University, Lanzhou, Gansu 730070, China
†Email: zhaosf@nwnu.edu.cn
‡J. R. Macdonald Laboratory, Physics Department,
Kansas State University, Manhattan, Kansas 66506-2604, USA

It is shown that measurement of alignment-dependent ionization probability and high-order harmonic generation (HHG) of molecules in an intense laser field can be used to probe the orbital symmetry of molecules. In this review, recent progress of molecular tunneling ionization (MO-ADK) model of Tong *et al.* [Phys. Rev. A 66, 033402 (2002)] is first reviewed. In particular, an efficient method to obtain wavefunctions of linear molecules in the asymptotic region was developed by solving the time-independent Schrödinger equation with B-spline functions, and molecular potential energy surfaces were constructed based on the density functional theory. The accurate wavefunctions are used to extract improved structure parameters in the MO-ADK model. The loss of accuracy of the MO-ADK model in the low intensity multiphoton ionization regime is also addressed by comparing with the molecular Perelomov–Popov–Terent'ev (MO-PPT) model, the single-active-electron time-dependent Schrödinger equation (SAE-TDSE) method, and the experimental data. Finally, how the orbital symmetry affects the HHG of molecules within the strong-field approximation (SFA) was reviewed.

1. Introduction

Probing molecular orbitals is one of the fundamental goals in physics and chemistry [1–5]. The conventional methods for studying the highest occupied molecular orbital (HOMO) are to use scanning tunneling microscopy (STM) [6, 7] and the angle-resolved photoelectron spectroscopy [8]. In 2004, based on the plane wave approximation, Itatani *et al.*, performed a complete tomographic reconstruction of the molecular *orbital wavefunction* of HOMO in N_2 from high harmonic spectra measured at various alignment angles [9]. Le *et al.*, examined the assumptions used in the tomographic procedure [10] and its underlying limitations. Since wavefunction in quantum mechanics is a complex function in general, it is a representation and not measurable. Thus the imaging of orbital wavefunction has to be taken with caution, and only within the confine of assumptions defined by the practitioners. The tomographic method has since been corrected by using scattering waves for the continuum states [11, 12], or by considering multielectron effects [13]. It has also been recently generalized to image the HOMO of asymmetric molecules [14]. The tomographic imaging is evidently based on the third recombination step of the three-step model of high-order harmonic generation (HHG) [15]. It is an indirect method and various approximations were made in the procedure. A more direct scheme, based on the first step (the ionization step), is to use the angle-dependent ionization rate (or probability) $P(\theta)$ of molecules, where θ is the angle between the molecular axis and the laser's polarization direction. Experimentally, the $P(\theta)$ has been obtained by ionizing a partially aligned ensemble of molecules [16–18, 20, 21]. Alternatively, the $P(\theta)$ can also be determined by measuring the molecular frame photoelectron angular distribution (MFPADs) [22–28] or by detecting the angular distribution of the emitted ionic fragments [29–34]. Theoretically, $P(\theta)$ can be calculated by solving the time-dependent Schrödinger equation (TDSE) of molecules, but often based on the single-active-electron (SAE) approximation [35–44] or by using the time-dependent density functional theory (TDDFT) [45–50]. Very recently, the time-dependent Hartree-Fock (TDHF) theory is also used to calculate the $P(\theta)$ of CO_2 [51]. Since these ab-initio calculations are rather time-consuming and still quite challenging for molecules, simpler theoretical models are quite desirable for interpreting experiments, such as the molecular Ammosov–Delone–Krainov model (MO-ADK) [52–55], the molecular strong-field approximation (MO-SFA) [56–60], the molecular Perelomov–Popov–Terent'ev (MO-PPT) model [61–63] and others [64–69].

Recall that the MO-ADK is a generalization of the atomic ADK model by Ammosov *et al.* [70] which was initially used to study the tunneling ionization of atoms. In the MO-ADK model [52], the static tunneling ionization rate is given analytically, and the rate depends on the molecular alignment angle, the instantaneous electric field of the laser, the structure parameters C_{lm} of the outermost molecular orbital and its ionization potential. Once the structure parameters of a specific molecule are available, one can easily calculate the orientation-dependent ionization rate with the MO-ADK. Moreover, the ionization probabilities and signals can also be obtained readily by including the temporal profile and spatial distribution of a focused laser beam in order to compare with experiments. Thus it is essential to determine and tabulate accurate structure parameters for several occupied orbitals of molecules. These structure parameters can be extracted directly from the molecular orbital wavefunction in the asymptotic region. In Tong *et al.* [52], the molecular wavefunctions were calculated originally with the multiple-scattering method [71]. At the present time, molecular wavefunctions are commonly calculated from quantum chemistry codes, such as GAUSSIAN [72], GAMESS [73] and others [74, 75]. Using the Hartree-Fock (HF) approximation and the conventional Gaussian bases, molecular wavefunctions calculated from these packages have been used to extract structure parameters of HOMO for several linear molecules [39, 47, 48, 76] and for some nonlinear polyatomic molecules [55, 77]. With structure parameters determined from these two methods, the MO-ADK model fits reasonably well for most experimental $P(\theta)$, except for CO_2 [17]. In Zhao *et al.* [78], it was found that the large discrepancies of $P(\theta)$ between the MO-ADK and the experimental data [17] can be attributed partly to the inaccurate structure parameters of CO_2. To determine the accurate structure parameters of molecules, accurate wavefunction of the ionizing orbital in the asymptotic region is required. In [54, 78–81], an efficient method was proposed to fix the asymptotic tail of the molecular wavefunction by solving the time-independent Schrödinger equation with B-spline functions, where the molecular potential was constructed numerically based on the density-functional theory (DFT). Accurate structure parameters of the HOMO and some inner orbitals (i.e., HOMO-1 and HOMO-2) for many linear molecules have been determined and tabulated [54, 78–81]. Using these improved structure parameters in the MO-ADK model, the $P(\theta)$ of CO_2, C_2H_2, H_2^+ and H_2 were improved significantly by comparing with those from more elaborate calculations and experimental data in deed. This method has also been used to extract structure parameters from molecular wavefunction calculated by propagating the TDSE in imaginary time [39]. Recently, Madsen *et al.* [65] demonstrated that wavefunction of the HOMO with the correct exponential behavior can also be obtained directly by solving the HF equations with the X2DHF code for diatomic molecules. These accurate wavefunctions have been used to determine the structure parameters for the MO-ADK model [82] and the weak-field asymptotic theory (WFAT) [65]. For triatomic molecules such as CO_2, OCS and H_2O, the possibility of obtaining the correct asymptotic tail of the HOMO wavefunction using optimized Gaussian basis sets from GAUSSIAN [72] or GAMESS [73] packages has been systematically studied [65, 66].

It is known that the alignment dependence of HHG is determined mostly by the orbital symmetry of molecules [9, 83]. In other words, it is possible to probe the HOMO orbital via the alignment-dependent HHG signals when contributions from inner orbitals (i.e., HOMO-1 and HOMO-2) are negligible. So far the alignment-dependent HHG has been studied theoretically by solving the TDSE of the simplest molecules such as H_2^+ [84–87] and HeH^{2+} [88, 89], or by performing the TDDFT calculations [90–93]. Since these two methods are rather time-consuming even for the single-molecule response of the HHG, most of the existing calculations for HHG from molecules were performed using the strong-field approximation (SFA) (to be called molecular Lewenstein) model [76, 94–101] which is a generalization of the atomic Lewenstein model [102]. In recent years, improvement on SFA has been proposed using the quantitative rescattering (QRS) theory [3, 103]. For large molecules, SFA can still be of interest in view of its simplicity and its reasonable accuracy.

The rest of this chapter is arranged as follows. In Sec. 2, the method of constructing one-electron potential of a linear molecule is summarized. This potential can be used to solve the time-independent Schrödinger equation with B-spline basis functions to fix the asymptotic tail of the molecular wavefunction. The basic equations of the MO-ADK, MO-PPT, MO-SFA and molecular Lewenstein models are then briefly reviewed. In Sec. 3, the improvement on the alignment dependence of ionization probability with the more accurate structure parameters is demonstrated. In this section, it is shown how to probe the orbital symmetry of molecules using the $P(\theta)$ and the alignment-dependent HHG signals by intense laser fields. A conclusion is given in Sec. 4.

2. Theoretical Methods

The theory part is separated into five subsections. First, it is shown on how to construct numerically one-electron potentials of linear molecules based on the DFT. Second, the method to improve the asymptotic tail of molecular wavefunction by solving the time-independent Schrödinger equation of linear molecules with B-spline functions and extract accurate structure parameters of molecules in the asymptotic region is given. Finally the basic equations of the MO-ADK, MO-PPT, MO-SFA and molecular Lewenstein models are reviewed, respectively.

2.1. *Construction of one-electron potentials of linear molecules*

The one-electron potentials are constructed numerically using the modified Leeuwen–Baerends (LBα) model [79, 80, 104, 105] where the electrostatic and exchange-correlation terms are included.

For linear molecules, based on the single-center expansion, the one-electron potential can be expressed as

$$V(r, \theta_e) = \sum_{l=0}^{l_{max}} v_l(r) P_l(\cos \theta_e). \tag{1}$$

Here, $v_l(r)$ is the radial component of the molecular potential and $P_l(\cos \theta_e)$ is the Legendre polynomial, θ_e is the angular coordinate of the active electron in the molecular frame. The radial potential is given by

$$v_l(r) = v_l^{nuc}(r) + v_l^{el}(r) + v_l^{xc}(r), \tag{2}$$

where the first two terms represent the electrostatic potential and the last term is the exchange-correlation interaction.

The electron-nucleus interaction $v_l^{nuc}(r)$ can be written as

$$v_l^{nuc}(r) = \sum_{i=1}^{N_a} v_l^i(r), \tag{3}$$

where i sums over all the N_a atoms in the molecule. By assuming the linear molecule is aligned along the z-axis, $v_l^i(r)$ is given by

$$v_l^i(r) = \begin{cases} -\left(\dfrac{r_<^i}{r_>^i}\right)^l \dfrac{Z_c^i}{r_>^i} & \text{if } z_i > 0 \\[3mm] -(-1)^l \left(\dfrac{r_<^i}{r_>^i}\right)^l \dfrac{Z_c^i}{r_>^i} & \text{if } z_i < 0. \end{cases} \tag{4}$$

Here, $r_<^i = \min(r, |z_i|)$ and $r_>^i = \max(r, |z_i|)$. Z_c^i and z_i are the nuclear charge and the z coordinate of the ith atom, respectively.

The partial Hartree potential $v_l^{el}(r)$ is written as

$$v_l^{el}(r) = \frac{4\pi}{2l+1} \int_0^\infty a_l(r') r'^2 \frac{r_<^l}{r_>^{l+1}} dr' \tag{5}$$

with $r_< = \min(r, r')$ and $r_> = \max(r, r')$. Here $a_l(r')$ is determined by

$$a_l(r') = \frac{2l+1}{2} \int_{-1}^1 \rho(r', \theta_e') P_l(\cos \theta_e') d(\cos \theta_e') \tag{6}$$

where $\rho(r', \theta_e')$ is the total electron density and

$$\rho(r', \theta_e') = \sum_{i=1}^{N_e} \frac{1}{2\pi} \int_0^{2\pi} |\Psi_i(r', \theta_e', \varphi_e')|^2 d\varphi_e' \tag{7}$$

with i runs over all the N_e electrons. The wavefunction of each molecular orbital can be calculated from quantum chemistry codes such as GAUSSIAN [72] and GAMESS [73].

In the LBα model, the partial exchange-correlation potential can be expressed as

$$v_l^{xc}(r) = \frac{2l+1}{2} \int_{-1}^{1} V_{xc,\sigma}^{LB\alpha}(r, \theta_e) P_l(\cos\theta_e) d(\cos\theta_e) \tag{8}$$

where

$$V_{xc,\sigma}^{LB\alpha}(r, \theta_e) = \alpha V_{x,\sigma}^{LDA}(r, \theta_e) + V_{c,\sigma}^{LDA}(r, \theta_e)$$
$$- \frac{\beta \chi_\sigma^2(r, \theta_e) \rho_\sigma^{1/3}(r, \theta_e)}{1 + 3\beta\chi_\sigma(r, \theta_e) \sinh^{-1}[\chi_\sigma(r, \theta_e)]} \tag{9}$$

with $\chi_\sigma(r, \theta_e) = |\nabla\rho_\sigma(r, \theta_e)|\rho_\sigma^{-4/3}(r, \theta_e)$ and $\rho_\sigma(r, \theta_e)$ is spin density. Here α and β are two empirical parameters. $V_{x,\sigma}^{LDA}(r, \theta_e)$ is the local density approximation (LDA) exchange potential

$$V_{x,\sigma}^{LDA}(r, \theta_e) = -\left[\frac{6}{\pi}\rho_\sigma(r, \theta_e)\right]^{1/3}. \tag{10}$$

In the present LBα calculations, the LDA correlation potential $V_{c,\sigma}^{LDA}(r, \theta_e)$ is calculated by using the Perdew–Wang representation for the correlation functionals [106]

$$V_{c,\sigma}^{LDA}(r, \theta_e) = \varepsilon_c(r_s, \zeta) - \frac{r_s}{3}\frac{\partial\varepsilon_c(r_s, \zeta)}{\partial r_s} - (\zeta - \text{sgn}\,\sigma)\frac{\partial\varepsilon_c(r_s, \zeta)}{\partial\zeta}, \tag{11}$$

where r_s and ζ are the density parameter and the relative spin polarization, respectively, and $\varepsilon_c(r_s, \zeta)$ is the correlation energy. Note that $\text{sgn}\,\sigma$ is 1 for $\sigma = \uparrow$ and -1 for $\sigma = \downarrow$.

2.2. Calculation of molecular wavefunction with correct asymptotic tail by solving the time-independent Schrödinger equation

Using the molecular potentials constructed in the previous subsection, the molecular wavefunction with the correct asymptotic behavior can be obtained by solving the following time-independent Schrödinger equation for linear molecules [54, 78–81]

$$H_{el}\Psi_n^{(m)}(\vec{r}) \equiv \left[-\frac{1}{2}\nabla^2 + V(r, \theta_e)\right]\Psi_n^{(m)}(\vec{r}) = E_n^{(m)}\Psi_n^{(m)}(\vec{r}), \tag{12}$$

where m is z component of the electronic orbital momentum and n is the orbital index. Because of the cylindrical symmetry, the wavefunction $\Psi_n^{(m)}(\vec{r})$ can be written as

$$\Psi_n^{(m)}(\vec{r}) = \frac{1}{\sqrt{2\pi}} e^{im\chi_e} \psi(r, \xi). \tag{13}$$

Here, $\xi = \cos\theta_e$ and χ_e is the angular coordinate of the active electron in the molecular frame. The wavefunction $\psi(r, \xi)$ can be expanded by B-spline functions as [80, 107]

$$\psi(r, \xi) = \sum_{i=1}^{N_r} \sum_{j=1}^{N_\xi} C_{ij} B_i(r)(1 - \xi^2)^{|m|/2} B_j(\xi), \tag{14}$$

where $B_i(r)$ and $B_j(\xi)$ are radial and angular B-spline functions, respectively. By substituting Eqs. (1), (13), and (14) into Eq. (12) and projecting onto the basis set $B_{i'}(r)(1 - \xi^2)^{|m|/2} B_{j'}(\xi)$, we obtain the following matrix equation

$$HC = ESC, \tag{15}$$

with

$$H_{i'j',ij} = \int_0^{r_{max}} \int_{-1}^1 B_{i'}(r)(1 - \xi^2)^{|m|/2} B_{j'}(\xi) H_{el}$$
$$\times B_i(r)(1 - \xi^2)^{|m|/2} B_j(\xi) r^2 dr d\xi, \tag{16}$$

and

$$S_{i'j',ij} = \int_0^{r_{max}} B_{i'}(r) B_i(r) r^2 dr$$
$$\times \int_{-1}^1 B_{j'}(\xi)(1 - \xi^2)^{|m|} B_j(\xi) d\xi, \tag{17}$$

where E and C are energy matrix and coefficient matrix, respectively. The eigenfunctions and eigenvalues for a given m can be obtained by diagonalizing Eq. (15).

For linear molecules, based on the single-center expansion, wavefunctions can also be expanded as

$$\Psi_n^{(m)}(\vec{r}) = \sum_l F_{lm}(r) Y_{lm}(\theta_e, \chi_e), \tag{18}$$

where $Y_{lm}(\theta_e, \chi_e)$ is the spherical harmonic functions. The radial wavefunction can be calculated by

$$F_{lm}(r) = \int \Psi_n^{(m)}(\vec{r}) Y_{lm}^*(\theta_e, \chi_e) \sin \theta_e d\theta_e d\chi_e. \tag{19}$$

Then accurate structure parameters C_{lm} can be determined by matching these radial functions to the form

$$F_{lm}(r) = C_{lm} r^{(Z_c/\kappa)-1} e^{-\kappa r} \tag{20}$$

where Z_c is the asymptotic charge, $\kappa = \sqrt{2I_p}$, and I_p is the ionization energy.

In the molecular frame, the angular distribution of the asymptotic electron density for the active electron can be written as

$$\rho_1(\theta_e, \chi_e) = \int_{r_1}^{\infty} \left| \Psi_n^{(m)}(r, \theta_e, \chi_e) \right|^2 r^2 dr \tag{21}$$

r_1 is the starting point of the fitting range. The θ_e-dependent electron density is given by

$$\rho(\theta_e) = \frac{1}{2\pi} \int_0^{2\pi} \rho_1(\theta_e, \chi_e) d\chi_e. \tag{22}$$

2.3. The MO-ADK and MO-PPT models

According to the MO-ADK model [52, 55, 77], the cycle-averaged ionization rate is given by

$$w_{MO-ADK}(F, \vec{R}) = \left[\frac{3F}{\pi k^3} \right]^{1/2} \sum_{m'} \frac{B^2(m')}{2^{|m'|} |m'|! \, \kappa^{2Z_c/\kappa-1}} \left(\frac{2\kappa^3}{F} \right)^{2Z_c/\kappa-|m'|-1} e^{-2\kappa^3/3F}, \tag{23}$$

where F is the peak field strength and $\vec{R} \equiv (\phi, \theta, \chi)$ is the Euler angles of the molecular frame with respect to the laboratory frame. Note that θ is the angle between the Z and z axes, ϕ and χ denote rotations around the Z axis and the z axis, respectively. For linear molecules, $B(m')$ can be expressed as

$$B(m') = \sum_l C_{lm} D_{m',m}^l(\vec{R}) Q(l, m'). \tag{24}$$

For nonlinear molecules, m is no longer a good quantum number and thus $B(m')$ is written as

$$B(m') = \sum_{lm} C_{lm} D_{m',m}^l(\vec{R}) Q(l, m'), \tag{25}$$

with

$$Q(l, m') = (-1)^{(m'+|m'|)/2} \sqrt{\frac{(2l+1)(l+|m'|)!}{2(l-|m'|)!}},$$ (26)

and $D^l_{m',m}(\vec{R})$ is the Wigner rotation matrix

$$D^l_{m',m}(\vec{R}) = e^{-im'\phi} d^l_{m',m}(\theta) e^{-im\chi}.$$ (27)

Based on the MO-PPT model [61, 63], the cycle-averaged ionization rate can be calculated analytically by

$$w_{MO-PPT}(F, \omega, \vec{R}) = \left(\frac{3F}{\pi k^3}\right)^{1/2} \sum_{m'} \frac{B^2(m')}{2^{|m'|}|m'|!} \frac{A_{m'}(\omega, \gamma)}{\kappa^{2Z_c/\kappa-1}} (1+\gamma^2)^{|m'|/2+3/4}$$

$$\times \left(\frac{2\kappa^3}{F}\right)^{2Z_c/\kappa-|m'|-1} e^{[-(2\kappa^3/3F)g(\gamma)]},$$ (28)

where γ is the Keldysh parameter and ω is the angular frequency of the laser pulse. The coefficients $A_{m'}(\omega, \gamma)$ can be found in Refs. [62, 108, 109]. $g(\gamma)$ is given by

$$g(\gamma) = \frac{3}{2\gamma} \left[(1 + \frac{1}{2\gamma^2}) \sinh^{-1}\gamma - \frac{\sqrt{1+\gamma^2}}{2\gamma} \right].$$ (29)

With the cycle-averaged ionization rates of molecules, one can calculate the total ionization probability by a laser pulse by

$$P(I, \vec{R}) = 1 - \exp(-\int_{-\infty}^{+\infty} w_m(F, \vec{R}) dt),$$ (30)

where m stands for the MO-ADK or the MO-PPT model.

To compare with experimental data, the ionization signal of molecules has to be calculated by

$$S(\vec{R}) \propto \int P(I, \vec{R}) 2\pi r \, dr \, dz = \int_0^{I_0} P(I, \vec{R}) \left[-\frac{\partial V}{\partial I}\right] dI$$ (31)

Here, I_0 is the peak intensity at the focal point and the volume element takes the form $-dV/dI \propto (2I + I_0)(I_0 - I)^{1/2} I^{-5/2}$. The spatial component of the electric field is assumed to be a Gaussian beam in our simulations.

2.4. *Calculation of the orientation-dependent ionization probability of molecules with the MO-SFA model*

Based on the MO-SFA [80, 103], the total ionization probability of molecules by a laser pulse can be written as

$$P(\vec{R}) = \int \left| f(\vec{p}, \vec{R}) \right|^2 d^3 p, \tag{32}$$

with

$$f(\vec{p}, \vec{R}) = i \int_{-\infty}^{+\infty} \langle \vec{p} + \vec{A}(t) | \vec{r} \cdot \vec{E}(t) | \Phi_0(\vec{r}) \rangle e^{-iS(\vec{p},t)} dt, \tag{33}$$

where $\vec{p}$ is the momentum of the emitted electron, $\vec{E}(t)$ and $\vec{A}(t)$ are the electric field and the vector potential, respectively. The action $S(\vec{p}, t)$ is given by

$$S(\vec{p}, t) = \int_t^\infty dt' \left\{ \frac{\left[\vec{p} + \vec{A}(t') \right]^2}{2} + I_p \right\}. \tag{34}$$

In Eq. (33), the ground-state wavefunction $\Phi_0(\vec{r})$ can be obtained from GAUSSIAN [72] or GAMESS [73] and the continuum state is approximated by a Volkov state.

2.5. *The molecular Lewenstein model for HHG from molecules*

In earlier work [94, 95], the Lewenstein model [102] was first generalized to diatomic molecules. It is then modified by others [76, 96–101]. According to the SFA, the parallel component of the induced dipole moment of a molecule driven by a linearly polarized laser field can be written as [110]

$$x(t) = i \int_0^\infty d\tau \left(\frac{\pi}{\varepsilon + i\tau/2} \right)^{3/2}$$
$$\times [\sin\theta \cos\chi d_x^*(t) + \sin\theta \sin\chi d_y^*(t) + \cos\theta d_z^*(t)]$$
$$\times [\sin\theta \cos\chi d_x(t - \tau) + \sin\theta \sin\chi d_y(t - \tau) + \cos\theta d_z(t - \tau)]$$
$$\times E(t - \tau) \exp[-i S_{st}(t, \tau)] a^*(t) a(t - \tau) + c.c., \tag{35}$$

with ε being a positive regularization constant. Here, $d_x(t)$, $d_y(t)$, and $d_z(t)$ are the x, y, and z components of the transition dipole moment between the ground state and the continuum state. The quasiclassical action at the stationary points for the electron propagating in the laser field is given by

$$S_{st}(t, \tau) = \int_{t-\tau}^{t} \left(\frac{[p_{st}(t, \tau) + A(t')]^2}{2} + I_p \right) dt', \tag{36}$$

where the canonical momentum at the stationary points is expressed as

$$p_{st}(t, \tau) = -\frac{1}{\tau} \int_{t-\tau}^{t} dt' A(t'). \tag{37}$$

In Eq. (35), $a(t)$ is introduced to account for the ground-state depletion and

$$a(t) = \exp\left[-\frac{1}{2} \int_{-\infty}^{t} w(t') dt' \right], \tag{38}$$

with the ionization rate $w(t')$ obtained from the MO-ADK model [52, 54, 55]. Again the ground state wavefunction of the molecule is calculated using the standard quantum chemistry program like GAUSSIAN [72] and the continuum state is described approximately with a plane wave. We note that Eq. (35) can be reduced to Eq. (4) in Ref. [95] if a linear molecule aligned along the z axis is exposed to a laser field, linearly polarized on the y-z plane.

3. Results and Discussions

3.1. *The one-electron potentials for* Cl_2

The model potentials of H_2 and N_2 were proposed in Refs. [41, 111], respectively. However, most of the effective potentials are calculated numerically based on the DFT [38, 39, 45–50, 54, 78–80, 90–93]. The one-electron potentials for linear molecules are created numerically following the procedure described in Sec. 2.1. Figure 1 shows partial wave expansions ($v_l^{nuc}(r)$, $v_l^{el}(r)$, $v_l^{xc}(r)$, and $v_l(r)$) of the effective potential $V(r, \theta_e)$ (see Eqs. (1) and (2)) for Cl_2.

3.2. *Extracting structure parameters for several highly occupied orbitals of linear molecules*

Once the wavefunctions with the correct asymptotic tail and the corresponding orbital binding energies of linear molecules are obtained by solving Eq. (12), accurate structure parameters can be extracted from these molecular wavefunctions in the asymptotic region. It has been confirmed that the calculated LBα orbital binding energies are in good agreement with the experimental data [45, 46, 50, 79, 80]. Figure 2 shows the radial wavefunctions of H_2^+ and CO_2 for the first three partial waves and compared to those obtained directly from the GAUSSIAN. In the small-r

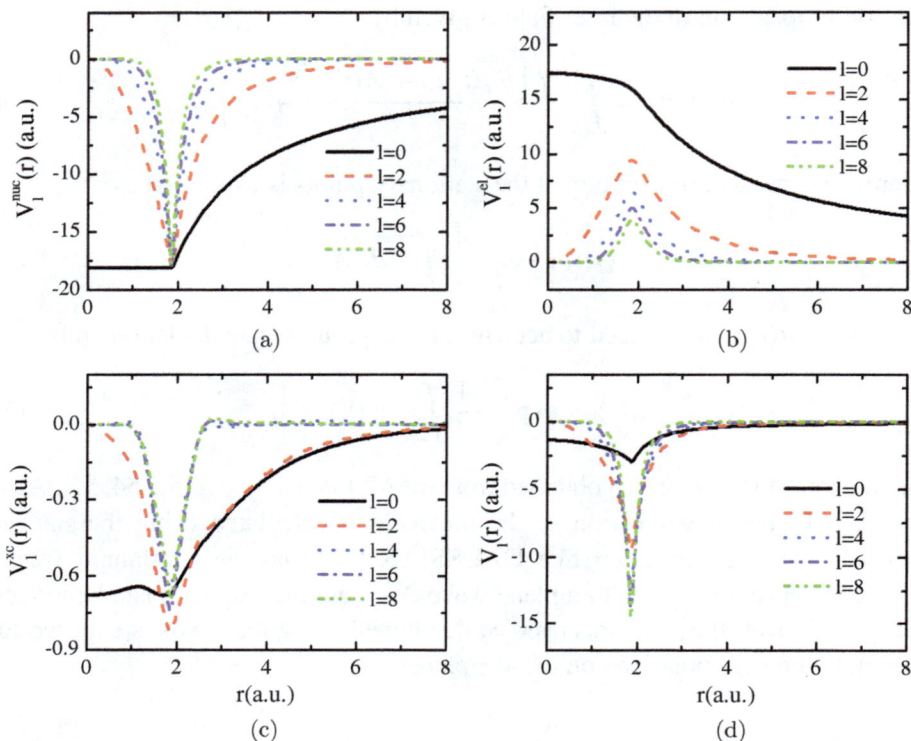

Fig. 1. (Color online) (a) Partial wave decomposition of the electron-nucleus potential; (b) the same for electron-electron repulsion potential; (c) for the exchange-correlation potential; (d) the total potential for each partial wave. We take $l_{max} = 80$ for Cl_2, but only low $l = 0, 2, 4, 6, 8$ terms are shown for clarity. Adapted from [80]. © (2013) by Taylor & Francis.

region, the present radial function agrees quite well with those from the GAUSSIAN for each partial wave. In the large-r region, one can see clearly the present calculated radial function displays the exponential decay form of Eq. (20). However, those from the GAUSSIAN exhibit oscillations and drop much faster like a Gaussian function. Note that conventional Gaussian bases are used in our present GAUSSIAN calculations. To determine the structure parameters of molecules, i.e., comparison of the asymptotic behavior of the calculated radial function to the correct asymptotic behavior [see Eq. (20)], the method proposed in Ref. [57] was followed. In Fig. 3, the structure parameters of H_2^+ at the equilibrium distance are obtained by fitting the calculated radial function to the correct one. The structure parameters C_{00}, C_{20} and C_{40} are 4.52, 0.62, 0.03, respectively. Using this method, we determined and tabulated structure parameters of the HOMO and of some inner orbitals for 37 linear molecules [54, 78–81].

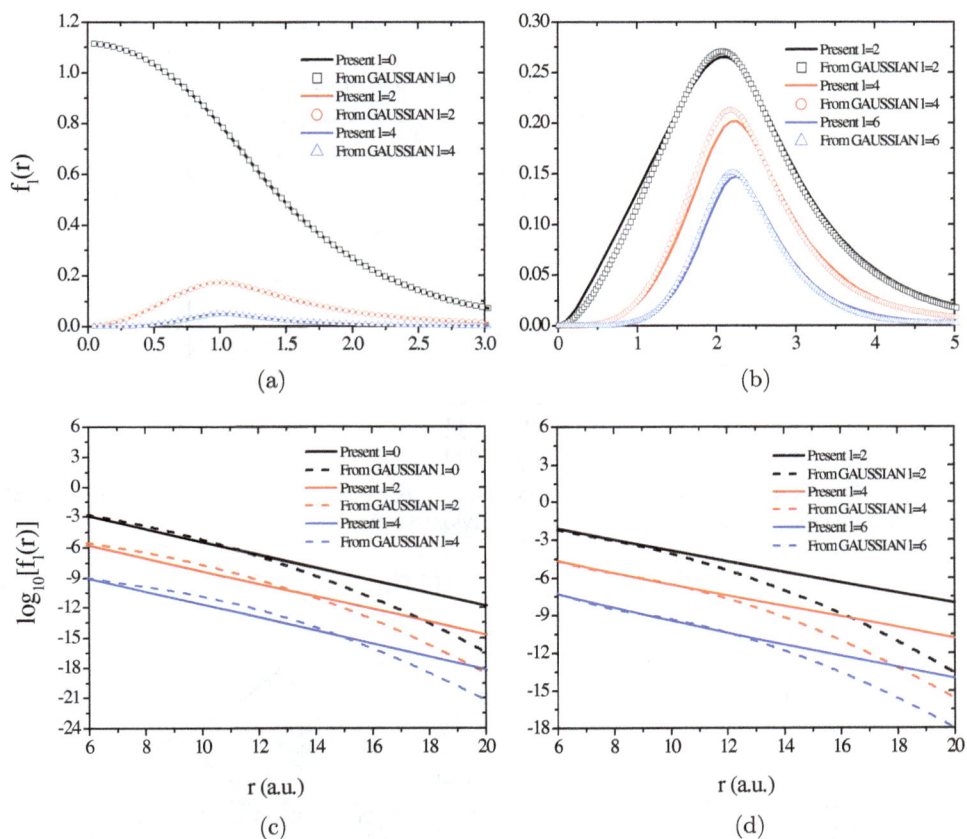

Fig. 2. (Color online) Partial wave radial function in the small-r region: (a) H_2^+; (b) CO_2 and in the large-r region: (c) H_2^+; (d) CO_2. For clarity, in (c) and (d), the radial functions for the last two partial waves are divided by 10^2 and 10^4, respectively. Figs. 2(b) and (d) are adapted from [78]. © (2009) by the American Physical Society.

3.3. Comparison of alignment dependent ionization probabilities between the MO-ADK model and other more elaborate calculations

Using the improved coefficients tabulated in Refs. [39, 54, 78–82], one can now calculate alignment dependent ionization probabilities for several selected molecules that have also been carried out by other theoretical methods. The comparison of these results is shown in Fig. 4. For simplicity, all the probabilities are normalized to 1.0 at the peak. For H_2^+ and H_2, one can see the MO-ADK results using the improved C_{lm} exhibit stronger angular dependence than the old ones (see Figs. 4(a) and 4(b)). The present MO-ADK results are in good agreement with those from TDSE [37, 40] for H_2^+ and with those from TDDFT [47] for H_2, respectively.

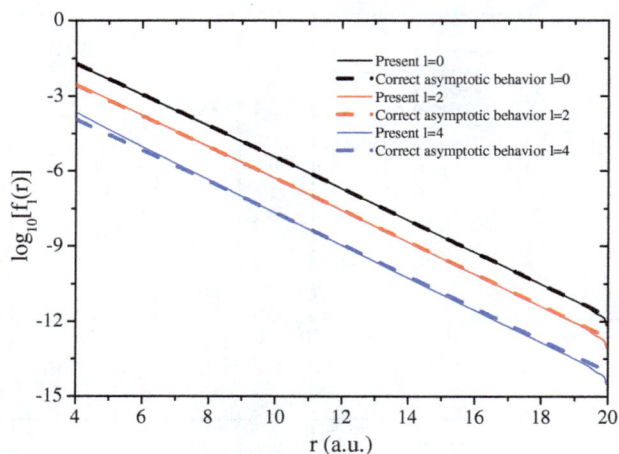

Fig. 3. (Color online) Comparison of the asymptotic behavior of the present calculated radial wavefunction (solid) to the correct asymptotic behavior (dashed) for H_2^+.

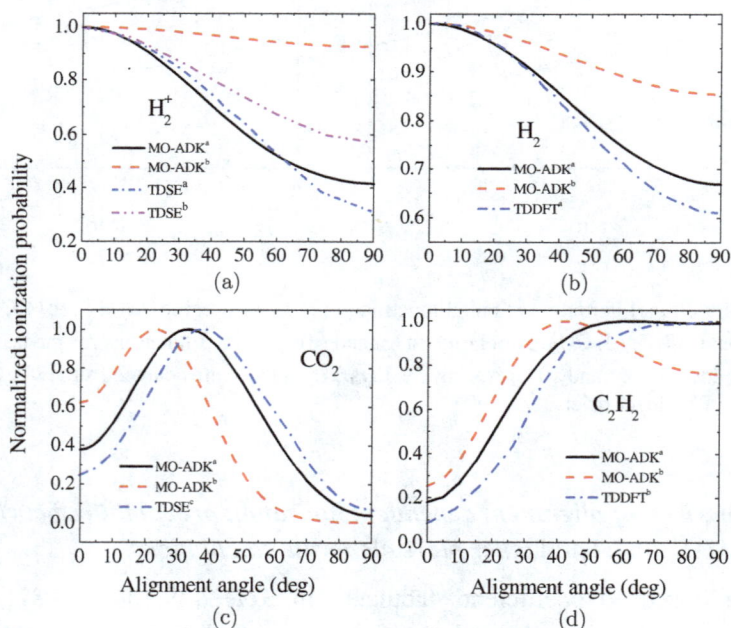

Fig. 4. (Color online) Normalized alignment-dependent ionization probability. (a) H_2^+ at 5×10^{14} W/cm^2; (b) H_2 at 1×10^{14} W/cm^2; (c) CO_2 at 1.1×10^{14} W/cm^2; (d) C_2H_2 at 5×10^{13} W/cm^2. MO-ADK[a] denotes MO-ADK results using the present improved C_{lm} coefficients [54, 78], MO-ADK[b] stands for MO-ADK results using the original C_{lm} coefficients determined from the multiple scattering theory [52, 76], TDSE[a] from Kamta *et al.* [40], TDSE[b] from Kjeldsen *et al.* [37], TDSE[c] from Petretti *et al.* [35], TDDFT[a] from Chu [47], and TDDFT[b] from Otobe *et al.* [49]. Fig. 4(a) is adapted from [54]. © (2010) by the American Physical Society.

For CO_2, the present MO-ADK peaks at 35° instead of 24° from the MO-ADK using the old coefficients. The improved MO-ADK result is much closer to the peak at 40° predicted by TDSE [35]. For C_2H_2, the older MO-ADK result has a peak at 45° and gives a minimum at 90°, while the TDDFT result [49] shows a peak at 90°. The new MO-ADK result agrees well with that from the TDDFT.

3.4. *Comparison with experiments*

It has been confirmed that the MO-ADK results using old coefficients agree reasonably well with the experimental data for N_2 and O_2 [16, 17, 31, 54]. For CO_2, The larger discrepancies of alignment-dependent ionization probabilities between the older MO-ADK results and the experimental data were found in Ref. [17]. The discrepancy brought out by the experiment [17] attracted a number of more accurate theoretical calculations such as the TDSE [39, 112], TDHF [51], and TDDFT [45]. For H_2, the original MO-ADK underestimates the experimental ratio of ionization rate for molecules aligned parallel vs. perpendicular with respect to the molecular axis [22, 31], while the present MO-ADK overestimates the experimental ratio [54]. The correct ratios have been obtained by solving the TDSE with a model potential at different laser intensities [41]. In Fig. 5(a), the normalized ionization probabilities of CO_2 from several theoretical methods with the experimental data for laser intensity $I = 1.1 \times 10^{14}$ W/cm^2 are compared. One can see the peak positions determined from the TDSE and the TDDFT are much closer to the experimental one than the MO-ADK result. However, so far all the theoretical results available fail to predict the narrow ionization distribution reported in the experiment. For laser intensity $I = 0.3 \times 10^{14}$ W/cm^2, the experimental data show a very broad angular distribution, consistent with all the theoretical results (see Fig. 5(b)). The reasonable agreement of the peak positions between these theoretical calculations and the experimental measurement can be observed in Fig. 5(b).

3.5. *Alignment dependence of ionization rates from HOMO, HOMO-1, and HOMO-2 orbitals*

In recent years, strong-field phenomena involving inner orbitals of molecules have been studied widely [23, 27, 28, 31, 35, 36, 45, 54, 79, 80, 113–122]. Since tunneling ionization is the first fundamental step to all rescattering processes including the HHG, it is important to investigate at what orientation (or alignment) angles the contributions from inner orbitals have to be considered by comparing the $P(\theta)$ of the HOMO with those from inner orbitals of molecules. Clearly the $P(\theta)$ of inner orbitals can also be calculated easily by the MO-ADK model using the corresponding experimental ionization potentials and structure parameters tabulated in Refs. [54, 79, 80]. Figure 6 shows the orientation-dependent ionization rates from the

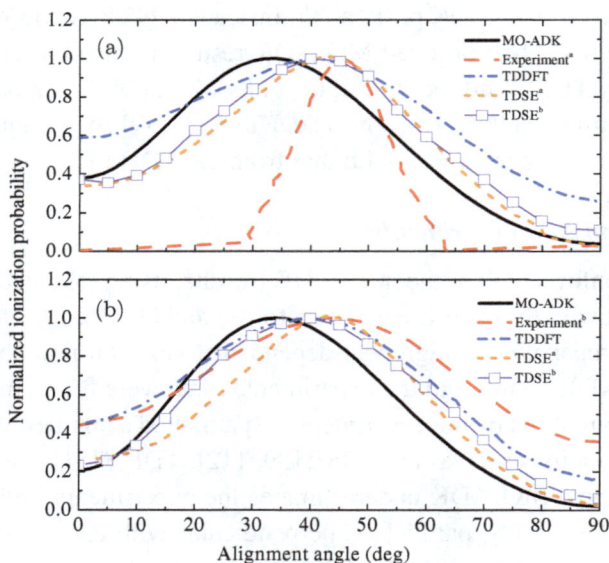

Fig. 5. (Color online) Normalized alignment-dependent ionization probabilities of CO_2. (a) Laser intensity is 1.1×10^{14} W/cm^2; (b) Laser intensity is 0.3×10^{14} W/cm^2. Note that 0.5×10^{14} W/cm^2 and 0.56×10^{14} W/cm^2 were used in Refs. [39, 45], respectively. TDDFT from Ref. [45], TDSE^a from Ref. [39], TDSE^b from Ref. [112], Experiment^a from Ref. [17], and Experiment^b from Ref. [18].

HOMO, HOMO-1, and HOMO-2 for N_2, CO_2, and Cl_2 and those from HOMO and HOMO-1 for the HBr molecule, respectively. Note that the angular dependence of ionization rates, $P(\theta)$, reflects vividly the shape of molecular orbitals. For the $P(\theta)$, a σ_g (or σ_u) orbital tends to have a peak at 0° and a minimum at 90°, a π orbital has the peak at 90° and minimum at 0° and 180°, a π_g orbital gives the peak near 45° and minimum at 0° and 90°, and a π_u orbital demonstrates a peak at 90° and a minimum at 0°. In Figs. 6(a) and 6(b), one can see that the contributions of ionization from HOMO-1 near 90° for N_2 and from HOMO-2 near 0° for CO_2 should be taken into account. Indeed, the contributions from the HOMO-1 (HOMO-2) to the HHG of N_2 (CO_2) have been reported widely [115–122]. For Cl_2, the contributions of ionization from the HOMO-1 near 90° and from the HOMO-2 near 0° are comparable to those from the HOMO (see Fig. 6(c)). For HBr, the ionization rates of the HOMO-1 are much higher than those of the HOMO near 180°. The significant contributions of ionization from the HOMO-1 near 0° and 180° can be seen in Fig. 6(d).

3.6. *Probing the shape of the ionizing molecular orbitals with the orientation-dependent ionization rates*

It has been confirmed that the orientation dependent ionization rates can reflect vividly the orbital symmetry of molecules [16, 17, 22, 28, 40, 54, 55]. In Fig. 7, the orientation dependent ionization rate $P(\theta)$ with the angle-dependent asymptotic

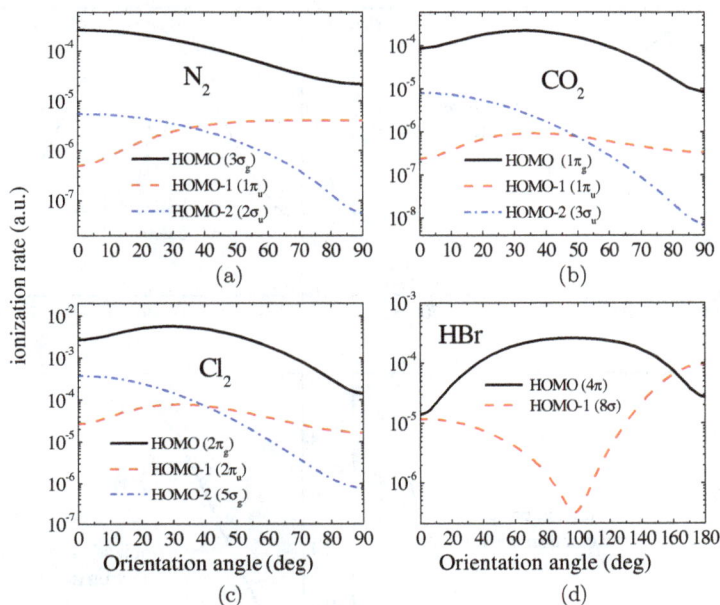

Fig. 6. (Color online) Orientation-dependent ionization rates of HOMO, HOMO-1, and HOMO-2 for N_2, CO_2, and Cl_2 and of HOMO and HOMO-1 for HBr. (a) N_2 at laser intensity of 1.5×10^{14} W/cm^2; (b) CO_2 at 1.1×10^{14} W/cm^2; (c) Cl_2 at 6.9×10^{13} W/cm^2; (d) HBr at 7.2×10^{13} W/cm^2. Figs. 6(a) and (b) are adapted from [54]. © (2010) by the American Physical Society. Figs. 6(c) and (d) are adapted from [80]. © (2013) by Taylor & Francis.

electron density $\rho(\theta_e)$ of the HOMO orbital for N_2, O_2, CO, and HBr are compared, respectively. Here, the MO-ADK model is used to calculate the $P(\theta)$ of these four linear molecules. One can see that the $P(\theta)$ follows well the shape of $\rho(\theta_e)$, as shown in Fig. 7. Note that the structure parameters C_{lm} of CO are taken from Ref. [80]. For the planar H_2O molecule, the molecule lies on the y-z plane, with the O atom along the z axis. The isocontour plot of the HOMO wavefunction is shown in Fig. 8(c). Clearly the HOMO orbital contains a nodal plane (i.e., y-z plane). The angular dependence of electron density is quite similar to that of ionization rate (see Figs. 8(a) and 8(b)). By averaging the electron density $\rho_1(\theta_e, \chi_e)$ and ionization rate $P(\theta, \chi)$ over χ_e or χ, the θ dependent ionization rate agrees very well with the electron density, as shown in Fig. 8(d). Thus it is possible to probe directly the electron density of the molecular orbital from which the electron is tunnel ionized using the corresponding alignment-dependent ionization rates when the ionization contributions from other occupied orbitals can be ignored.

3.7. *Examination of the validity of the MO-ADK and MO-PPT models*

Next the MO-ADK and MO-PPT models are examined by comparing them to the SAE-TDSE calculations and the experimental results. It has been confirmed that

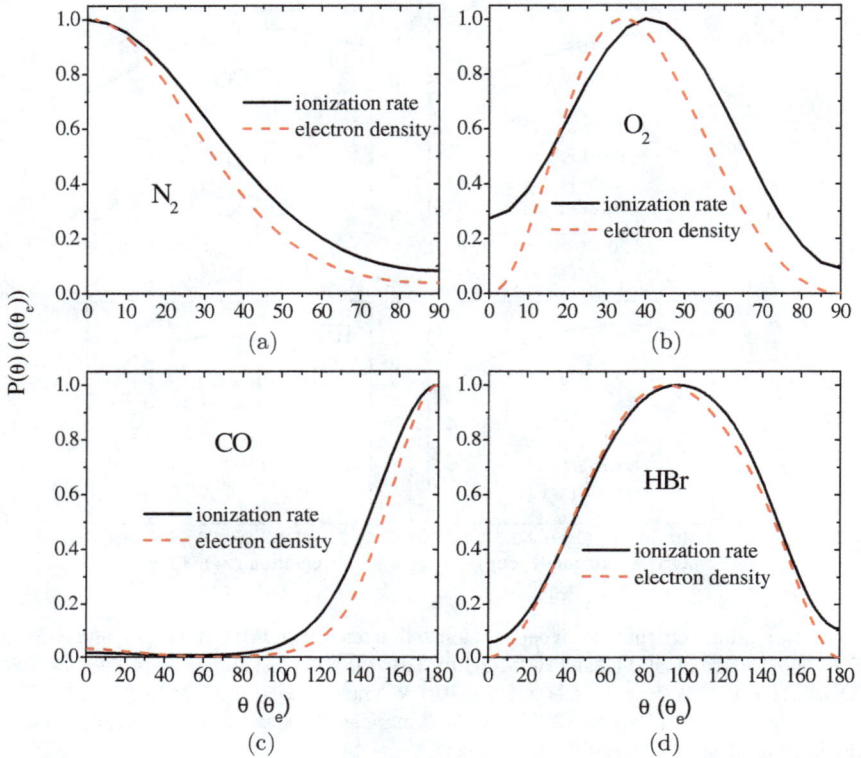

Fig. 7. (Color online) Comparison of the orientation-dependent ionization rate (solid line) to the angular distribution of asymptotic electron density (dashed line). The ionization rates are obtained from the MO-ADK model. (a) N_2 at laser intensity of 1.0×10^{14} W/cm^2; (b) O_2 at 1.0×10^{14} W/cm^2; (c) CO at 1.9×10^{14} W/cm^2; (d) HBr at 7.2×10^{13} W/cm^2.

the MO-ADK model works well in the tunneling ionization regime. This model has also been empirically modified to study the ionization of the H_2^+ molecule in the over-the-barrier ionization (TBI) regime [123]. However, the MO-ADK model is not valid in the multiphoton ionization regime [52, 62, 63]. Figure 9 compares the present calculated ionization probabilities of H_2 from the MO-ADK and MO-PPT models with those SAE-TDSE results using the Hartree-Fock functionals [43]. All the ionization probabilities from the MO-ADK and MO-PPT models are normalized to those of the SAE-TDSE at the saturation laser intensity of 2.29×10^{14} W/cm^2. One can see that the MO-ADK fits quite well the SAE-TDSE in the tunneling ionization region (i.e., $\gamma < 1$), while it underestimates remarkably the ionization probabilities in the multiphoton ionization regime indeed (i.e., $\gamma > 1$). The MO-PPT agrees very well with the SAE-TDSE in the whole range covering from the multiphoton to tunneling ionization regimes. In Fig. 10, ionization signals of NO, Cl_2, N_2 and O_2 obtained from the MO-ADK and MO-PPT models with the experimental data

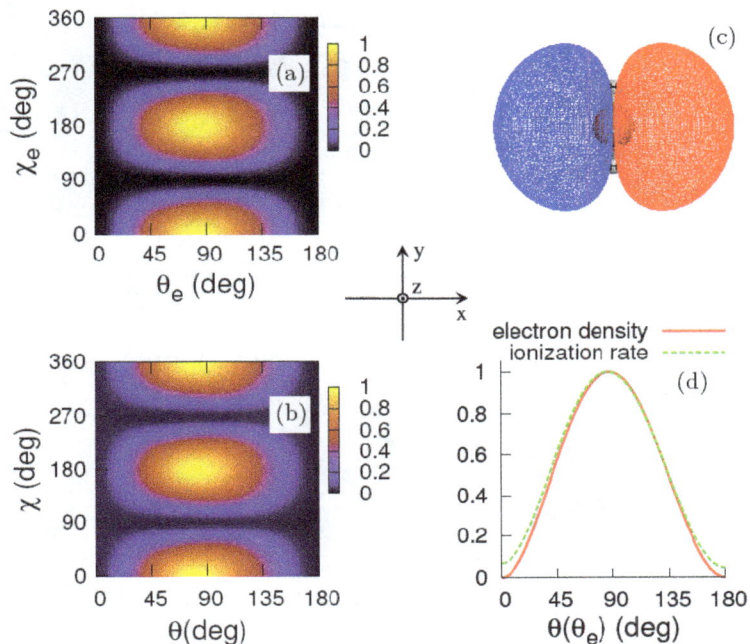

Fig. 8. (Color online) (a) Angular distribution of the normalized asymptotic electron density for H_2O. (b) Normalized alignment-dependent ionization rate of H_2O at laser intensity of 8×10^{13} W/cm^2. (c) The isocontour plot of the HOMO wavefunction for H_2O. The sign of the HOMO wavefunction is indicated by different colors, i.e., red denotes positive sign and blue stands for negative sign. (d) Comparison of the normalized χ_e or χ averaged electron density and ionization rate for H_2O. The x, y, z axes of the molecular frame are also shown. Adapted from [55]. © (2011) by IOP Publishing.

[52, 61, 124] are compared. Clearly the MO-PPT fits well with the experimental results, while the MO-ADK deviates seriously from the experimental data in the multiphoton ionization region.

3.8. *Probing the molecular orbital with the alignment-dependent HHG signals*

Finally, the possibility for probing the molecular orbital using the alignment-dependent HHG signals from molecules fixed in space is investigated. Here, HHG signals at several alignment angles from the molecular Lewenstein model are shown. In Fig. 11(a), the angular dependence of the yields of the 35th, 39th, and 43rd harmonics with angle-dependent asymptotic electron density for N_2 are compared. The HHG yield of each of the $(2n + 1)$-th harmonic is obtained by integrating over the intensity within the energy between the $2n$-th and $(2n + 2)$-th order. For simplicity, all the HHG yields and electron densities in Fig. 11 are normalized to 1.0 at the peak. For O_2, direct comparison of the alignment-dependent yields of

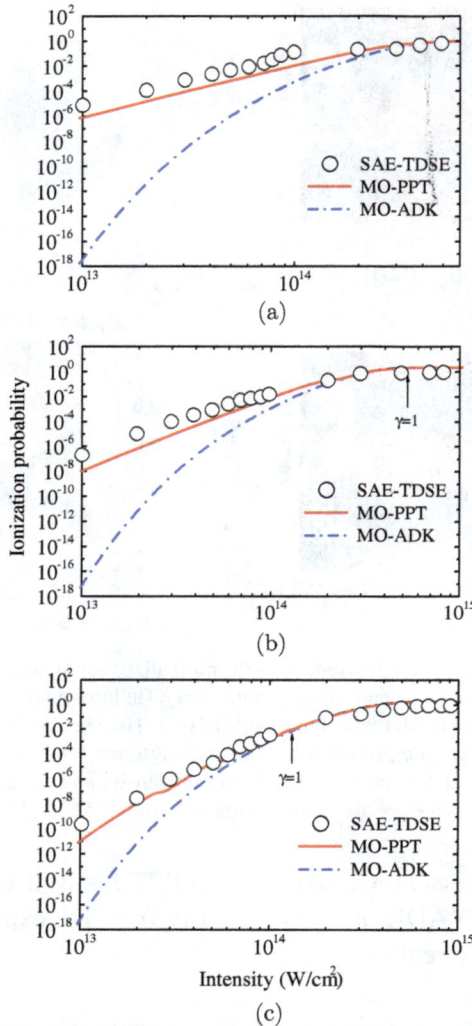

Fig. 9. (Color online) Ionization probabilities of the H_2 molecule as a function of laser peak intensity at central wavelengths of (a) 266 nm; (b) 400 nm; and (c) 800 nm. The laser field is a cosine square pulse with 36 cycles, 24 cycles and 12 cycles for 266 nm, 400 nm and 800 nm, respectively. SAE-TDSE from Ref. [43]. Adapted from [63]. © (2014) by Elsevier.

the 23rd, 27th, and 31st to the electron density is also shown in Fig. 11(b). In the present simulations, a Gaussian pulse with laser intensity of 3×10^{14} W/cm^2 for N_2 and 2×10^{14} W/cm^2 for O_2 was used, respectively. The central wavelength and the pulse duration are chosen to be 800 nm and 30 fs, respectively. In Figs. 11(a) and 11(b), the angular dependence among the different harmonics does not change much for N_2 and O_2, respectively. It is emphasized that the alignment dependence

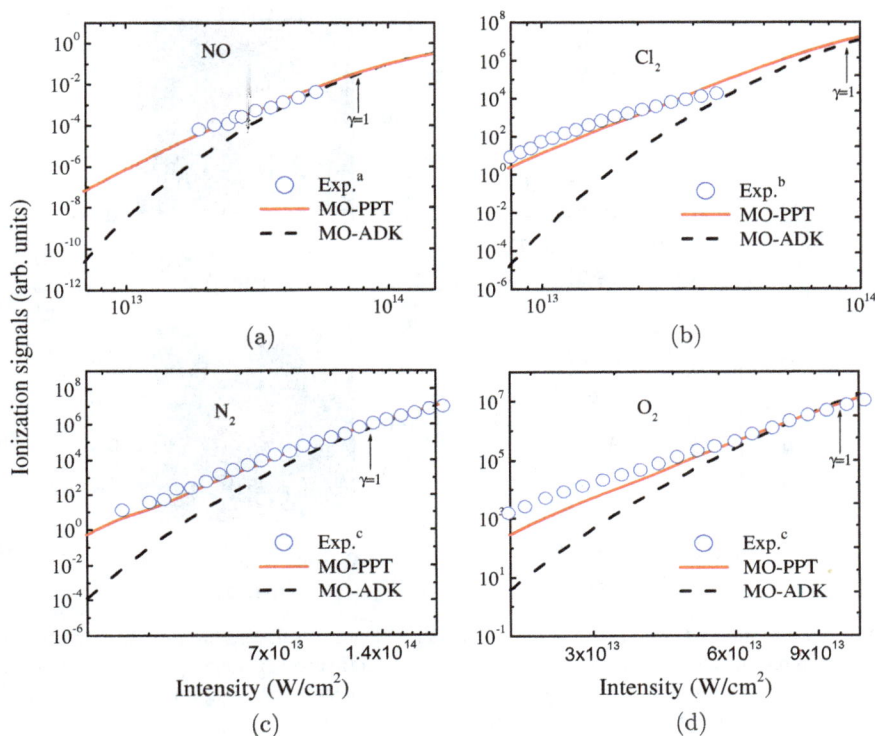

Fig. 10. (Color online) Ionization signals as a function of laser peak intensity. (a) NO at laser central wavelength $\lambda = 800$ nm and pulse duration (FWHM) $\tau = 25$ fs; (b) Cl_2 at $\lambda = 790$ nm and $\tau = 25$ fs; (c) N_2 at $\lambda = 800$ nm and $\tau = 30$ fs; and (d) O_2 at $\lambda = 800$ nm and $\tau = 30$ fs. The laser field is a Gaussian pulse in the calculations. Exp.[a] from Ref. [52], Exp.[b] from Ref. [61] and Exp.[c] from Ref. [124]. Adapted from [63]. © (2014) by Elsevier.

of HHG yields is determined mostly by the orbital symmetry within the molecular Lewenstein model. For example, the HOMO of N_2 (O_2) is a σ_g (π_g) orbital, thus both the HHG yields and electron density have a peak near $\theta = 0°$ ($\theta = 45°$) (see Figs. 11(a) and 11(b)). For the planar H_2O molecule, the alignment dependence of the 37th harmonic follows closely the angular distribution of electron density (see Figs. 11(c) and 11(d)). A Gaussian pulse with the laser intensity of 0.6×10^{14} W/cm^2, central wavelength of 1200 nm, and pulse duration of 25 fs was used in the calculation for H_2O. Therefore, the alignment-dependent HHG yields can be used to probe the molecular orbital from which the electron is removed if the contributions from other occupied orbitals to the HHG can be ignored. Note that these results are not surprising since in the molecular Lewenstein model only the initial wavefunction of the molecule enters into the theory. The original tomographic paper [9] was also based on the molecular Lewenstein model.

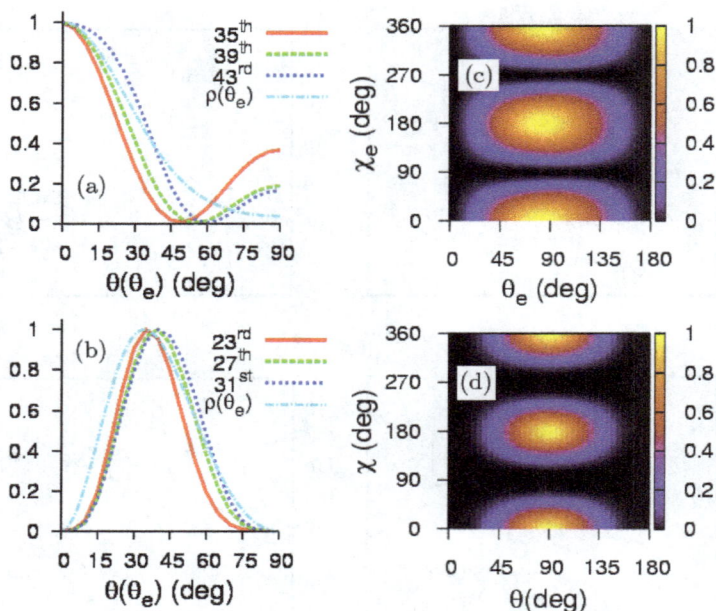

Fig. 11. (Color online) Comparison of alignment dependence of selected high-order harmonic yields in the plateau region and angular distribution of the asymptotic electron density for (a) N_2 and (b) O_2. (c) Angular distribution of electron density for the H_2O molecule; (d) Alignment dependence of the 37th harmonic from H_2O. For the laser parameters used, see text.

More accurate treatment of HHG like the QRS model [3, 103, 125] requires that the recombining electrons be described by scattering wavefunctions instead of plane waves used in the Lewenstein model. In addition, experimental HHG spectra are due to the coherent emission of light from all molecules in the medium and thus propagation effects should be included [126]. For HHG due to simpler molecules, such as N_2, O_2 and CO_2, accurate calculations based on the QRS model, including multiple orbital contributions as well as propagation effect, has been carried out and the results have been compared well with experiments [121, 127–129]. In fact, HHG from large polyatomic molecules have also been carried out within the QRS model [130–132]. Such calculations are quite tedious. In such situations, molecular Lewenstein model would offer a convenient qualitative theory for interpreting experimental observations.

4. Conclusions

In this chapter, it is shown that molecular orbital in a molecule can be directly probed using the alignment-dependent ionization probabilities or HHG yields from molecules exposed to an intense laser field. The ionization probabilities can be calculated easily with simple models like MO-ADK, MO-PPT, MO-SFA *et al.* In the MO-ADK model, the static ionization rate depends on the structure parameters (i.e., expansion coefficients) of molecules. An efficient method to obtain the molecular wavefunction with the correct asymptotic behavior by solving the time-independent Schrödinger equation with the B-spline functions has been described where the one-electron potential is constructed numerically based on the DFT. These correct wavefunctions are used to extract accurate structure parameters of molecules in the asymptotic region. The failure of the MO-ADK model in the multiphoton regime was also addressed. Interestingly the MO-PPT model was able to fit both the TDSE results and the available experimental data covering from the multiphoton to tunneling ionization regimes. Based on the MO-ADK model, it is shown that tunneling ionization rates of inner orbitals are comparable to that of the HOMO at some alignment angles. Indeed, contributions from inner orbitals to strong-field phenomena have been observed experimentally [23, 27, 28, 31, 113–119]. Moreover, it is also demonstrated that the molecular orbital from which the electron is removed can be probed using alignment-dependent ionization probabilities or HHG signals by an intense laser field when the contributions from inner occupied orbitals can be neglected.

Acknowledgments

The authors thank Prof. X. M. Tong, Prof. T. F. Jiang, Prof. Zengxiu Zhao, Dr. Anh-Thu Le, Dr. Cheng Jin and Dr. Junliang Xu for their contributions on these researches. S. -F. Z would like to thank Dr. Shi-Lin Hu for providing their ionization probabilities obtained from the SAE-TDSE for H_2. CDL was supported in part by Chemical Sciences, Geosciences and Biosciences Division, Office of Basic Energy Sciences, Office of Science, U.S. Department of Energy. S. -F. Z and X. X. Z was also supported by the National Natural Science Foundation of China under Grant Nos. 11044007, 11164025, 11264036, 11465016 and the Specialized Research Fund for the Doctoral Program of Higher Education of China under Grant Nos. 20096203120001, and 20116203120001.

References

1. W. H. E. Schwarz, *Angew. Chem. Int. Ed.* 45, 1508 (2006).
2. A. D. Bandrauk and M. Ivanov (eds.), *Quantum Dynamic Imaging* (New York: Springer, 2011).
3. C. D. Lin, A. -T. Le, Z. Chen, T. Morishita, and R. R. Lucchese, *J. Phys. B: At. Mol. Opt. Phys.* **43**, 122001 (2010).
4. M. Lein, *J. Phys. B: At. Mol. Opt. Phys.* **40**, R135 (2007).
5. S. Haessler, J. Caillat, and P. Salières, *J. Phys. B: At. Mol. Opt. Phys.* **44**, 203001 (2011).
6. J. Repp and G. Meyer, *Phys. Rev. Lett.* **94**, 026803 (2005).
7. A. Bellec, F. Ample, D. Riedel, G. Dujardin, and C. Joachim, *Nano Lett.* **9**, 144 (2009).

8. P. Puschnig, S. Berkebile, A. J. Fleming, G. Koller, K. Emtsev, T. Seyller, J. D. Riley, C. Ambrosch-Draxl, F. P. Netzer, and M. G. Ramsey, *Science* **326**, 702 (2009).
9. J. Itatani, J. Levesque, D. Zeidler, H. Niikura, H. Pépin, J. C. Kieffer, P. B. Corkum, and D. M. Villeneuve, *Nature* (London) **432**, 867 (2004).
10. V. -H. Le, A. -T. Le, R. H. Xie, and C. D. Lin, *Phys. Rev. A* **76**, 013414 (2007).
11. C. Vozzi, M. Negro, F. Calegari, G. Sansone, M. Nisoli, S. De Silvestri, and S. Stagira, *Nat. Phys.* **7**, 822 (2011).
12. Y. Li, X. Zhu, P. Lan, Q. Zhang, M. Qin, and P. Lu, *Phys. Rev. A* **89**, 045401 (2014).
13. S. Patchkovskii, Z. X. Zhao, T. Brabec, and D. B. Villeneuve, *Phys. Rev. Lett.* **97**, 123003 (2006).
14. Y. J. Chen, L. B. Fu, and J. Liu, *Phys. Rev. Lett.* **111**, 073902 (2013).
15. P. B. Corkum, *Phys. Rev. Lett.* **71**, 1994 (1993).
16. I. V. Litvinyuk, K. F. Lee, P. W. Dooley, D. M. Rayner, D. M. Villeneuve, and P. B. Corkum, *Phys. Rev. Lett.* **90**, 233003 (2003).
17. D. Pavičić, K. F. Lee, D. M. Rayner, P. B. Corkum, and D. M. Villeneuve, *Phys. Rev. Lett.* **98**, 243001 (2007).
18. I. Thomann, R. Lock, V. Sharma, E. Gagnon, S. T. Pratt, H. C. Kapteyn, M. M. Murnane, and W. Li, *J. Phys. Chem. A* **112**, 9382 (2008).
19. V. Kumarappan, L. Holmegaard, C. Martiny, C. B. Madsen, T. K. Kjeldsen, S. S. Viftrup, L. B. Madsen, and H. Stapelfeldt, *Phys. Rev. Lett.* **100**, 093006 (2008).
20. R. Itakura, H. Hasegawa, Y. Kurosaki, A. Yokoyama, and Y. Ohshima, *J. Phys. Chem. A* **114**, 11202 (2010).
21. J. L. Hansen, L. Holmegaard, J. H. Nielsen, H. Stapelfeldt, D. Dimitrovski, and L. B. Madsen, *J. Phys. B: At. Mol. Opt. Phys.* **45**, 015101 (2012).
22. A. Staudte, S. Patchkovskii, D. Pavičić, H. Akagi, O. Smirnova, D. Zeidler, M. Meckel, D. M. Villeneuve, R. Dörner, M. Yu. Ivanov, and P. B. Corkum, *Phys. Rev. Lett.* **102**, 033004 (2009).
23. H. Akagi, T. Otobe, A. Staudte, A. Shiner, F. Turner, R. Dörner, D. M. Villeneuve, and P. B. Corkum, *Science* **325**, 1364 (2009).
24. M. Magrakvelidze, F. He, S. De, I. Bocharova, D. Ray, U. Thumm, and I. V. Litvinyuk, *Phys. Rev. A* **79**, 033408 (2009).
25. J. Wu, M. Meckel, S. Voss, H. Sann, M. Kunitski, L. Ph. H. Schmidt, A. Czasch, H. Kim, T. Jahnke, and R. Dörner, *Phys. Rev. Lett.* **108**, 043002 (2012).
26. J. B. Williams, C. S. Trevisan, M. S. Schöffler, T. Jahnke, I. Bocharova, H. Kim, B. Ulrich, R. Wallauer, F. Sturm, T. N. Rescigno, A. Belkacem, R. Dörner, Th. Weber, C. W. McCurdy, and A. L. Landers, *Phys. Rev. Lett.* **108**, 233002 (2012).
27. J. Wu, L. Ph. H. Schmidt, M. Kunitski, M. Meckel, S. Voss, H. Sann, H. Kim, T. Jahnke, A. Czasch, and R. Dörner, *Phys. Rev. Lett.* **108**, 183001 (2012).
28. H. Liu, S. -F. Zhao, M. Li, Y. Deng, C. Wu, X. X. Zhou, Q. Gong, and Y. Liu, *Phys. Rev. A* **88**, 061401 (R) (2013).
29. A. S. Alnaser, S. Voss, X. M. Tong, C. M. Maharjan, P. Ranitovic, B. Ulrich, T. Osipov, B. Shan, Z. Chang, and C. L. Cocke, *Phys. Rev. Lett.* **93**, 113003 (2004).
30. A. S. Alnaser, C. M. Maharjan, X. M. Tong, B. Ulrich, P. Ranitovic, B. Shan, Z. Chang, C. D. Lin, C. L. Cocke, and I. V. Litvinyuk, *Phys. Rev. A* **71**, 031403(R) (2005).
31. P. von den Hoff, I. Znakovskaya, S. Zherebtsov, M. F. Kling, and R. de Vivie-Riedle, *Appl. Phys. B* **98**, 659 (2010).
32. X. Liu, C. Wu, Z. Wu, Y. Liu, Y. Deng, and Q. Gong, *Phys. Rev. A* **83**, 035403 (2011).
33. H. Chen, L. Fang, V. Tagliamonti, and G. N. Gibson, *Phys. Rev. A* **84**, 043427 (2011).
34. J. McKenna, A. M. Sayler, B. Gaire, Nora G. Johnson, K. D. Carnes, B. D. Esry, and I. Ben-Itzhak, *Phys. Rev. Lett.* **103**, 103004 (2009).
35. S. Petretti, Y. V. Vanne, A. Saenz, A. Castro, and P. Decleva, *Phys. Rev. Lett.* **104**, 223001 (2010).

36. M. Spanner and S. Patchkovskii, *Phys. Rev. A* **80**, 063411 (2009).
37. T. K. Kjeldsen, L. A. A. Nikolopoulos, and L. B. Madsen, *Phys. Rev. A* **75**, 063427 (2007).
38. M. Abu-samha and L. B. Madsen, *Phys. Rev. A* **80**, 023401 (2009).
39. M. Abu-samha and L. B. Madsen, *Phys. Rev. A* **81**, 033416 (2010).
40. G. Lagmago Kamta and A. D. Bandrauk, *Phys. Rev. A* **74**, 033415 (2006).
41. Y. J. Jin, X. M. Tong, and N. Toshima, *Phys. Rev. A* **83**, 063409 (2011).
42. B. Zhang, J. Yuan, and Z. X. Zhao, *Phys. Rev. A* **85**, 033421 (2012).
43. M. Awasthi, Y. V. Vanne, A. Saenz, A. Castro, and P. Decleva, *Phys. Rev. A* **77**, 063403 (2008).
44. S. Petretti , A. Saenz, A. Castro, and P. Decleva, *Chem. Phys.* **414**, 45 (2013).
45. S. K. Son and Shih-I. Chu, *Phys. Rev. A* **80**, 011403(R) (2009).
46. D. A. Telnov and Shih-I. Chu, *Phys. Rev. A* **79**, 041401(R) (2009).
47. X. Chu, *Phys. Rev. A* **82**, 023407 (2010).
48. X. Chu and M. McIntyre, *Phys. Rev. A* **83**, 013409 (2011).
49. T. Otobe and K. Yabana, *Phys. Rev. A* **75**, 062507 (2007).
50. S. K. Son and Shih-I Chu, *Chem. Phys.* **366**, 91 (2009).
51. B. Zhang, J. Yuan, and Z. X. Zhao, *Phys. Rev. Lett.* **111**, 163001 (2013).
52. X. M. Tong, Z. X. Zhao, and C. D. Lin, *Phys. Rev. A* **66**, 033402 (2002).
53. Z. X. Zhao, X. M. Tong, and C. D. Lin, *Phys. Rev. A* **67**, 043404 (2003).
54. S. -F. Zhao, C. Jin, A. -T. Le, T. F. Jiang, and C. D. Lin, *Phys. Rev. A* **81**, 033423 (2010).
55. S. -F. Zhao, J. Xu, C. Jin, A. -T. Le, and C. D. Lin, *J. Phys. B: At. Mol. Opt. Phys.* **44**, 035601 (2011).
56. J. Muth-Böhm, A. Becker, and F. H. M. Faisal, *Phys. Rev. Lett.* **85**, 2280 (2000).
57. T. K. Kjeldsen and L. B. Madsen, *J. Phys. B: At. Mol. Opt. Phys.* **37**, 2033 (2004).
58. D. B. Milošević, *Phys. Rev. A* **74**, 063404 (2006).
59. B. Zhang and Z. X. Zhao, *Phys. Rev. A* **82**, 035401 (2010).
60. Y. J. Chen and B. Zhang, *J. Phys. B: At. Mol. Opt. Phys.* **45**, 215601 (2012).
61. E. P. Benis, J. F. Xia, X. M. Tong, M. Faheem, M. Zamkov, B. Shan, P. Richard, and Z. Chang, *Phys. Rev. A* **70**, 025401 (2004).
62. Y. Z. Fu, S. -F. Zhao, and X. X. Zhou, *Chin. Phys. B* **21**, 113101 (2012).
63. S. -F. Zhao, L. Liu, and X. X. Zhou, *Opt. Commun.* **313**, 74 (2014).
64. L. B. Madsen, O. I. Tolstikhin, and T. Morishita, *Phys. Rev. A* **85**, 053404 (2012).
65. L. B. Madsen, F. Jensen, O. I. Tolstikhin, and T. Morishita, *Phys. Rev. A* **87**, 013406 (2013).
66. L. B. Madsen, F. Jensen, O. I. Tolstikhin, and T. Morishita, *Phys. Rev. A* **89**, 033412 (2014).
67. R. Murray, M. Spanner, S. Patchkovskii, and M. Yu. Ivanov, *Phys. Rev. Lett.* **106**, 173001 (2011).
68. G. A. Gallup and I. I. Fabrikant, *Phys. Rev. A* **81**, 033417 (2010).
69. B. Zhang and Z. X. Zhao, *Chin. Phys. Lett.* **27**, 043301 (2010).
70. M. V. Ammosov, N. B. Delone, and V. P. Krainov, Zh. Eksp. Teor. Fiz. 91, 2008 (1986) [(*Sov. Phys. JETP* **64**, 1191 (1986)].
71. D. Dill and J. L. Dehmer, *J. Chem. Phys.* **61**, 692 (1974).
72. M. J. Frisch, G. W. Trucks, H. B. Schlegel, G. E. Scuseria, M. A. Robb, J. R. Cheeseman, J. A. Montgomery Jr., T. Vreven, K. N. Kudin, J. C. Burant, J. M. Millam, S. S. Iyengar, J. Tomasi, V. Barone, B. Mennucci, M. Cossi, G. Scalmani, N. Rega, G. A. Petersson, H. Nakatsuji, M. Hada, M. Ehara, K. Toyota, R. Fukuda, J. Hasegawa, M. Ishida, T. Nakajima, Y. Honda, O. Kitao, H. Nakai, M. Klene, X. Li, J. E. Knox, H. P. Hratchian, J. B. Cross, V. Bakken, C. Adamo, J. Jaramillo, R. Gomperts, R. E. Stratmann, O. Yazyev, A. J. Austin, R. Cammi, C. Pomelli, J. W. Ochterski, P. Y. Ayala, K. Morokuma, G. A. Voth, P. Salvador, J. J. Dannenberg, V. G. Zakrzewski, S. Dapprich, A. D. Daniels, M. C. Strain, O. Farkas, D. K. Malick, A. D. Rabuck, K. Raghavachari, J. B. Foresman, J. V. Ortiz, Q. Cui, A. G. Baboul, S. Clifford, J. Cioslowski, B. B. Stefanov, G. Liu, A. Liashenko, P. Piskorz, I. Komaromi, R. L. Martin, D. J. Fox, T. Keith, M. A. Al-Laham, C. Y. Peng, A. Nanayakkara, M. Challacombe,

P. M. W. Gill, B. Johnson, W. Chen, M. W. Wong, C. Gonzalez, and J. A. Pople, GAUSSIAN 03, Revision C.02 (Gaussian Inc. Pittsburgh, PA, 2003).

73. M. W. Schmidt, K. K. Baldridge, J. A. Boatz, S. T. Elbert, M. S. Gordon, J. H. Jensen, S. Koseki, N. Matsunaga, K. A. Nguyen, S. Su, T. L. Windus, M. Dupuis, and J. A. Montgomery Jr., *J. Comput. Chem.* **14**, 1347 (1993).

74. H. -J. Werner and P. J. Knowles, MOLPRO, Version 2002.6, A Package of Ab Initio Programs, Birmingham, UK, 2003.

75. J. Kobus, L. Laaksonen, and D. Sundholm, *Comput. Phys. Commun.* **98**, 346 (1996).

76. A. -T. Le, X. M. Tong, and C. D. Lin, *J. Mod. Opt.* **54**, 967 (2007).

77. T. K. Kjeldsen, C. Z. Bisgaard, L. B. Madsen, and H. Stapelfeldt, *Phys. Rev. A* **71**, 013418 (2005).

78. S. -F. Zhao, C. Jin, A. -T. Le, T. F. Jiang, and C. D. Lin, *Phys. Rev. A* **80**, 051402 (R) (2009).

79. S. -F. Zhao, C. Jin, A. -T. Le, and C. D. Lin, *Phys. Rev. A* **82**, 035402 (2010).

80. J. P. Wang, S. -F. Zhao, C. R. Zhang, W. Li, and X. X. Zhou, *Mol. Phys.* **112**, 1102 (2014).

81. X. J. Li, S. -F. Zhao, and X. X. Zhou, *Commun. Theor. Phys.* **58**, 419 (2012).

82. T. K. Kjeldsen and L. B. Madsen, *Phys. Rev. A* **71**, 023411 (2005).

83. R. Torres, N. Kajumba, Jonathan G. Underwood, J. S. Robinson, S. Baker, J. W. G. Tisch, R. de Nalda, W. A. Bryan, R. Velotta, C. Altucci, I. C. E. Turcu, and J. P. Marangos, *Phys. Rev. Lett.* **98**, 203007 (2007).

84. M. Lein, N. Hay, R. Velotta, J. P. Marangos, and P. L. Knight, *Phys. Rev. Lett.* **88**, 183903 (2002).

85. G. Lagmago Kamta and A. D. Bandrauk, *Phys. Rev. A* **71**, 053407 (2005).

86. G. Lagmago Kamta and A. D. Bandrauk, *Phys. Rev. A* **80**, 041403 (R) (2009).

87. D. A. Telnov and Shih-I Chu, *Phys. Rev. A* **76**, 043412 (2007).

88. X. B. Bian and A. D. Bandrauk, *Phys. Rev. Lett.* **105**, 093903 (2010).

89. X. B. Bian and A. D. Bandrauk, *Phys. Rev. A* **86**, 053417 (2012).

90. D. A. Telnov and Shih-I Chu, *Phys. Rev. A* **80**, 043412 (2009).

91. X. Chu and G. C. Groenenboom, *Phys. Rev. A* **87**, 013434 (2013).

92. E. P. Fowe and A. D. Bandrauk, *Phys. Rev. A* **84**, 035402 (2011).

93. E. F. Penka, E. Couture-Bienvenue, and A. D. Bandrauk, *Phys. Rev. A* **89**, 023414 (2014).

94. X. X. Zhou, X. M. Tong, Z. X. Zhao, and C. D. Lin, *Phys. Rev. A* **71**, 061801 (R) (2005).

95. X. X. Zhou, X. M. Tong, Z. X. Zhao, and C. D. Lin, *Phys. Rev. A* **72**, 033412 (2005).

96. A. -T. Le, X. M. Tong, and C. D. Lin, *Phys. Rev. A* **73**, 041402 (R) (2006).

97. J. P. Marangos, C. Altucci, R. Velotta, E. Heesel, E. Springate, M. Pascolini, L. Poletto, P. Villoresi, C. Vozzi, G. Sansone, M. Anscombe, J.-P. Caumes, S. Stagira, and M. Nisoli, *J. Mod. Opt.* **53**, 97 (2006).

98. S. Odžak and D. B. Milošević, *Phys. Rev. A* **79**, 023414 (2009).

99. C. B. Madsen and L. B. Madsen, *Phys. Rev. A* **74**, 023403 (2006).

100. C. B. Madsen and L. B. Madsen, *Phys. Rev. A* 76, 043419 (2007).

101. Y. J. Chen and B. Hu, *Phys. Rev. A* 80, 033408 (2009).

102. M. Lewenstein, Ph. Balcou, M. Yu. Ivanov, A. L'Huillier, and P. B. Corkum, *Phys. Rev. A* 49, 2117 (1994).

103. A. -T. Le, R. R. Lucchese, S. Tonzani, T. Morishita, and C. D. Lin, *Phys. Rev. A* 80, 013401 (2009).

104. P. R. T. Schipper, O. V. Gritsenko, S. J. A. van Gisbergen, and E. J. Baerends, *J. Chem. Phys.* **112**, 1344 (2000).

105. Shih-I Chu, *J. Chem. Phys.* **123**, 062207 (2005).

106. J. P. Perdew and Y. Wang, *Phys. Rev. B* **45**, 13244 (1992).

107. X. B. Bian, L. Y. Peng, and T. Y. Shi, *Phys. Rev. A* **77**, 063415 (2008).

108. A. M. Perelomov, V. S. Popov, and M. V. Terent'ev, *Sov. Phys. JETP* **23**, 924 (1966).

109. F. A. Ilkov, J. E. Decker, and S. L. Chin, *J. Phys. B: At. Mol. Opt. Phys.* **25**, 4005 (1992).

110. S. -F. Zhao, C. Jin, R. R. Lucchese, A. -T. Le, and C. D. Lin, *Phys. Rev. A* **83**, 033409 (2011).

111. Y. -M. Lee, T. -F. Jiang, Z. -Y. Su, and J. -S. Wu, *Comput. Phys. Commun.* **182**, 140 (2011).
112. S. L. Hu, Z. X. Zhao, and T. Y. Shi, *Chin. Phys. Lett.* **30**, 103103 (2013).
113. C. Wu, H. Zhang, H. Yang, Q. Gong, D. Song, and H. Su, *Phys. Rev. A* **83**, 033410 (2011).
114. I. Znakovskaya, P. Von den Hoff, S. Zherebtsov, A. Wirth, O. Herrwerth, M. J. J. Vrakking, R. de Vivie-Riedle, and M. F. Kling, *Phys. Rev. Lett.* **103**, 103002 (2009).
115. B. K. McFarland, J. P. Farrell, P. H. Bucksbaum, and M. Gühr, *Science* **322**, 1232 (2008).
116. O. Smirnova, Y. Mairesse, S. Patchkovskii, N. Dudovich, D. Villeneuve, P. B. Corkum, and M. Yu. Ivanov, *Nature* (London) **460**, 972 (2009).
117. H. J. Wörner, J. B. Bertrand, P. Hockett, P. B. Corkum, and D. M. Villeneuve, *Phys. Rev. Lett.* **104**, 233904 (2010).
118. Z. Diveki, A. Camper, S. Haessler, T. Auguste, T. Ruchon, B. Carré, P. Salières, R. Guichard, J. Caillat, A. Maquet, and R. Taïeb, *New J. Phys.* **14**, 023062 (2012).
119. J. Li, P. Liu, H. Yang, L. Song, S. Zhao, H. Lu, R. Li, and Z. Xu, *Opt. Express* **21**, 7599 (2013).
120. A. -T. Le, R. R. Lucchese, and C. D. Lin, *J. Phys. B: At. Mol. Opt. Phys.* **42**, 211001 (2009).
121. C. Jin, J. B. Bertrand, R. R. Lucchese, H. J. Wörner, P. B. Corkum, D. M. Villeneuve, A. -T. Le, and C. D. Lin, *Phys. Rev. A* **85**, 013405 (2012).
122. J. Heslar, D. Telnov, and Shih-I Chu, *Phys. Rev. A* **83**, 043414 (2011).
123. X. M. Tong and C. D. Lin, *J. Phys. B: At. Mol. Opt. Phys.* **38**, 2593 (2005).
124. C. Guo, M. Li, J. P. Nibarger, and G. N. Gibson, *Phys. Rev. A* **58**, R4271 (1998).
125. T. Morishita, A. -T. Le, Z. Chen, and C. D. Lin, *Phys. Rev. Lett.* **100**, 013903 (2008).
126. C. Jin, A. -T. Le, and C. D. Lin, *Phys. Rev. A* **79**, 053413 (2009).
127. C. Jin, A. -T. Le, and C. D. Lin, *Phys. Rev. A* **83**, 023411 (2011).
128. C. Jin, H. J. Wörner, V. Tosa, A. -T. Le, J. B. Bertrand, R. R. Lucchese, P. B. Corkum, D. M. Villeneuve, and C. D. Lin, *J. Phys. B: At. Mol. Opt. Phys.* **44**, 095601 (2011).
129. C. Jin, A. -T. Le, and C. D. Lin, *Phys. Rev. A* **83**, 053409 (2011).
130. A. -T. Le, R. R. Lucchese, M. T. Lee, and C. D. Lin, *Phys. Rev. Lett.* **102**, 203001 (2009).
131. M. C. H. Wong, A. -T. Le, A. F. Alharbi, A. E. Boguslavskiy, R. R. Lucchese, J. -P. Brichta, C. D. Lin, and V. R. Bhardwaj, *Phys. Rev. Lett.* **110**, 033006 (2013).
132. A. -T. Le, R. R. Lucchese, and C. D. Lin, *Phys. Rev. A* **88**, 021402 (R) (2013).

Chapter 7

**High-order Harmonic Generation Driven by Sub-Cycle
Shaped Laser Field**

Yinghui Zheng, Zhinan Zeng*, Pengfei Wei, Jing Miao,
Ruxin Li, and Zhizhan Xu

*State Key Laboratory of High Field Laser Physics,
Shanghai Institute of Optics and Fine Mechanics,
Chinese Academy of Sciences, Shanghai 201800, China*
*zhinan_zeng@mail.siom.ac.cn

High-order harmonic generation can be described by the semiclassical three-step model, in which an electron is freed, accelerated away from an atom or molecule by a strong oscillating laser field, and then, upon reversal of the field, careened back into its parent ion. The shaped laser field has been proved to be an effective tool to control the three-step process and consequently to achieve the high intensity harmonic generation or an isolated attosecond pulse generation by changing the relative phase, intensity ratio, polarization, etc, between the pulses of shaped laser field. High-order harmonic and attosecond pulse generation driven by a shaped laser field synthesized with two or three laser pulses of controlled related phase are reviewed.

1. Introduction

One of the most extensively studied topics in the field of strong-field processes in last two decades is high-order harmonic generation (HHG) in atoms or molecules driven by an intense laser field [1–4]. It provides a powerful technique to generate a single attosecond pulse [5] or a train of attosecond pulses [6] by converting light from the infrared into the extreme ultraviolet (XUV) [7, 8] and x-ray regime [9]. More recently, HHG in gases has been widely used as a probe to study the ultrafast electronic structure [10–12], electronic dynamics [13, 14], multichannel dynamics in strong-field ionization [15] and the tomographic image of molecular orbitals [16, 17].

One of the most important applications of the HHG is to produce a subfemtosecond or attosecond XUV pulse. Generation and application of the attosecond pulses are actively investigated in this decade. The duration of the experimentally achievable isolated attosecond pulse (IAP) has been shortened down to 67 as [6]. So far, the first demonstrated scheme to produce single attosecond pulses requires phase stabilized few-cycle laser pulses and a subsequent spectral filtering of cut-off harmonics [1, 18]. An alternative approach for the generation of IAP is based on the temporal confinement of the HHG by polarization-gating [3, 19, 20]. Another method to generate single attosecond pulses is the application of a two-color field. Two-color HHG using relatively long pulses has been studied both theoretically [21] and experimentally [22, 23], and enhancement of harmonic intensity, extension of harmonic cutoff frequency and appearance of even-order harmonics were observed.

The high-order harmonic source usually emits in the form of XUV frequency comb in the plateau region with different harmonics at similar intensities, due to the non-perturbative nature of the HHG process. Therefore, besides the IAP generation, the selective generation of a single high-order harmonic emission from the harmonic comb as an intense monochromatic coherent source is also very important for many applications, such as the seeding of an XUV free electron laser for obtaining fully coherent output [24].

HHG is a strongly nonlinear response of a medium to the driving laser field. HHG can be understood by a three-step model [25]: an electron is first pulled out from an atom or a molecule, is accelerated by the driving laser field, and then recollides with its parent ion core, followed by harmonic photon emission. By controlling the laser field on subcycle timescale carefully, the motion of the electron can be accurately steered to generate an IAP, to enhance the conversion efficiency of the HHG, to selectively generate a single high-order harmonic emission and so on. A robust technique to control the waveform of the driving laser field is to produce a shaped laser field by synthesizing two or three laser pulses with controlled related phase.

In this chapter, we address the recent progress at SIOM: a theoretical prediction of the generation of two XUV attosecond pulses with tunable delay between them using two orthogonally-polarized linearly chirped laser pulses; and experimentally demonstrate a robust scheme to select a single high-order harmonic among the harmonic comb by using a driving laser field with sub-cycle waveform control, which is synthesized by the fundamental 800 nm laser pulse and two controlling laser pulses at 400 nm and 267 nm with perpendicular polarizations.

2.1. *Two pulses scheme*

As mentioned above, the main methods applied experimentally for IAP generation involve using few-cycle laser pulse [1, 26] (amplitude gating), time-varying ellipticity of the driving laser field (polarization gating) [20, 27, 28], and rapid ionization of the medium (ionization gating [29, 30]). Besides, several techniques are based on combination of one of these methods with using second harmonic in conjunction to the fundamental field [31, 32]. In the presence of the second harmonic the period of the recollision process is the full optical cycle [33, 34] (but not the half cycle as in the case of one-color field), thus using of the second harmonic provides additional time for "closing" of the gate. Combining the two-color and time-gating schemes, driving laser pulse can be further extended to be multi-cycle and such technique is known as double optical gating (DOG) [35].

Recently, in the work of J. Miao *et al.* [36], we suggest and investigate theoretically a new polarization gating technique allowing generation of a pair of attopulses with controllable delay. This technique is based on using two orthogonally polarized linearly chirped laser pulses. We calculate the XUV emission in such field using the Lewenstein model [37] and find that a pair of attosecond pulses can be generated under proper choice of the field parameters. The time delay between these two XUV pulses is controlled by simply changing the delay between the two driving fields.

The laser pulses used in our simulation are linearly chirped. Note that experimentally the linear chirp can be easily controlled by changing the distance between the gratings [38]. To simplify the simulation, the pulses have the same chirp, but one is negative and the other positive, respectively. Two crossed linearly polarized fields, E_x and E_y, can be written as below, which have the same spectral amplitude with the initial pulse,

$$E_x = E_0 \exp[-(t + T_d/2)^2 \times a] \sin[\omega(t + T_d/2) + b(t + T_d/2)^2], \qquad (1)$$

$$E_y = E_0 \exp[-(t - T_d/2)^2 \times a] \sin[\omega(t - T_d/2) - b(t - T_d/2)^2], \qquad (2)$$

where T_d is the relative delay between the two pulses, ω is the angular frequency, $a = 1/(\tau_p N)^2$, $b = a(N^2 - 1)^{1/2}$ and N is a parameter describing the chirp, τ_p is the duration of the initial pulse. In our simulations we use the initial laser pulse with the peak intensity of 8×10^{13} W/cm^2. The initial pulse duration is 25 fs and $N = 3$, which means that the durations of both chirped laser pulses are $N\tau_p = 75$ fs. We use the Lewenstein model [37] in our simulations.

Using 800 nm laser field, we calculate the HHG spectra as a function of the delay T_d between the laser pulses. Selecting a single harmonic, we apply inversed Fourier transform to study its temporal behavior. Figure 1(a) shows the temporal

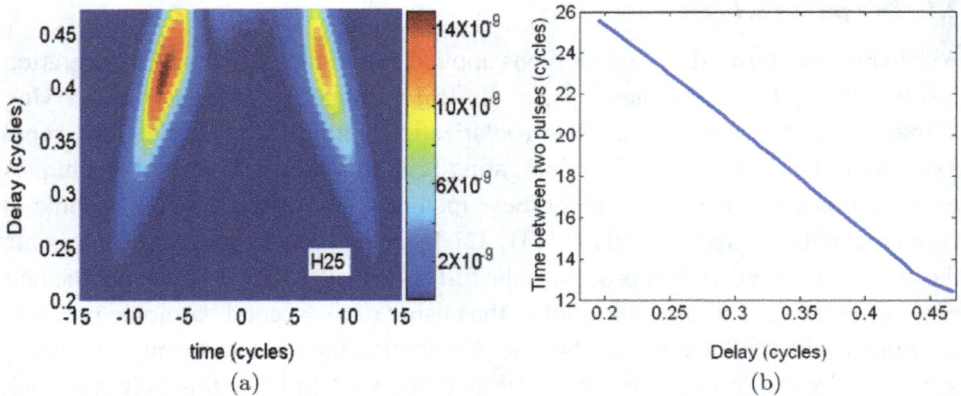

Fig. 1. (a) temporal profiles of H25 intensity vs delay between two orthogonally polarized linearly chirped pulses. (b) The time interval between two XUV pulses vs delay between the two laser pulses. The driving wavelength is 800 nm, the peak laser intensity is 8×10^{13} W/cm^2, the initial pulse duration is 25 fs and $N = 3$, thus the durations of both chirped pulses are 75 fs.

profiles of H25 changing with the delay between the two laser pulses. When the delay is about 0.225 cycles (0.6 fs), the time interval between two XUV pulses is about 24 cycles. When the delay is adjusted between 0.15 cycles and 0.45 cycles, the time interval between two femtosecond XUV pulses varies from about 12 cycles to 26 cycles (Fig. 1(b)). Note that this behavior is typical for other harmonics as well.

With this kind of gating, we can also produce the double attopulses with controllable delay. Figure 2(a) shows the attopulses with the harmonics above 19th order generated by the same laser pulse as used in Fig. 1. In each gate several bursts appear. This is because each gate is too long, namely longer than the laser half-cycle. The gate duration can be decreased using the shorter driving pulse, or providing generation conditions when the threshold ellipticity is lower. The second option can be realized using the mid-infrared laser pulse and higher order harmonics, because the threshold ellipticity decreases when the harmonic order increases [20] and the driving wavelength increases [39]. Figure 2(b) shows the generation of the double attopulses with the harmonics above 141st order using 2 μm/75 fs laser field. From the figure we can see, that the time interval between the two pulses can be controlled.

Note, that the exact emission time of every attosecond pulse is defined by the instantaneous generating field. Here we deal with the attosecond pulses from the cut-off region, so they are emitted close to time when the field is zero. Changing the position of the gate, we do not change these time instants. That is why the emission time for every attosecond pulse in Fig. 2 does not change with the delay,

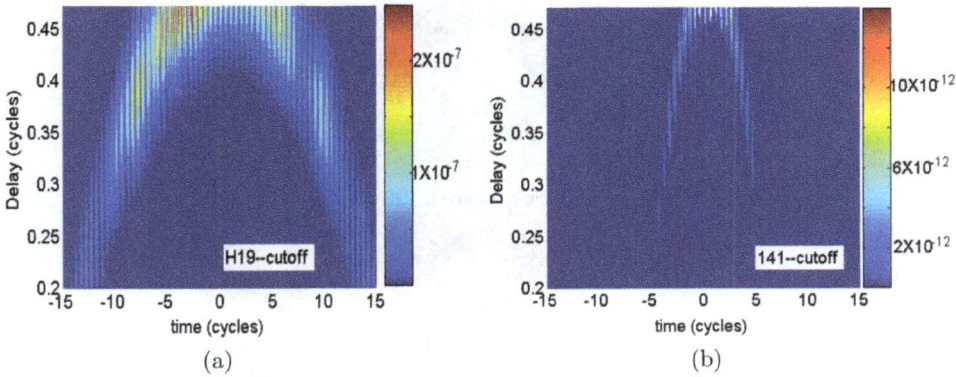

Fig. 2. (a) Attopulses generated by the 800 nm driving field; the other field parameters are the same as for Fig. 1. The harmonics above 19th order are chosen to produce the attopulse. (b) The same for the 2 μm, 75 fs chirped laser pulse under $N = 3$, the other parameters are the same as for the panel (a). Note that when the delay T_d is changed from 1 fs (0.15 cycles) to 3 fs (0.45 cycles), the intensity of the linear field in the gate changes almost linearly from 5×10^{13} W/cm^2 to 9×10^{13} W/cm^2.

but the efficiency of the emission depends (continuously) on the gate position. Under certain delays the emission times miss the gate, and two weak attopulses are generated for every window. For the "usual" polarization gating similar behavior was studied in Refs. [20, 40]. However, in these papers the gate position was fixed, and the emission times were varied changing the carrier-envelope phase of the laser pulse. Thus, in conditions of Fig. 2(b) two IAP are generated, separated with the integer number of half-cycles. This number is controlled with the delay T_d, besides, under certain delays a pair of weak attopulses is generated inside every gate.

In order to understand the physical mechanisms in the generation of these two XUV pulses, we calculate the ellipticity of the electric field and the rotation angle of the polarization ellipse. The ellipticity of the field is presented in Fig. 3. One can clearly see that the region of low ellipticities in this figure corresponds to the region where the attopulses are generated (see Fig. 2). In Fig. 3 we see that when the delay is increased, the two gates get closer to each other, and finally for T_d equal to a half-cycle they overlap. In this case the resulting gate is longer and several attopulses are generated within it, see Fig. 2. Thus, one can conclude that the suggested gating scheme is not suitable for generating a pair of attopulses with very short delay between them. The used delays T_d are much less than the pulse duration. So the generating pulse given by Eq. (1) has almost the same duration $N\tau_p$, as the initial chirped pulse. If the time interval between the two gates exceeds this duration, the gates take place at the edges of the pulse, so intensity within the gate is low. Thus, this duration gives the upper limit of the time separating the attosecond pulses achieved with the suggested technique.

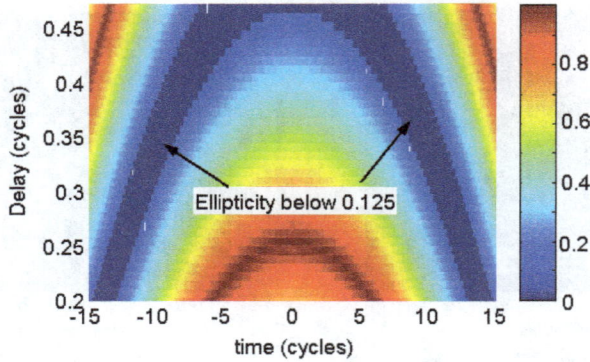

Fig. 3. The ellipticity modulus of the field vs the delay between two laser pulse. The parameters of the laser pulse are the same as for Fig. 1. The region where the ellipticity is lower than the threshold ellipticity for H25 (0.125) is shown in the figure.

In our similation, the gate duration is very close to the one provided by the usual ellipticity gating technique [41]. Thus, the main difference is that in the suggested technique we have a pair of gates taking place in the time instants. Finally, we would like to discuss the limitations of the experimental applicability of the suggested scheme caused by the medium ionization. The ionization decreases the XUV generation efficiency due to the ground state depletion, and also affects phase-matching of the generation. This limitation is important for all polarization gating schemes: laser intensity outside the gate is enough to provide some ionization, although the rate is less than inside the gate because of the elliptical (or circular) polarization. In the suggested scheme this limitation is especially important if the generation of the pair of attopulses with similar intensities is required. The effect of the ionization on the phase-matching depends on the medium density and propagation distance. In general, the role of phase-matching can be minimized using a dilute and thin generating target (certainly, this would limit the total XUV generation efficiency). In this case the difference of the attosecond pulse intensities is only due to the ground state depletion. In our simulations we have found 35% ground state depletion using the ADK ionization rate for 800 nm, 10^{14} W/cm^2 laser pulse under T_d equal to quarter-cycle. Corresponding attopulse intensity decrease can be seen in Fig. 1(a), Fig. 2(a). However, this suggested double gating for XUV generation with controllable delay can be realized by using orthogonally-polarized linearly chirped laser pulses, and it is quite easy for the experimental realization: the chirp can be introduced by changing the distance between the gratings or passing through the dispersion media, thus no nonlinear process is required. This will make possible using this technique, in particular, for high energy driving laser pulse, e.g., the laser pulse from the PW laser system. Generation of the pair of intense attosecond pulses with easily controllable delay can be very useful for forthcoming application of attopulses in XUV pump – XUV probe experiments.

2.2. *Multi-color field scheme*

The high-order harmonic source usually emits in the form of extreme ultraviolet (XUV) frequency comb in the plateau region with different harmonics at similar intensities, due to the non-perturbative nature of the HHG process. On one hand, the harmonic emission in the form of broad-band supercontinuum is pursed for the isolated attosecond pulse [6, 42, 43] by using a few-cycle or a multi-color laser field. For example, Bandulet *et al.* [44] reported a gating technique to produce an isolated attosecond pulse by combining the 800 nm pulse with other two noncommensurate infrared pulses. On the other hand, the selective generation of a single high-order harmonic emission from the harmonic comb as an intense monochromatic coherent source is also very important for many applications, such as the seeding of an XUV free electron laser for obtaining fully coherent output [24]. In order to select a single high-order harmonic emission from the harmonic comb, the intra-atomic phase matching scheme was proposed [45, 46]. Although the intensity of the target harmonic (27th) was maximized with 8 times enhancement, the adjacent harmonic peaks were not suppressed efficiently. The intensity contrast ratio between the target harmonic and the adjacent harmonics was enhanced only by a factor of 4, indicating that the adjacent harmonics were also enhanced simultaneously.

In the work of P. Wei *et al.* [47], we report a robust scheme to control the intra-atomic phase matching by using a driving laser field with sub-cycle waveform control. The laser field is synthesized with the fundamental, second and third harmonics of an 800 nm multi-cycle Ti:Sapphire laser pulse. The required optical waveform for the successful intra-atomic phase matching is obtained by shaping (both in amplitude and in polarization) the laser field on the sub-cycle time scale through simply adjusting the relative time delay among the laser pulses of different colors and polarizations. We experimentally demonstrate that a specific harmonic is selectively enhanced while the adjacent harmonics are dramatically suppressed, with both the harmonic intensity and the selectivity (the intensity contrast ratio between the target harmonic and the adjacent harmonics) significantly improved.

The schematic of experimental setup is shown in Fig. 4(a). A commercial Ti:Sapphire femtosecond laser (Coherent Inc.) is used to produce 2 mJ laser pulses at 800 nm center wavelength with 45 fs pulse duration at a repetition rate of 1 kHz. The output pulses are directed into the vacuum chamber for HHG. In the vacuum chamber, the laser beam is collimated and passes through sequentially a focusing lens (with a focal length of 500 mm), a 0.3 mm-thick SH-BBO crystal (used for the second harmonic generation, the polarization of the generated 400 nm pulse

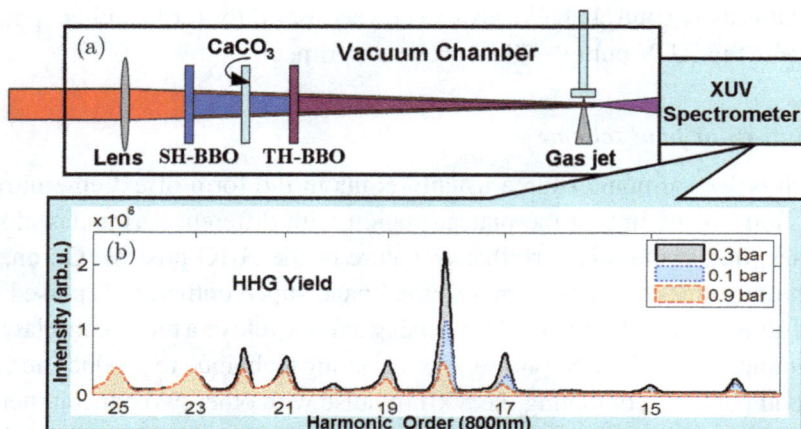

Fig. 4. (a) Schematic of the experimental setup. (b) Harmonic spectra obtained with different gas pressures.

is perpendicular to that of the fundamental 800 nm pulse), a 0.4 mm-thick CaCO3 crystal (used for controlling the time delay between the 400 nm and 800 nm pulses), and a TH-BBO crystal (0.1 mm thickness, type II phase matching, for the third harmonic generation, the polarization of the generated 267 nm pulse is parallel to that of the fundamental 800 nm pulse), and is then focused into a continuous argon gas jet emitted from a nozzle with 0.2 mm inner diameter. The measured energy ratio among the 800 nm, 400 nm, and 267 mm pulses is about 100: 14: 5, with a total laser intensity of 2.0×10^{14} W/cm^2. The generated high-order harmonics are detected by a home-made flat-field grating spectrometer equipped with a soft-x-ray CCD (Princeton Instruments, SX 400). A 500 nm thick aluminum foil is used in the spectrometer to block the driving laser.

Using the above-mentioned three-color laser field, we first optimize the high-order harmonic signal by adjusting the argon gas pressure. The gas pressure is adjusted from 0.1 to 1.0 bar, and the measured harmonic yield is first increased and then decreased. The optimum stagnation pressure of argon gas is about 0.3 bar for obtaining the maximum harmonic yield, which is shown in Fig. 4(b). One can see that several harmonics, especially the 18th harmonic, are selectively enhanced when the gas pressure is optimized, due to that the harmonic emission can be macroscopic phase-matched and the spectral width of the harmonic emission is confined to only a few harmonic orders [48]. By the way, we also measure the HHG with the lower intensity of the 400 nm and 267 nm pulses through tilting the BBO crystal to the phase-mismatching position, and find that the harmonic yield is decreased and the intensity of even-order harmonics are reduced particularly.

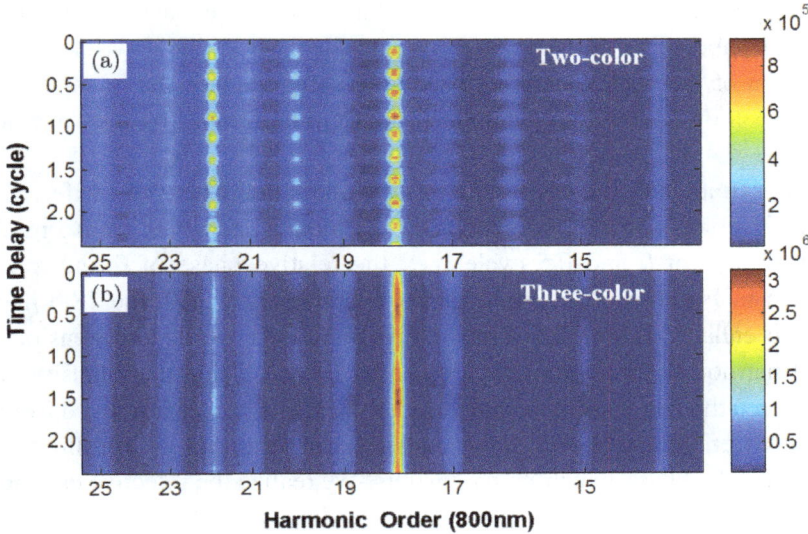

Fig. 5. Harmonic spectra driven by (a) the two-color laser field and (b) the three-color laser field as a function of the time delay, respectively.

The two-color scheme has been proved to be an effective way to enhance the harmonic yield by about two orders of magnitude [22]. For comparison, we measure the harmonic yields by using the traditional two-color scheme with the orthogonally polarized 800 nm/400 nm pulses (just remove the TH-BBO crystal, the measured energy ratio between 800 nm and 400 nm pulses is about 100: 17). The measured harmonic yields are shown in Fig. 5(a) as a function of the time delay between the two-color pulses. The measured harmonic yields by using the above three-color scheme are also shown in Fig. 5(b) as a function of the time delay between the 800 nm and 400 nm fields. The modulation period of the generated harmonics in the two-color case is 1/4 cycle (the optical cycle of 800 nm pulse, the same hereafter), which is determined by the evolution period of the phase difference between the driving laser pulse at 800 nm and the controlling laser pulse at 400 nm. In the three-color case, the harmonic intensity is modulated with a period of about one cycle under the experimental condition since the time delay of the 267 nm pulse is correlated to the time delay between the 800 nm and 400 nm pulses.

For a more detailed comparison, we extract the maximum harmonic intensities at the optimum time delay from Fig. 5(a) and Fig. 5(b), which are shown in Fig. 6(a), together with the harmonic spectrum obtained by using only the fundamental 800 nm pulse. Compared to the harmonic generation in the fundamental 800 nm pulse, the addition of the 400 nm pulse of perpendicular polarization can enhance the conversion efficiency by about 2 orders of magnitude [22], and the further addition of a weak 267 nm pulse of parallel polarization further enhances the conversion efficiency by about 5 times. Moreover, compared to the traditional two-color scheme, the addition of the weak 267 mm pulse can greatly enhance the intensity contrast ratio between the 18th harmonic and the adjacent harmonics by over one order of magnitude.

In order to identify the role of the relative phase delay between different colors in the three-color scheme, we extract the harmonic intensities from Fig. 5(b) at the time delays of 0 and 0.5 cycle (i.e., the relative phase of 0 and π), which are shown in Fig. 6(b). Compared to the relative phase of 0, the 18th harmonic intensity is enhanced by 43.7% at the relative phase of π, while the intensities of the adjacent harmonics are at least suppressed by 65.4%. Then, the intensity contrast ratio between the 18th harmonic and the adjacent harmonics is improved from 3.8 to 15.9. Moreover, the specific harmonic emission can be tuned by controlling the laser intensity and the phase-matching parameters. We realize the selective enhancement

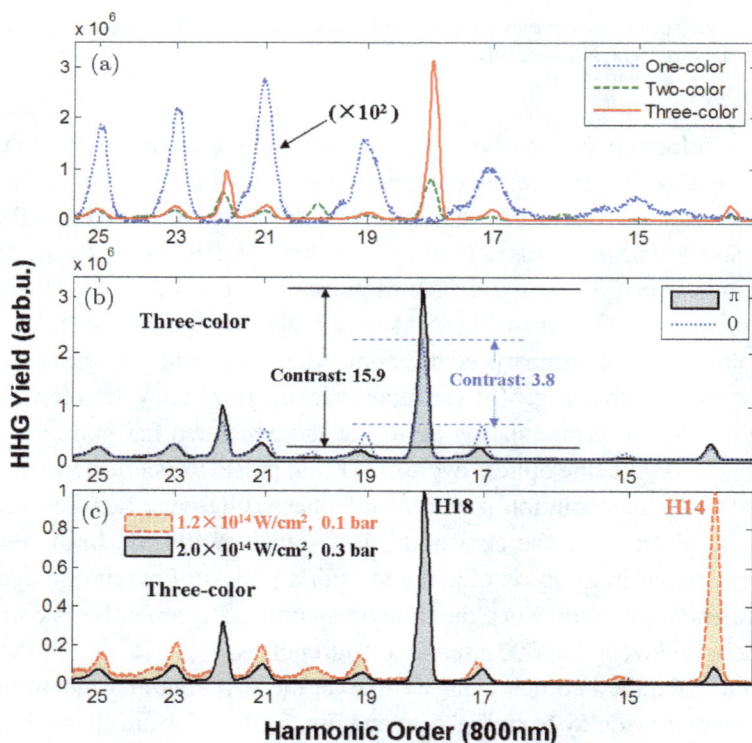

Fig. 6. (a) Comparison of the maximum harmonic yields driven by different laser fields. (b) Comparison of the harmonic yields driven by the three-color laser field with two different relative phases. (c) Comparison of the harmonic yields driven by the three-color laser field with different intensity and density conditions.

of the 14th harmonic at the laser intensity of 1.2×10^{14} W/cm^2 and the gas pressure of 0.1 bar, as shown in Fig. 6(c) together with the selective enhancement of the 18th harmonic at the laser intensity of 2.0×10^{14} W/cm^2 and the gas pressure of 0.3 bar.

The selective enhancement of the harmonic emission can be attributed to the phase matching effects. The optimization of phase matching [49] includes both the macroscopic and intra-atomic phase matching. For the macroscopic phase matching, the effective spectral range and the relative weight of the individual harmonic order strongly depend on the species of gas and its pressure, the interaction geometry, the intensity and diameter of the laser beam, and etc. By properly choosing these parameters, the harmonic emission can be phase-matched and confined to a few orders [48]. For the intra-atomic phase matching, the emission of different trajectories in each half-cycle should be constructively interfered [46]. For the driving laser pulse with a finite duration, the pulse envelope slightly tailors the electric field of each half-cycle, leading to the phase difference of the harmonic emissions from each half-cycle. For an 800 nm laser pulse with the intensity of 10^{14} W/cm^2, the phase of the harmonic emission is several tens of radian and a small change of the electric field will affect the phase of the harmonic greatly.

Shaping the laser field waveform by using the combination of multi-color laser pulses [50, 51] can lead to not only a broadened XUV supercontinuum for isolated attosecond pulse generation but also the confinement of the harmonic emission in a small spectral region for narrow-bandwidth XUV radiation [52] and even a single-order harmonic emission as a monochromatic XUV source in the present work. In our waveform shaping with three-color laser field, the fundamental 800 nm laser pulse is the main driving laser field for the HHG, the two controlling pulses of different colors with perpendicular polarization help to improve the harmonic yield and spectral selectivity simultaneously. The addition of the 400 nm pulse with perpendicular polarization can enhance the conversion efficiency by about two orders of magnitude, and the further addition of a weak 267 nm pulse with parallel polarization can enhance the contrast ratio by about one order of magnitude. The above three-color combination is the basic configuration for controlling the electron trajectory in both the perpendicular and parallel directions. Although a finer control is possible by using the laser field of more than three colors, the improvement would be very limited and it is difficult to implement experimentally. As we know, the ability of the waveform shaping with multi-color field is dependent on the spectrum-spanning range. Such three-color laser field spans from 800 nm to 267 nm, and the addition of the fourth harmonic will extend only from 267 nm to 200 nm.

Therefore, the improvement in intensity contrast can be mainly attributed to the intra-atomic phase matching realized with the sub-cycle waveform controlled driving laser field consisting of a main driving laser pulse and two controlling laser pulses of perpendicular polarizations, with a total spectrum-spanning range of 1.6 octaves from 267 nm to 800 nm. In fact, the maximum intensity contrast ratio in our three-color scheme is reached by a factor of 15.9, which is much better than the optimized results obtained by the pulse-shaping technique [45]. Our results can be understood as follows: First, the harmonic emission is selectively enhanced and confined to only a few harmonic orders by controlling the macroscopic phase matching conditions. Second, the target harmonic (18th) is further enhanced and the adjacent harmonics (16th, 17th, 19th, and 20th) are greatly suppressed by controlling the intra-atomic phase matching conditions. The photon number of 18th harmonic in each shot is estimated using the method [53] to be larger than 9.2×10^6. A conversion efficiency of about 1.0×10^{-6} is expected if the loss of spectrometer slit is taken into account. This scheme is promising for the development of compact monochromatic coherent XUV sources.

In the subsequent work, we optimized the macroscopic phase matching from a long gas cell with loosely focusing optics with a three-color scheme so as to further improve the intensity contrast ratio between the target and the adjacent harmonics [54]. When compared with single harmonic emission from a continuous gas jet, the intensity of the selected single harmonic emission from the long gas cell is more intense by as much as 1–2 orders of magnitude. Simultaneously, the contrast ratio (i.e., the spectral purity) is also increased by several times.

The main differences between the setup used in this work and that of Ref. [47] are the focusing optics and the length of the lasing medium, while the other experimental conditions are all the same. Because one promising technique for further enhancement of harmonic emission involves the use of loosely focusing optics and therefore a much longer medium, we selected a longer focus lens with a focal length of 1100 mm (while the focal length used in Ref. [47] is 500 mm) and a much longer medium with a cell length of 25 mm (while the medium used in Ref. [47] is a continuous gas jet emitted from a nozzle with an inner diameter of 0.2 mm). Also, a diaphragm (with diameter of approximately 10 mm) is used in front of the lens to further control the quantum trajectories by modifying the laser beam profile in the HHG process. Therefore, we can optimize the harmonic signal by modifying the laser beam profile, selecting the focusing position, adjusting the argon gas pressure, and controlling the time delays among the different colors.

Using the three-color scheme, we first optimized the single harmonic emission by modifying the laser beam profile and selecting the focusing position. The experimentally measured harmonic emissions from the long gas cell are shown as functions of the focusing position in Fig. 7, and the corresponding emissions from the gas jet are also shown for comparison. We see that the 18th harmonic is selectively

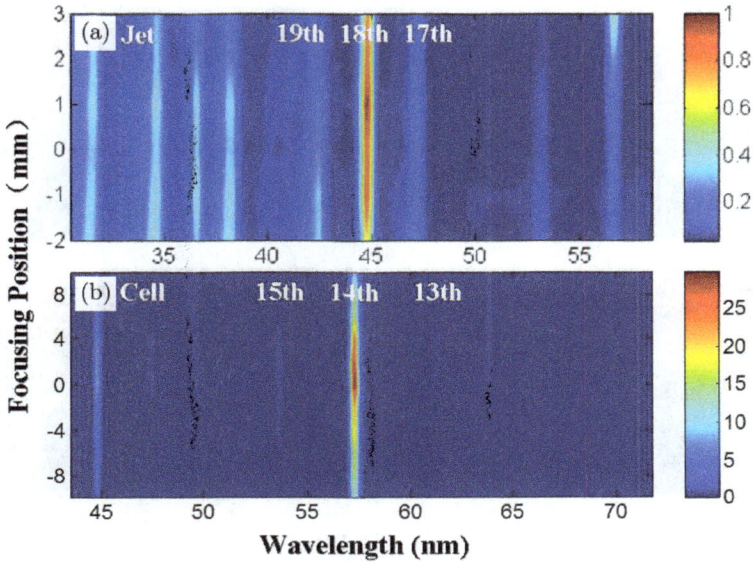

Fig. 7. Experimentally measured harmonic emissions as functions of focusing position in (a) the jet case, and (b) the cell case.

enhanced in the jet case, while the 14th harmonic is selectively enhanced in the cell case. This selectivity can be attributed to the fact that stronger laser intensity corresponds to a higher harmonic order for the selective enhancement (both the jet and the cell are pumped by 2 mJ laser pulses at a center wavelength of 800 nm with 45 fs pulse duration; the cell uses the loosely focusing optics, and therefore has the lower laser intensity and the lower selected harmonic order). We can also see that the peak intensity of the selected harmonic from the cell is approximately 30 times more intense than that of the jet, which can be attributed to the use of the loosely focusing optics in the cell case and the fact that the much longer medium provides a much higher conversion efficiency.

Then, we further optimized the single harmonic emission by controlling the time delays among the different colors. The experimentally measured harmonic emissions as functions of the time delay between the 800 nm and 400 nm laser pulses are shown in Fig. 8. We see that, because the time delay is optimized, the selected harmonic is enhanced, while most of the other harmonics are suppressed and even disappear at some particular time delays. This is because of intra-atomic phase matching. We can also see that the depth of the time modulation in the jet case is deeper than that in the cell case, which can be attributed to the fact that the propagation effect can consume the time modulation (i.e., the interaction medium in the cell case is much longer than that in the jet case, and the longer medium corresponds to more effective

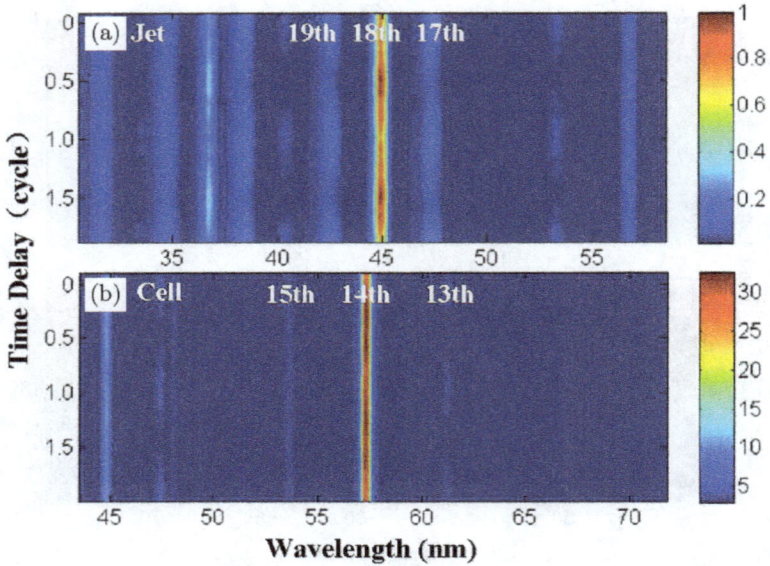

Fig. 8. Experimentally measured harmonic emissions as functions of the time delay between the 800 nm and 400 nm laser pulses in (a) the jet case, and (b) the cell case.

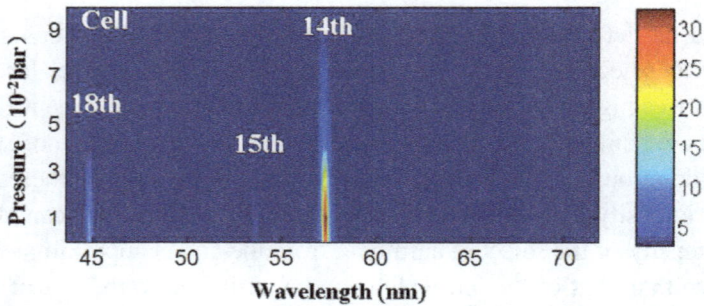

Fig. 9. Experimentally measured harmonic emission as a function of gas pressure in the cell case. The selected single harmonic is optimized by adjustment of the argon gas pressure at the optimum time delay.

propagation), similar to the relationship between the carrier-envelope-phase (CEP) effect and the duration of the laser pulse.

To take advantage of the time modulation and the effective propagation, we further optimized the selected harmonic emission at the optimum time delay by adjusting the gas pressure. The experimentally measured harmonic emission as a function of gas pressure is shown in Fig. 9. We see that the measured harmonic yield first increases and then decreases. The optimum pressure to obtain the maximum harmonic yield is approximately 1×10^{-2} bar. We can also see that the selected

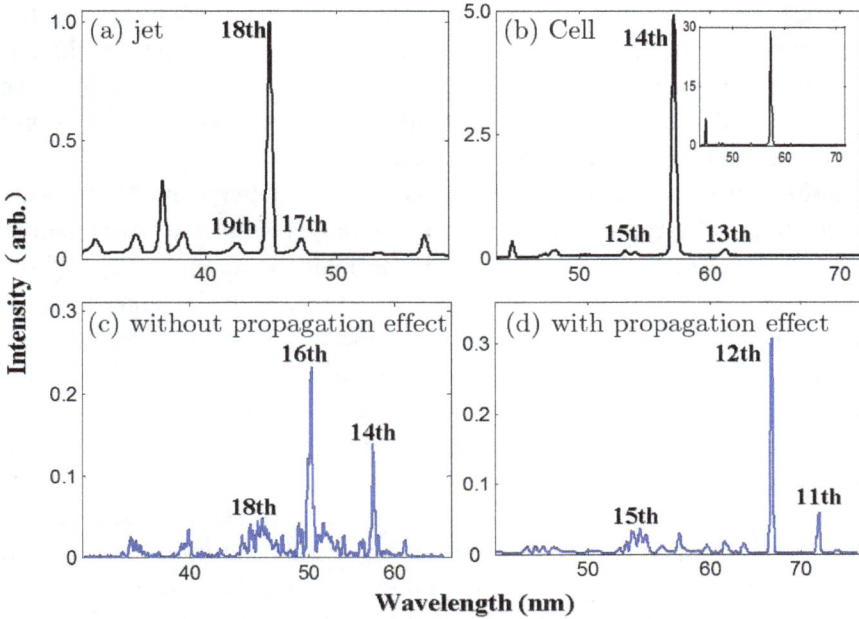

Fig. 10. Top: Experimentally extracted results for the optimum single-harmonic emissions in (a) the jet case, and (b) the cell case. Bottom: Theoretically calculated results for the optimum single-harmonic emissions (c) without, and (d) with the propagation effect.

harmonics are almost reduced to only one (i.e., the rest of the harmonics have almost disappeared) when the gas pressure is increased to 4×10^{-2} bar.

For a more detailed comparison, we extracted the purest harmonic emission at the gas pressure of 4×10^{-2} bar from Fig. 9, which is shown in Fig. 10(b) (the inset shows the maximum harmonic emission at the gas pressure of 1×10^{-2} bar), together with the optimum emission of the selected harmonic in the jet case, which is shown in Fig. 10(a). Using the jet case as a standard, we then normalized the harmonic intensity in the cell case. In the jet case (shown in Fig. 10(a)), the peak intensity is 1, the contrast ratio between the selected harmonic and the adjacent harmonics is 15.9, and the contrast ratio between the selected harmonic and all remaining harmonics is 3.2. In the cell case with gas pressure of 4×10^{-2} bar (the purest harmonic emission, as shown in Fig. 10(b)), the peak intensity is approximately 5 (i.e., when compared with the jet case, the intensity is increased by five times), the contrast ratio between the selected harmonic and the adjacent harmonics is 41.4, and the contrast ratio between the selected harmonic and all remaining harmonics is 15.3. In the cell case with a gas pressure of 1×10^{-2} bar (the maximum harmonic emission, as shown in the inset of Fig. 10(b)), the peak intensity is approximately 30 (i.e., when compared with the jet case, the intensity is increased by 30 times), the contrast ratio between the selected harmonic and the adjacent harmonics is 48.5, and the contrast ratio between the selected harmonic and all remaining harmonics is 4.2. Therefore, when compared with the single harmonic emission in the jet case, both the harmonic intensity and the spectral purity have obviously been enhanced by use of the long gas cell and the long-focus lens.

Figures 10(c) and 10(d) show the calculated results without and with the propagation effect, respectively. We see that only one or two even harmonics dominate the spectra, while almost all other harmonics are dramatically suppressed. Although there are some differences between the theoretical and experimental results, it has been demonstrated that only one or two even harmonics can be selectively generated, while the remaining harmonics are dramatically suppressed. The simulation results that include the propagation effect, as shown in Fig. 10(d), indicate that only one harmonic peak (the 12th) is selected, while the simulation results without the propagation effect, as shown in Fig. 10(c), indicate that two harmonic peaks (the 14th and the 16th) are selected. While the calculated results have not yet fully reproduced the experimental results and more time is required to find the appropriate parameters to optimize the 14th harmonic, the numerical simulation does show better results for the single harmonic when the propagation effect is considered. In addition, these simulation results indicate another reason why the selected harmonic in the cell case is better than that in the jet case, in that the longer interaction medium corresponds to a better propagation effect.

In general, HHG is regarded as the best coherent source available in the ultraviolet and XUV regions of the spectrum [55–57]. Our selective generation of an almost pure single harmonic should be more coherent than the general HHG in the form of a frequency comb, because the monochromaticity of such a selected harmonic is much higher and leads to better coherence. Therefore, as an intense monochromatic coherent source, the pure single harmonic is likely to be highly important for many applications, including the seeding of XUV free electron lasers, nanolithography, XUV interferometry, and biological imaging.

HHG is a strongly nonlinear response of a medium to the laser field. By shaping the electric field, the ionization, the electron motion and electron-ion recollision of the HHG process can be controlled, and thus the IAP generation and conversion efficiency of HHG, etc, can be manipulated. In this chapter, we review a theoretical prediction of the generation of two XUV attosecond pulses with tunable delay between them using two orthogonally-polarized linearly chirped laser pulses and a experimental demonstration of a robust scheme to select a single high-order harmonic among the harmonic comb by using a driving laser field with sub-cycle waveform control. The sub-cycle shaped laser field has promoted the development of the lightwave electronics [58].

Acknowledgements

This work was supported by the National Natural Science Foundation of China (Grants No. 11127901, No. 61221064, No. 11134010, No. 11227902, No. 11222439, and No. 11274325), the 973 Project (Grant No. 2011CB808103), and Shanghai Commission of Science and Technology (Grant No. 12QA1403700).

References

1. M. Hentschel, R. Kienberger, Ch. Spielmann, G. A. Reider, N. Milosevic, T. Brabec, P. Corkum, U. Heinzmann, M. Drescher, and F. Krausz, *Nature (London).* **414**, 509 (2001).
2. C. Gohle, T. Udem, M. Herrmann, J. Rauschenberger, R. Holzwarth, H. A. Schuessler, F. Krausz, and T. W. Hänsch, *Nature (London).* **436**, 234 (2005).
3. G. Sansone, E. Benedetti, F. Calegari, C. Vozzi, L. Avaldi, R. Flammini, L. Poletto, P. Villoresi, C. Altucci, R. Velotta, S. Stagira, S. D. Silvestri, and M. Nisoli, *Science.* **314**, 443 (2006).
4. E. Goulielmakis, M. Schultze, M. Hofstetter, V. S. Yakovlev, J. *Science.* **320**, Gagnon, M. Uiberacker, A. L. Aquila, E. M. Gullikson, D. T. Attwood, R. Kienberger, F. Krausz, and U. Kleineberg, *Science.* **320**, 1614 (2008).
5. O. Smirnova, V. S. Yakovlev, and M. Ivanov, *Phys. Rev. Lett.* **94**, 213001 (2005).
6. K. Zhao, Q. Zhang, M. Chini, Y. Wu, X. Wang, and Z. Chang, *Opt. Lett.* **37**, 3891 (2012).
7. N. A. Papadogiannis, B. Witzel, C. Kalpouzos, and D. Charalambidis, *Phys. Rev. Lett.* **83**, 4289 (1999).
8. P. M. Paul, E. S. Toma, P. Breger, G. Mullot, F. Augé, Ph. Balcou, H. G. Muller, and P. Agostini, *Science.* **292**, 1689 (2001).
9. T. Popmintchev, M. Chen, D. Popmintchev, P. Arpin, S. Brown, S. Ališauskas, G. Andriukaitis, T. Balčiunas, O. D. Mücke, A. Pugzlys, A. Baltuška, B. Shim, S. E. Schrauth, A. Gaeta, C. H-García, L. Plaja, A. Becker, A. J-Becker, M. M. Murnane, and H. C. Kapteyn, *Science.* **336**, 1287 (2012).
10. J. Itatani, J. Levesque, D. Zeidler, H. Niikura, H. Pépin, J. C. Kieffer, P. B. Corkum, and D. M. Villeneuve, *Nature (London).* **432**, 867 (2004).
11. B. K. McFarland, J. P. Farrell, P. H. Bucksbaum, and M. Gühr, *Science.* **322**, 1232 (2008).
12. O. Smirnova, Y. Mairesse, S. Patchkovskii, N. Dudovich, D. Villeneuve, P. Corkum, and M. Yu. Ivanov, *Nature (London).* **460**, 972 (2009).
13. A. L. Cavalieri, N. Müller, Th. Uphues, V. S. Yakovlev, A. Baltuška, B. Horvath, B. Schmidt, L. Blümel, R. Holzwarth, S. Hendel, M. Drescher, U. Kleineberg, P. M. Echenique, R. Kienberger, F. Krausz, and U. Heinzmann, *Nature (London).* **449**, 1029 (2007).
14. D. Shafir, H. Soifer, B. D. Bruner, M. Dagan, Y. Mairesse, S. Patchkovskii, M. Yu. Ivanov, O. Smirnova, and N. Dudovich, *Nature (London).* **485**, 343 (2012).
15. Y. Mairesse, J. Higuet, N. Dudovich, D. Shafir, B. Fabre, E. Mével, E. Constant, S. Patchkovskii, Z. Walters, M. Yu. Ivanov, and O. Smirnova, *Phys. Rev. Lett.* **104**, 213601 (2010).
16. S. Haessler, J. Caillat, W. Boutu, C. G-Teixeira, T. Ruchon, T. Auguste, Z. Diveki, P. Breger, A. Maquet, B. Carré, R. Taïeb, and P. Salières, *Nat. Phys.* **6**, 200 (2010).
17. C. Vozzi, M. Negro, F. Calegari, G. Sansone, M. Nisoli, S. De. Silvestri, and S. Stagira, *Nat. Phys.* **7**, 822 (2011).
18. R. Kienberger, E. Goulielmakis, M. Uiberacker, A. Baltuška, V. Yakovlev, F. Bammer, A. Scrinzi, Th. Westerwalbesloh, U. Kleineberg, U. Heinzmann, M. Drescher, and F. Krausz, *Nature.* **427**, 817 (2004).
19. C. Altucci, C. Delfin, L. Roos, M. B. Gaarde, A. L'Huillier, I. Mercer, T. Starczewski, and C. -G. Wahlström, *Phys. Rev. A.* **58**, 3934 (1998).

20. I. J. Sola, E. Mével, L. Elouga, E. Constant, V. Strelkov, L. Poletto, P. Villoresi, E. Benedetti, J. -P. Caumes, S. Stagira, C. Vozzi, G. Sansone, and M. Nisoli, *Nat. Phys.* **2**, 319 (2006).
21. H. Eichmann, A. Egbert, S. Nolte, C. Momma, B. Wellegehausen, W. Becker, S. Long, and J. K. Mclver, *Phys. Rev. A.* **51**, R3414 (1995).
22. I. J. Kim, C. M. Kim, H. T. Kim, G. H. Lee, Y. S. Lee, J. Y. Park, D. J. Cho, and C. H. Nam, *Phys. Rev. Lett.* **94**, 243901 (2005).
23. T. T. Liu, T. Kanai, T. Sekikawa, and S. Watanabe, *Phys. Rev. A.* **73**, 063823 (2006).
24. G. Lambert, T. Hara, D. Garzella, T. Tanikawa, M. Labat, B. Carre, H. Kitamura, T. Shintake, M. Bougeard, S. Inoue, Y. Tanaka, P. Salieres, H. Merdji, O. Chubar, O. Gobert, K. Tahara, and M.-E. Couprie, *Nat. Phys.* **4**, 296 (2008).
25. P. B. Corkum, *Phys. Rev. Lett.* **71**, 1994 (1993).
26. A. Baltuska, Th. Udem, M. Uiberacker, M. Hentschel, E. Goulielmakis, Ch. Gohle, R. Holzwarth, V. S. Yakovlev, A. Scrinzi, T. W. Hänsch, and F. Krausz, *Nature.* **421**, 611 (2003).
27. P. B. Corkum, N. H. Burnett, and M. Y. Ivanov, *Opt. Lett.* **19**, 1870 (1994).
28. V. T. Platonenko, and V. V. Strelkov, *J. Opt. Soc. Am. B.* **16**, 435 (1999).
29. V. V. Strelkov, A. F. Sterjantov, N. Yu. Shubin, and V. T. Platonenko, *J. Phys. B.* **39**, 577 (2006).
30. P. Lan, P. Lu, W. Cao, Y. Li, and X. Wang, Phys. Rev. A. **76**, 051801 (2007).
31. Z. Zeng, Y. Cheng, X. Song, R. Li, and Z. Xu, *Phys. Rev. Lett.* **98**, 203901 (2007).
32. X. Feng, S. Gilbertson, H. Mashiko, H. Wang, S. D. Khan, M. Chini, Y. Wu, K. Zhao, and Z. Chang, *Phys. Rev. Lett.* **103**(18), 183901 (2009).
33. T. Pfeifer, L. Gallmann, M. J. Abel, D. M. Neumark, and S. R. Leone, *Opt. Lett.* **31**, 975 (2006).
34. J. Mauritsson, P. Johnsson, E. Gustafsson, A. L'Huillier, K. J. Schafer, and M. B. Gaarde, *Phys. Rev. Lett.* **97**, 013001 (2006).
35. Z. Chang, *Phys. Rev. A.* **76**, 051403 (2007).
36. Jing Miao, Zhinan Zeng, Peng Liu, Yinghui Zheng, Ruxin Li, Zhizhan Xu, V. T. Platonenko, and V. V. Strelkov, *Opt. Express.* **20**, 5196 (2012).
37. M. Lewenstein, Ph. Balcou, M. Yu. Ivanov, A. L'Huillier, and P. B. Corkum, *Phys. Rev. A.* **49**, 2117 (1994).
38. D. G. Lee, J.-H. Kim, K.-H. Hong, and C. H. Nam, *Phys. Rev. Lett.* **87**, 243902 (2001).
39. S. D. Khan, Y. Cheng, M. Möller, K. Zhao, B. Zhao, M. Chini, G. G. Paulus, and Z. Chang, *Appl. Phys. Lett.* **99**, 161106 (2011).
40. G. Sansone, E. Benedetti, J. P. Caumes, S. Stagira, C. Vozzi, M. Nisoli, V. Strelkov, I. Sola, L. B. Elouga, A. Zaïr, E. Mével, and E. Constant, *Phys. Rev. A.* **80**, 063837 (2009).
41. Z. Chang, *Phys. Rev. A.* **70**, 043802 (2004).
42. E. J. Takahashi, P. Lan, O. D. Mucke, Y. Nabekawa, and K. Midorikawa, *Phys. Rev. Lett.* **104**, 233901 (2010).
43. E. J. Takahashi, P. Lan, and K. Midorikawa, Lasers and Electro-Optics 2012, San Jose, California, United States (IEEE Computer Society, Washington, DC, 2012).
44. H.-C. Bandulet, D. Comtois, E. Bisson, A. Fleischer, H. Pépin, J.-C. Kieffer, P. B. Corkum, and D. M. Villeneuve, *Phys. Rev. A.* **81**, 013803 (2010).
45. R. Bartels, S. Backus, E. Zeek, L. Misoguti, G. Vdovin, I. P. Christov, M. M. Murnane, and H. C. Kapteyn, *Nature (London).* **406**, 164 (2000).
46. I. P. Christov, R. Bartels, H. C. Kapteyn, and M. M. Murnane, *Phys. Rev. Lett.* **86**, 5458 (2001).
47. Pengfei Wei, Jing Miao, Zhinan Zeng, Chuang Li, Xiaochun Ge, Ruxin Li, and Zhizhan Xu, *Phys. Rev. Lett.* **110**, 233903 (2013).
48. S. Teichmann, P. Hannaford, and L. Van Dao, *Appl. Phys. Lett.* **94**, 171111 (2009).
49. T. Popmintchev, M.-C. Chen, P. Arpin, M. M. Murnane, and H. C. Kapteyn, *Nat. Photonics.* **4**, 822 (2010).
50. L. E. Chipperfield, J. S. Robinson, J.W. G. Tisch, and J. P. Marangos, *Phys. Rev. Lett.* **102**, 063003 (2009).

51. S. Huang *et al.*, *Nat. Photonics.* **5**, 475 (2011).
52. J. Yao *et al.*, *Phys. Rev. A.* **83**, 033835 (2011).
53. N. Yamaguchi, Z. Takahashi, Y. Nishimura, A. Sakata, K. Watanabe, Y. Okamoto, Y. Takemura, H. Azuma, and T. Hara, *J. Plasma Fusion Res.* **81**, 391 (2005).
54. Pengfei Wei, Zhinan Zeng, Jiaming Jiang, Jing Miao, Yinghui Zheng, Xiaochun Ge, Chuang Li, and Ruxin Li, *Appl. Phys. Lett.* **24**, 085302 (2014).
55. T. Ditmire, E. T. Gumbrell, R. A. Smith, J. W. G. Tisch, D. D. Meyerhofer, and M. H. R. Hutchinson, *Phys. Rev. Lett.* **77**, 4756 (1996)
56. M. Bellini, C. Lyngå, A. Tozzi, M. B. Gaarde, T. W. Hänsch, A. L'Huillier, and C.-G. Wahlström, *Phys. Rev. Lett.* **81**, 297 (1998).
57. R. A. Bartels, A. Paul, H. Green, H. C. Kapteyn, M. M. Murnane, S. Backus, I. P. Christov, Y. Liu, D. Attwood, and C. Jacobsen, *Science.* **297**, 376 (2002).
58. F. Krausz and M. Ivanov, *Rev. Mod. Phys.* **81**, 163 (2009).

Chapter 8

Imaging Ultra-fast Molecular Dynamics
in Free Electron Laser Field

Y. Z. Zhang*,† and Y. H. Jiang*,‡,§

*Shanghai Advanced Research Institute,
Chinese Academy of Sciences (CAS), Shanghai, 201210, China
‡Shanghaitech University, Shanghai, 201210, China
†zhangyz@sari.ac.cn
§jiangyh@sari.ac.cn

The free electron laser (FEL) provides the coherent, brilliant and ultrashort light pulse in short wavelength (extreme ultraviolet and X-ray) regimes, opening up possibilities to study ultra-fast molecular dynamics in photo-induced chemical reactions with new methodologies. In this chapter, we introduce the time-resolved pump-probe experiments on gas-phase targets with FEL facilities to image the nuclear and electronic motions in molecular reactions, which serve as a benchmark for further FEL applications like coherent diffraction imaging and coherent control of functional dynamics in complex molecular reactions.

1. Introduction

The Free Electron Laser (FEL) facilities — a rapid developing technology in advanced radiation sources in the last decade — provide the coherent, brilliant and ultrashort light pulse which, tunable from Terahertz to X-ray regime, opens up new opportunities to study the light-matter interaction in various scientific areas ranging from fundamental physics to chemistry and bio-molecular research. Due to the huge impact on a variety of fields in science, many countries are constructing or planning the large-size radiation sources. In mid-2005 the FEL at DESY, now known as Free electron LASer in Hamburg (FLASH) opened as a user facility [1]. Right now the FLASH offers the extreme ultraviolet (EUV) radiation of 28–295 eV with pulse duration of 30–300 fs carrying an average energy of $200\,\mu J$ per pulse [2]. Since FLASH operation, a number of new short wavelength FELs have begun operation. The first lasing of Linac Coherent Light Source (LCLS) in the USA was observed in April 2009 [3], which currently operates in the X-ray region between 7.1–9.5 keV, providing typically $100\,\mu J$ per pulse with pulse duration below 50 fs [4]. In Japan, the SPring-8 Compact SASE Source (SCSS), which served as a prototype of further XFEL conducted by RIKEN, was firstly accessible in EUV regime in 2008 [5]. And the successful operation of SACLA in Japanese Spring-8 at 1.2 nm was achieved [6], launching to user operation in March 2012. And the SACLA provides the photon energy range of 4.5–15 keV, with an average energy of $300\,\mu J$ and 4.5–31 fs pulse duration [7]. The seeded-FEL source FERMI at Trieste in Italy covering from 100 nm to 10 nm has recently been open to user applications [8]. The FLASH II, LCLS II, XFEL and SwissFEL are under construction, and will be coming soon. The Chinese FEL projects of Dalian Coherent Light Source (DCLS) in Dalian and soft X-ray FEL (SXFEL) in Shanghai have been funded and expected to operation in 2016.

The FEL radiation delivers brilliant, ultrashort and short-wavelength pulse, manifesting itself as a number of applications in various areas of science. The most far-reaching application might be the direct imaging of matter on the length of atomic scale, which is particularly helpful for structural determination in a broad range of disciplines. The indispensible superiority of FEL imaging techniques is that the intense FEL pulse allows to image a single non-crystallizable macromolecule within a femtosecond FEL pulse, so called coherent diffraction imaging [9, 10]. Comparatively, the conventional X-ray crystallographic imaging on synchrotron is only able to integrate weak signal from a periodic structure, which is not compatible with non-crystalline samples. Another unique application of FEL facilities is the observation of ultrafast processes in photo-induced chemical reaction. The characteristic of ultrashort pulse duration of FEL transfers the time-resolved experimental method introduced by femtosecond chemistry to the EUV/X-ray regime. In this kind of experiments, a single atom, molecule or extended liquid and condensed matter are excited by a FEL pulse or a ultrashort pulse from other laser sources, and the structural changes are probed as a function of evolution time by means of many kinds of spectroscopy including absorption or emission spectroscopy, as well as photoionization spectroscopy (photoion/photoelectron spectroscopy). Combining the coherent diffraction imaging and time-resolved experimental method, the ultimate goals are approaching that we are able to directly take snapshots of single biomolecule and follow their functional dynamics, nurturing our dream of producing a "molecular movie".

All the aforementioned applications are based on the fully understanding of behaviors of molecules in strong, short-wavelength light-field. The atomic, molecular physics in FEL field, which answers the question of how the matters behave in strong FEL field in the simplest forms, serves as a benchmark to study more complicated dynamics in complex molecular systems. Thus, the Atomic, Molecular and Optical Physics (AMO) research activities compose an important ingredient of experiments of FEL facilities, and AMO applications accomplished with FEL have been reviewed in a number of literatures [7, 11–17]. In this chapter, we will confine ourselves to the time-resolved (pump-probe) experiments of short-wavelength FELs on molecular gas-phase targets, and take a selection of very recent illustrative results, which concentrate on the electronic and nuclear structural changes in photo-induced chemical reactions and their correlated dynamics in these processes.

In pump-probe scheme, the molecules interact with both pump pulse and probe pulse, which is instinctively a nonlinear process. Comparatively, the tunable synchrotron radiations, in which the interactions are governed by single-photon processes due to the weak field, are not feasible to nonlinear behaviors and time-resolved experiments in molecules. Although the pump-probe experiments could be conducted by optical femtosecond laser, the optical femtosecond lasers are mostly manipulated in low photon energy (visible, near-infrared and IR regimes), only allowing to monitor the vibrational, rotational motion of molecules and to interact with valence electrons. The new generation of FEL radiation combines the advantages of both synchrotron radiation and femtosecond laser, delivering the short-wavelength, tunable radiation with high intensity, short pulse duration and high coherence, which readily excite high-lying electronic states and valence electrons and implement time-resolved experiments. Taking advantage of these unprecedented characteristics, some new phenomena of molecular responds in intense EUV/X-ray fields are firstly observed, which is impossible with other light sources. Exemplarily, multiphoton-multielectron excitation in molecule has been studied by inspecting the angular and energy dependent emission of ionic fragments for different charge states, for the first time proving that the sequential ionization is the dominate nonlinear process when the molecules were irradiated by the EUV/X-ray strong field [18–20]. The hard X-ray FEL from the LCLS opens a new field dominated by multiphoton and multiple core-hole dynamics in molecules. While the conventional nonlinear behaviors of matter only involve the valence electrons, the nonlinear effect of inner shell electrons involving the creation and decay of multiple core-holes was explored for the first time employing the strong short-wavelength FEL pulse [21, 22].

Besides the aforementioned new phenomena, the intense ultrashort FELs present unique strengths to conduct pump-probe experimental method, that is not available with tunable synchrotrons or intense femtosecond infrared laser. Firstly, the wavelength-tunable EUV/X-ray FEL pulse can excite the specific states of interest in molecules by single-photon absorption. And a probe pulse on the order of femtosecond will inspect the conformational changes or the dynamics at conical intersection by employing the absorption spectroscopic method or photoionization-based spectroscopy. Thus, the pump-probe method in EUV/X-ray regime is able to identify the adiabatic intermediate states and to disentangle the corresponding reaction pathways of photo-induced molecular reactions along the reaction coordinate. Secondly, when the laser energy is tuned to EUV/X-ray region, the strong-field effects can be neglected due to the decreasing ponderomotive energy. Comparatively, the strong laser field in visible/IR region may influence the molecular potential curves and alter the outcome of the chemical reaction, which leads to complicated results not to be easily understood. The XUV/X-ray time-resolved method intends to detect molecular dynamics along the unperturbed potential curves, facilitating the probe of dynamics "within molecules". Thirdly, the excitation in molecule induced by FEL light typically involves a sequence of single-photon absorptions, in which process the momentum distributions of reactive fragments reflect the electron density distributions of initial and final states during the photon-induced reaction. Thus, by recording the kinetic energy release (KER) and momentum distributions of reactive fragments as a function of reactive time, the instantaneous electron density distributions can be reconstructed, allowing to investigate the electronic dynamics in nuclear coordinate in chemical reactions. Finally, the very small wavelength radiation from FEL efficiently creates vacancies in strong-bound, inner-shell electron in molecule. Therefore, the intense short-wavelength FEL provides access to create nonlinear absorption and relaxation involving the inner-shell electrons in molecules, in contrast to intense optical lasers, which is only able to investigate the nonlinear behaviors of the outer valence electrons. Additionally, the short-wavelength FEL pulse has capability to excite inner-shell electrons in the specific constitute of a molecule without perturbing valence electrons. In this type of experiments we concentrate one atom in a molecule, and thereby isolate the molecular dynamics at specific site.

Exploiting aforementioned selling points of FEL, a number of pioneer experiments on molecular dynamics have been implemented. In the chapter, we briefly summarize the time-resolved experimental methods and apply them to some fundamental processes of molecular reactions. The paper is organized as follows. Section 1 describes the temporal characteristics of free electron laser, which determines the temporal resolution of time-resolved pump-probe experiments employing FEL facilities. In sec. 2, we select a number of experiments using the combination of XUV-pump — XUV-probe geometry with the multi-particle coincident measurement (Reaction Microscope, ReMi) that mainly operated on FLASH at Hamburg, exploring the molecular dynamics in real time upon the illumination of EUV-FEL. In sec. 3, we describe experiments implementing optical-pump — X-ray-probe scheme on LCLS at Stanford, aiming at the electron

rearrangement and structural change in photo-excited molecular reaction. We end
the article with a summary and an outlook.

2. The Temporal Characteristic and Resolution
   of Free Electron Laser

As of now, despite the recent development of seeding FEL technique providing
Fourier-transform-limited pulse, most of free electron lasers are operated in SASE
(Self-Amplified Spontaneous Emission) mode. In the SASE mode, the FEL facilities
are operated with significant shot-to-shot variations regardless of pulse-energy,
spectral distribution and temporal shape. Even in individual FEL pulse, there
are random internal structures which are very difficult to be determined with
experimental measurement and theoretical prediction. The Fig. 1(a) shows the shot-
to-shot pulse shapes simulated by partial-coherence method [23], considering the
random spectral phase of each frequency components and the realistic FEL pulse
duration. The major benefit of the partial-coherence method to construct a random
FEL pulse is that the complicated machine-related parameters are not need any
more to predict the pulse shape when comparing the experimental results with the
theoretical simulations. The Fig. 1(a) illustrates that the individual pulse profiles
significantly fluctuate from shot to shot. And there are complex internal structures,
i.e., several sharp spikes, in each pulse-envelop.

Because of the shot-to-shot fluctuations and abundant internal structures in
the FEL pulse-envelop, there is a problematic issue that how we measure and
describe the temporal characteristics of FEL pulse. In the atomic and molecular

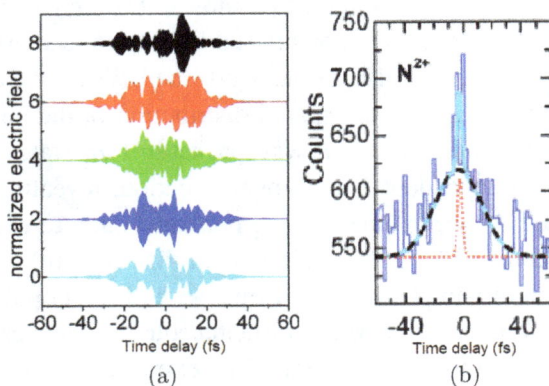

Fig. 1.   (a) shot-to-shot pulse shapes simulated by partial-coherence method. Adapted from Ref. [23].
(b) Experimental autocorrelation trace of N_2^+ fragments. There are two features that can be identified
from the result. The broad Gaussians basement with a FWHW of ~40 fs corresponds the average pulse
duration, whereas the narrow spike of ~4 fs on the top reflects the temporal coherence of EUV-pulse.
Adapted from Ref. [24].

experiment hosted by FEL, the average pulse duration and temporal coherence characteristics can be measured by EUV-pump — EUV-probe scheme (Fig. 3). Likewise the autocorrelation measurements of optical femtosecond laser using the nonlinear crystal, the nonlinear effect of gas-phase molecules in EUV region can be exploited to determine the temporal profile of FEL pulses [24]. In experiment, ion yields of nonlinear processes are measured as a function of pump-probe time delay, reflecting the temporal characteristics of FEL pulse. Fig. 1(b) shows the exemplary ionic yield of noncoincident N^{2+} fragments with 45 eV EUV photon energy along the time delay, i.e., nonlinear autocorrelation. The N^{2+} fragments dominantly come from the N_2^{3+} and N_2^{4+} dissociative states, respectively launched by sequential three-photon and four-photon absorption. The autocorrelation trace can be fitted by the combination of two Guassian distributions of ~40 fs and ~4 fs. The Guassian shape of ~40 fs can be easily interpreted as the statistical averages of the pulse duration. However, surprisingly, the sharp spike of ~4 fs on the top is much shorter than the average pulse duration of FEL, and the essential question is why the temporal resolution of order of few femtoseconds can be achieved by a pulse with pulse duration of 40 fs. The question can be explained by a scenario that the shape spikes in pulse-envelop lead to the temporal coherence of ~4 fs, that is only one tenth of FEL pulse duration, and the temporal resolution is mainly determined by the longitude coherence of FEL pulse instead of the pulse duration.

The noisy shape of FEL pulse can be utilized to resolve the evolution of molecular dynamics in sub-pulse-duration resolution, and this concept has been demonstrated in series of pump-probe experiments. The concept has been explained by the theoretical work in [25]. The nuclear wave-packet in D_2^+ $1s\sigma_g$ molecular potential curve was chosen as a prototype example to demonstrate the concept. The first photon promotes D_2 to D_2^+ $1s\sigma_g$ molecular potential curve, then the second photon launches the D_2^+ to a double charged repulsive state, while the KER from Coulomb explosion of D_2^{2+} repulsive state maps out the spatial distribution of the nuclear wavepacket in $1s\sigma_g$ molecular potential curve. The wavepacket in $1s\sigma_g$ state shown in Fig. 2(a) is calculated by time-dependent Schrödinger equation, revealing that the nuclear wavepacket oscillates between inner turning point and outer turning point on a time scale of approximately 20 fs. The simulations in Fig. 2(b), (c), (d) show the KER spectra as a function of pump-probe delay time with different pulse structures. The final KER distributions correspond the instantaneous internuclear distances when the probe pulse arrives the D_2^+ $1s\sigma_g$ state. The extracted internuclear wavepacket evolutions with different types of FEL pulse are shown in Fig. 2. The Fig. 2(b) presents the temporal nuclear evolution acquired by 30 fs pulse but with several sharp spikes in it, and the pulse structure is simulated by partial-coherence method. The result is compared with the dynamics employing bandwidth-limit 1.12 fs (shown

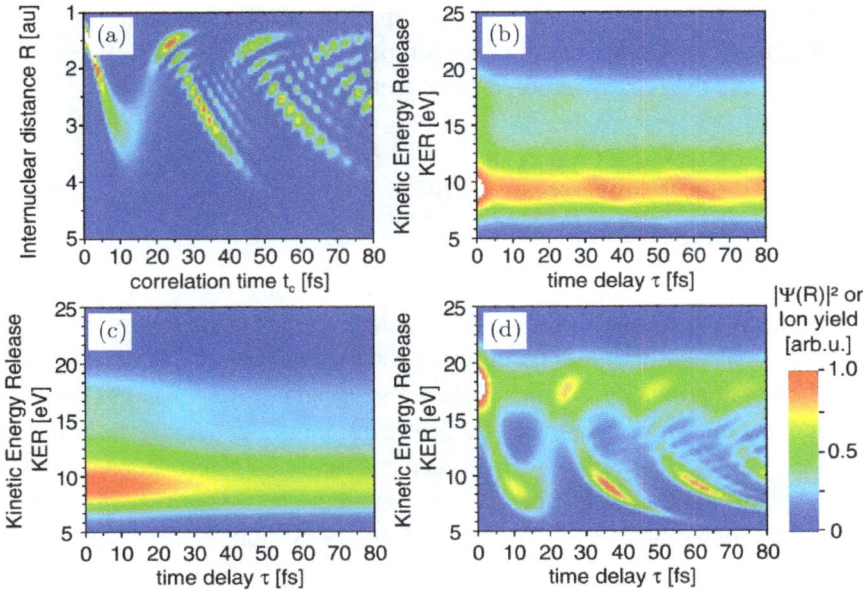

Fig. 2. (a) The nuclear wavepacket evolution of D_2^+ $1s\sigma_g$. The periodic structures indicate the nuclear wavepacket oscillation between the inner-point and outer-point of the potential curve. (b) The KER spectra as a function of pump-probe time delay using the 30 fs FEL pulse but with complex internal structures. (c) The time-resolved KER spectra with a 30 fs bandwidth-limited pulse. (d) The KER spectra with 1.12 fs pulse. Adapted from Ref. [25].

in Fig. 2(c)), which is not yet achievable in experiments with FEL facility until now. Interestingly, though it is not as clear as the result with very short pulse, the sub-pulse-duration dynamics of ∼20 fs can be easily retrieved. It can be explained that the time resolution is not limited by the pulse duration but the longitude-coherence of FEL pulse. The longitude-coherence is determined by the sharp spikes inside the Gaussian pulse (Fig. 1(a)), but not the bandwidth of the Gaussian pulse itself. The concept can be further proved by replacing the spiky pulse with 30 fs Fourier-transform-limited pulse, washing out dynamical evolution inside molecules (shown in Fig. 2(c)). The concept of the high temporal resolution employing the long spiky pulse has been utilized in a number of EUV-pump — EUV-probe experiments to obtain high temporal solution beyond the pulse duration.

In the optical-pump–X-ray-probe (OPXP) experiment, the situation is considerably different with the EUV-pump–EUV-probe geometry that individual FEL pulse is split up and recombined again at interaction volume. Unlike the EUV-pump–EUV-probe experiment, providing an inherent stable pulse-pair, the OPXP experiment need the synchronization of the optical femtosecond laser and FEL facility. The temporal jitter between the two radiation sources is commonly much larger than the pulse duration of FEL pulse, which would limit the temporal resolution in this kind of experiments. The improvement of the temporal resolution of OPXP relies on the effort to upgrade the FEL machine-related technique introduced in sec. 3.

3. The EUV Pump/Probe Measurement

3.1. *The split-mirror scheme*

The EUV-pump — EUV-probe experiments are usually based on the wavefront pulse segmentation to introduce the time delay, either by a back-reflecting focus mirror that is physically separated into two parts for the pulse-pair creation [26], or rely on the grazing incidence Mach-Zehnder geometry which provides the photon-energy-independent transmission [27]. The setup of back-reflecting splitting mirror is shown in Fig. 3, which combines the general Reaction Microscope (ReMi) [28] with the back-reflection split-mirror. The spherical multi-layer mirror is cut into two identical "half-mirror". While one half-mirror is mounted at a fixed position, the other one is movable along the FEL beam axis by means of a high-precision piezo-stage. Owing to the intrinsic time-stability between the two pulses in a pump-probe arrangement, the back-reflecting splitting mirror enables femtosecond time resolutions. During measurements the incoming FEL beam is equally distributed over both half-mirrors, and foci are merged inside a dilute and well localized supersonic gas-jet in the center of the reaction microscope. Ionic fragments are

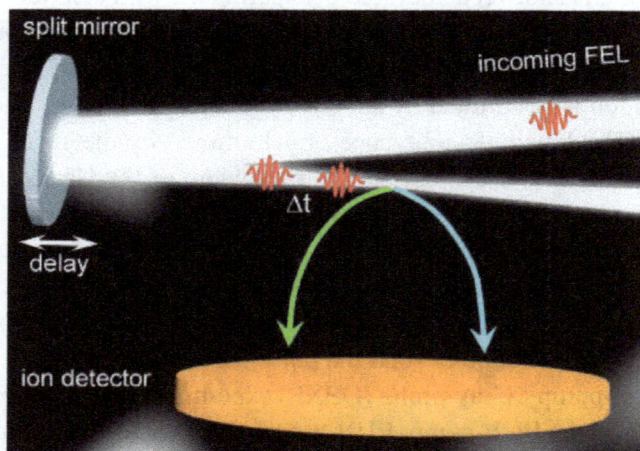

Fig. 3. Schematic drawing of the experimental setup with split-mirror stage and ion detection of the reaction microscope. The back-reflection split-mirror setup focuses the FEL beam as well as creates the pulse pair with various time delay Δt. The fragments can be detected by the time- and position-sensitive detector, allowing the 3D momentum reconstruction. Adapted from Ref. [29].

projected onto a time- and position-sensitive detector. From the time-of-flight and position of each individual fragment the initial momentum of ions can be reconstructed.

Employing the EUV-pump — EUV-probe scheme described above, a number of experiments are performed and some impressive outcomes are published for very first time. Nonlinear multiphoton excitation/relaxation in strong EUV field have been investigated by inspecting the angular and energy dependent emission of ionic fragments for different charge states [18]. Additionally, the wavepacket oscillation in the D_2^+ potential curve is chosen as a benchmark system to resolve *in situ* nuclear wavepacket, demonstrating a proof-of-principle time-resolved experiment in EUV regime [26]. The EUV-pump — EUV-probe also enable to follow the evolution of nuclear motions in bound and dissociative states and disentangle a series of reaction pathways in dissociative reaction in some diatomic molecules like D_2 [26], N_2 [29, 30] and O_2 [30]. In addition, the EUV-induced electronic and nuclear structural evolutions in excited molecular states beyond Born-Oppenheimer approximation are mapped out in fs resolution, providing instantaneous electronic and molecular structure duration the light-induced reaction. Here, we briefly introduce time-resolved experiments which employ EUV photons for both initiating and probing molecular dynamics of interested molecules. The experiments are mainly performed with the workstation located at FLASH in Hamburg.

3.2. Tracing the nuclear wave-packet in D_2^+ by EUV time-resolved experiment

The proof-of-principle pump-probe experiment in EUV is performed in D_2 molecule. The D_2 molecule is chosen due to its simplicity and availability of time-dependent theoretical calculation. The pump pulse of 38 eV photon energy creates the ground state of D_2^+ ($1s\sigma_g$) by removing the valence electron, and initiates vibrational wavepacket. By scanning the pump-probe delay, we select the different instance to ionize the second electron, resulting in a repulsive $D^+ + D^+$ dissociative state. With the ReMi described above, the $D^+ + D^+$ fragment can be captured coincidently, and the 3D momentum of ions can be reconstructed. In this process, the R-dependent nuclear wavepackets at the instant of the second pulse arrival are launched into the repulsive $D^+ + D^+$ state, leading to an R-dependent KER of fragments. Thus, the nuclear wavepacket in R-coordinate is projected into momentum space, which is so called Coulomb explosion imaging. The KER evolution with pump-probe delay time reflects the nuclear motion in D_2^+ $1s\sigma_g$ potential curve at different instant of the probe arrival. Hence, the stepwise sequential ionization allows to trace the nuclear wavepacket in real time domain with temporal limitation down to the pulse duration.

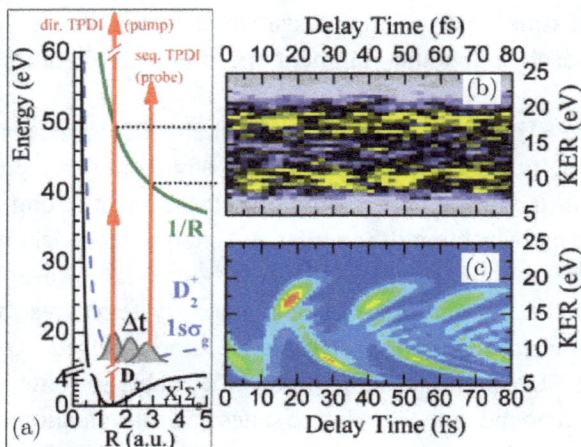

Fig. 4. (a) Illustration of the EUV-pump — EUV-probe experiment in D_2 molecule. The first pump launched the wavepacket to ground state of $D_2{}^+$ ($1s\sigma_g$). The probe pulse arrives at different time delay to sequentially ionize the $D_2{}^+$ to a repulsive $D^+ + D^+$ state. (b) KER spectrum of coincident $D^+ + D^+$ fragments as a function of delay time. (c) The theoretical results corresponding to experimental results in (b). Adapted from Ref. [26].

The experimental KER spectrum as a function of delay time (Fig. 4(b)) carries information of nuclear wavepacket in $D_2{}^+$ ground state, which is quantitatively defined by the shape of $D_2{}^+$ $1s\sigma_g$ potential curve. In EUV-pump — EUV-probe experiment, the nuclear wavepacket is populated by the pump pulse and propagate freely in the potential well (Fig. 4(a)). The probe pulse further ionizes the second electron, and projects the R-dependent wavepacket to a dissociative state, i.e., Coulomb explosion imaging. The KERs of D^+ fragments at different pump-probe delay time are proportional to $1/R$, reflecting the instantaneous nuclear wavepacket at the moment of the second ionization event. Thus the ionization on out turning point with large R leads to small explosion energy, i.e., small KER, while the inner point corresponds to large KER. There are two pronounced structures center at 10 eV and 18 eV. The high energetic structure might be attributed to the direct two-photon double-ionization pathway that the ground state simultaneously absorbs two photons to the repulsive $1/R$ Coulomb state. Also, the high energetic structure may come from the sequential ionization pathway that the second-step absorption happens at the moment that the wavepacket reaches the inner turning point. The relative contributions of the direct or sequential ionization can be elucidated by advanced many-particle quantum calculation [31].

Since the nuclear wavepacket of $D_2{}^+$ ($1s\sigma_g$) prepared by first pulse will oscillate between the classical inner turning point and outer turning point, the KER distributions of coincident $D^+ + D^+$ channel should show the periodic structure with pump-probe delay time. When the statistical events between 6 eV and 12 eV are projected onto the delay axis, a periodic ionic yield of 22 ± 4 fs will be clearly observed. The corresponding theoretical calculation by numerically solving the time-dependent Schrödinger equation is shown in Fig. 4(c). The theoretical picture gives clear periodic structures than experimental result probably due to the limitation of pulse duration and statistical background. The theory gives the ionic yield period of 23.8 fs in low energy region, which well agrees with 22 ± 4 fs in the experimental result. The proof-of-principle experiment in time-resolved evolution of D_2^+ ($1s\sigma_g$) nuclear wavepacket opens a variety of possibilities including *in situ* imaging out the nuclear vibrational motion induced by photo-excitation on temporal resolution of the order of few femtosecond.

3.3. *Nuclear wavepacket dynamics in excited states of* N_2

Nitrogen is one of the most common diatomic molecules which is abundant in atmosphere and universe. Accordingly, N_2 is widely studied by synchrotron facilities, since that priori spectroscopic knowledge on N_2 is collected by a variety of experimental techniques. In FEL facility, the multi-photon multi-ionization of N_2 in EUV regime is investigated by the single-pulse experiment [18, 32], and the ion-fragment angular distributions are recorded, allowing to investigate the symmetries of initial state and final state in electronic transition. In pump-probe experiment, N_2 molecule is chosen to investigate the time-resolved process in multi-photon multi-ionization and subsequent relaxation. Based on the sequential ionization, the relaxation processes along various excited states in N_2 were temporally followed, and the electronic dynamics interwined with nuclear motion can be explored.

In aforementioned autocorrelation experiment in N_2 shown in Fig. 1(b), the temporal pulse profile can be extracted from EUV-pump — EUV-probe measurement. Besides that, the sequential ionization induced by pump-probe pulse-pair can be used to detect and disentangle the nuclear wavepacket along different reaction pathways via Coulomb explosion imaging. The identification with reaction pathways provides the basement to further investigate the electronic and nuclear dynamics in pathways of interest. Here, the first EUV pulse removed one electron in a neutral molecule, and excited N_2 molecule onto several excited states (bound states and dissociative states) via a Franck-Condon transition, simultaneously initializing the nuclear-wavepacket in a number of ionic species on potential energy curves. The time-delayed EUV pulse launched the evolving nuclear wavepacket onto Coulomb explosion state, leading to different KER distributions of ionic fragments related to R-dependent nuclear wavepacket. The exemplary KER spectrum of coincidence $N^+ + N^+$ fragments as a function of time delay at 38 eV photon energy is shown in Fig. 5, which can be obtained by reconstructing the 3D momentum and restricting

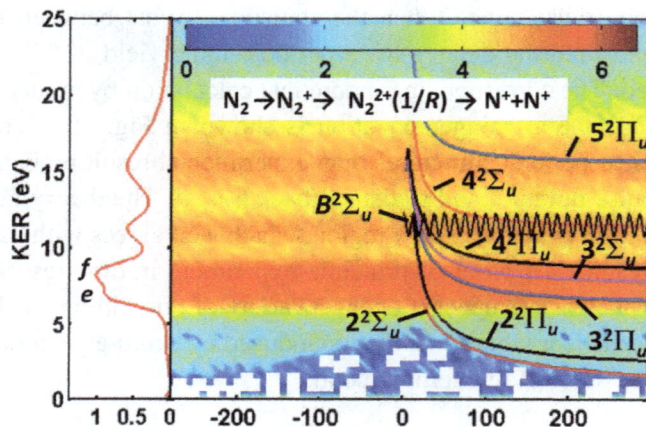

Fig. 5. The KER spectrum of $N^+ + N^+$ fragments with pump-probe time delay. Left side is the delay-integrated KER spectrum. The solid lines are the classical simulations. Adapted from Ref. [30].

the momentum conservation conditions. The spectrum is symmetric with respect to zero delay due to the alternative pump-probe order of two identical pulse-replicas. There are two types of features to be observed. One is the delay-dependent band below 5 eV, in which high energetic KER distribution at small delay decreases as the increasing pump-probe time delay and converges to an asymptotic energy. It can be explained that the intermediate state upon the first photon is a dissociative state and the second pulse launches the dissociative state to higher charged repulsive states at increasing internuclear distances during the dissociation, resulting in smaller and smaller Coulombic explosion energies. The spectrum above 5 eV manifests a series of time-independent structures. In contrast to the time-dependent band, the constant bands come from the pathways that the intermediate states are bound states. In bound states, the vibrational wavepackets are restricted into a potential well, from which the further ionization will lead to KER distribution in a narrow band. The experimental results can be reproduced and explained by the classical simulation, depicted with solid lines in Fig. 5. The descending curves, whose KERs decrease with increasing time delay, correspond to the time-dependent structures. The oscillate lines resemble the time-independent structure with constant KER band. The period should be consistent with nuclear vibrational period in potential well of the bound state. In experiment, the oscillation is too rapid to resolve with FEL pulse nowadays which is so far limited around 10 fs. The pump-probe KER spectrum is of great value because it manifests and separates different reaction pathways. Relying on the time-dependent KER spectra, the pathways of interest can be isolated from a series of populated intermediate states, allowing us to interrogate the specific reaction pathway in molecular dynamics.

3.4. *Ultrafast photoisomerization of acetylene cation*

In aforementioned N_2 experiment, we present that pump-probe method can be utilized to trace the nuclear motion in simple diatomic molecules and to disentangle a series of dissociative reaction pathways. Now we turn to more complicated nuclear motion in the chemical reaction in polyatomic molecules. Photo-induced isomerization is one of the most elementary chemical reaction occurring in photosynthesis, eye vision as well as viral infection (HIV isomerization) [33–35]. The acetylene molecule (C_2H_2) is probably the simplest molecule undergoing isomerization, serving as a prototype for studying the structural rearrangement during isomerization. The excitation of an acetylene molecule by a EUV pulse of 38 eV photon energy opens several reaction channels including deprotonation ($C_2H_2 \xrightarrow{nh\nu} H^+ + C_2H^+$ and $C_2H_2 \xrightarrow{nh\nu} H^+ + C_2H^{2+}$), symmetric break-up ($C_2H_2 \xrightarrow{nh\nu} CH^+ + CH^+$) and isomerization ($C_2H_2 \xrightarrow{nh\nu} CH_2^+ + C^+$), and the corresponding molecular dynamics were recorded via Coulomb explosion imaging by EUV-pump — EUV-probe method [36]. Here, we highlight the isomerization channel that the proton in $[HCCH]^+$ cation moves from the end of C atom to the other side, forming vinylidene $[H_2CC]^+$ structure (Fig. 6(a) and (b)).

Upon pumping by the first EUV photon, the single charged state $A^2\Sigma_g$ is populated with emitting the first electron e_1. The transition pathway for isomerization

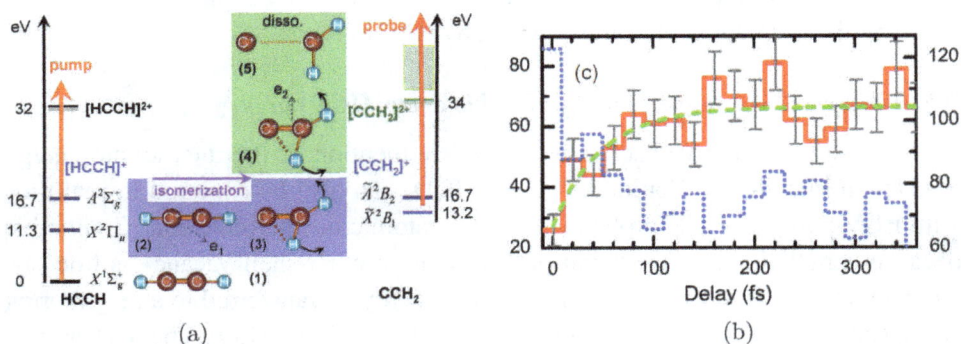

Fig. 6. (a) Schematic of acetylene-vinylidene isomerization. The pump pulse ionizes the first electron and initializes the molecular reaction. During the isomerization reaction, the H atom migrates from the end of the C atom to the other C atom. The second probe pulse arrives at different instant of the reaction and produces coincident $CH_2^+ + C^+$ fragments, which serves as a signature for the H-atom transfer. (b) Yield of the $CH_2^+ + C^+$ coincident fragments (red line) as a function of the pump-probe delay. The exponential fitting (green dash line) indicates that the ultrafast process of the isomerization is estimated within 52 ± 15 fs. Adapted from Ref. [37].

from the acetylene to the vinylidene cation occurs:

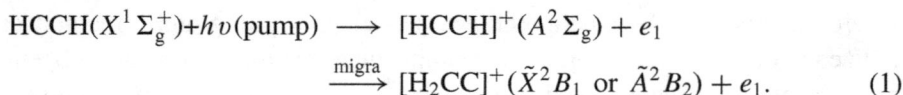

$$HCCH(X^1\Sigma_g^+)+h\upsilon(\text{pump}) \longrightarrow [HCCH]^+(A^2\Sigma_g) + e_1$$

$$\xrightarrow{\text{migra}} [H_2CC]^+(\tilde{X}^2B_1 \text{ or } \tilde{A}^2B_2) + e_1. \qquad (1)$$

On the $A^2\Sigma_g$ potential curve, the $[HCCH]^+$ is unstable, and transform to vinylidene $[H_2CC]^+$ $\tilde{A}^2B_1$ or $\tilde{X}^2B_1$ state via the isomerization process. The molecular rearrangement can be mapped by further ionizing the vinylidene $[H_2CC]^+$ with the probe pulse, and the coincident fragments $CH_2^+ + C^+ \leftarrow [H_2CC]^+$ are recorded as a function of time delay. The probing process can be expressed by

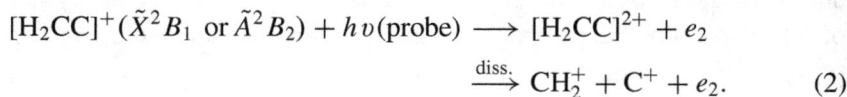

$$[H_2CC]^+(\tilde{X}^2B_1 \text{ or } \tilde{A}^2B_2) + h\upsilon(\text{probe}) \longrightarrow [H_2CC]^{2+} + e_2$$

$$\xrightarrow{\text{diss.}} CH_2^+ + C^+ + e_2. \qquad (2)$$

In Fig. 6 (b), the ion yield of $CH_2^+ + C^+$ channel displays a clear time-dependence, indicating the migration of H atom in the acetylene cation from one side of C atom to the other C atom. By fitting the data with exponential function, the isomerization time has been estimated to 52 ± 15 fs [37], and supported by the theoretical calculation [38]. Recently, by recording triple coincidences in molecular rearrangement, the atomic-scale structural changes during the isomerization can be visualized in fs resolution by three-body Coulomb explosion imaging. By observing the acetylene-vinylidene isomerization of C_2H_2, we prove the capability of the FEL facility to temporally probe the nuclear rearrangement in photo-induced chemical reaction. With the further development of the FEL and corresponding coincident detection techniques, the structural changes in chemical reaction can be mapped in real time, realizing the ultimate goal of molecular movie.

3.5. *Time-resolved interatomic Coulombic decay (ICD) in Ne₂*

Besides the nuclear wavepacket as previously mentioned, the interatomic energy transfer in interatomic Coulombic decay (ICD) can also be explored in real time with EUV-pump — EUV-probe method. Interatomic or intermolecular Coulombic decay is a radiationless relaxation process that an inner-shell vacancy is filled by an outer-valence electron, while the released energy is transferred to a neighboring atom, simultaneously emitting an electron into the continuum [39]. The ICD process may occur in a various environment like clusters, He droplets, fullerenes and aqueous solutions. In 2003, ICD was experimentally confirmed in large neon clusters [40], and Jahnke *et al.* performed an unambiguous experiment on neon dimers by a kinematically complete experiment [41] in 2004. By means of EUV-pump — EVU-probe experiment, the direct time-resolved measurement of ICD lifetime is performed for the first time at FEL facility [42]. As shown in Fig. 7, the pump pulse

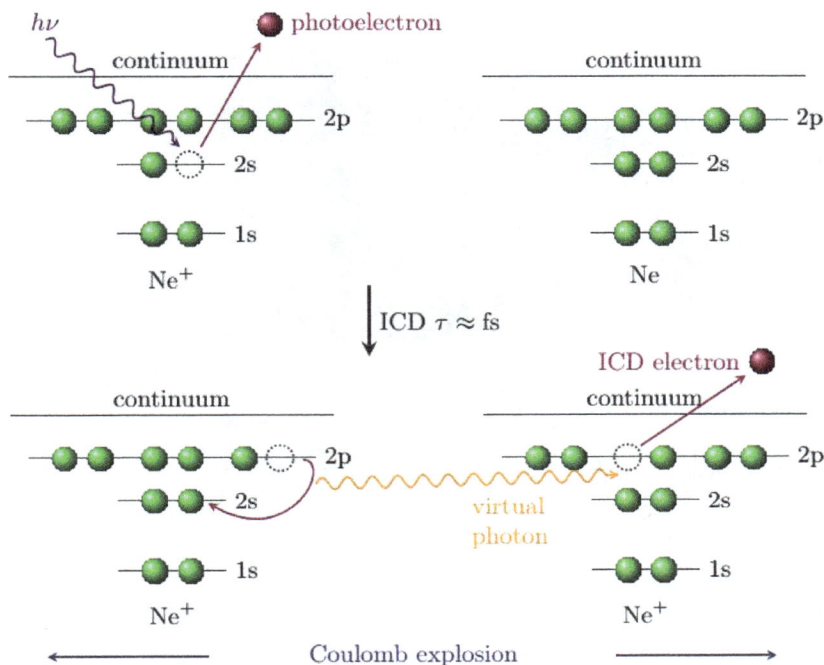

Fig. 7. Schematic of the ICD process in Neon dimer. A EUV photon ionizes $2s$ inner-valence electron in one of Ne_2 dimer. The hole is filled by $2p$ electron and the excess energy is transferred to the neighboring Ne atom via virtual photon, leading to $2p$ electron ionization. Adapted from Ref. [43].

of 58.2 eV photon energy creates a $2s$ inner-valence vacancy ($2^2\Sigma_g^+$ or $2^2\Sigma_u^+$ state) in either of the two Ne atoms. The hole is filled by a $2p$ electron and the excess energy is transferred to the neighboring Ne atom, which removes a $2p$ electron from a neutral Ne atom. Finally, a repulsive Ne^+ $(2p^{-1})$ + Ne^+ $(2p^{-1})$ is produced. The lifetime of the $2s$ inner-valence vacancy, i.e. lifetime of ICD, can be determined by pump-probe approach. During the ICD process, the wavepacket of $2^2\Sigma_g^+$ or $2^2\Sigma_u^+$ state in Ne_2^+ dimer decays into one of the various repulsive Ne^+ $(2p^{-1})$ + Ne^+ $(2p^{-1})$ states via ICD, and the subsequent probe pulse removes another electron in either Ne^+ ion from two Ne^+ fragments, launching to highly charged Coulombic explosion channels Ne^{2+} $(2p^{-2})$ + Ne^+ $(2p^{-1})$. Experimentally, the coincidence Ne^{2+} + Ne^+ can be detected as a function of time-delay. In comparison, without ICD process, the wavepacket of $2^2\Sigma_g^+$ or $2^2\Sigma_u^+$ state in Ne_2^+ propagates along the potential curve of $2^2\Sigma_g^+$ or $2^2\Sigma_u^+$ state, and finally dissociates into Ne^+ $(2s^{-1})$ + Ne. The probe pulse will create the Ne^+ + Ne^+ or Ne + Ne^{2+} ions instead of Ne^{2+} + Ne^+ fragments in ICD channel. Hence, the ion yield in Ne^{2+} $(2p^{-2})$ + Ne^+ $(2p^{-1})$ is expected to increase with pump-probe time delay, reflecting lifetime of ICD process.

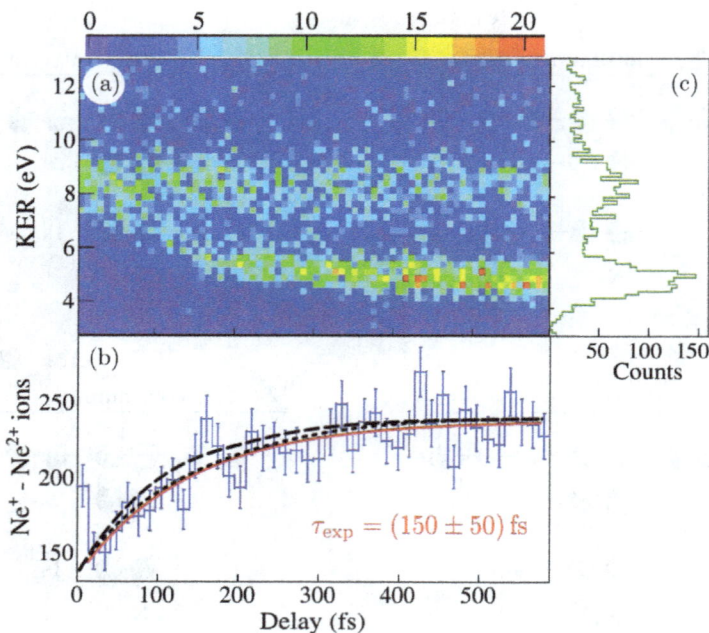

Fig. 8. (a) Delay-dependent KER spectrum of $Ne^+ + Ne^{2+}$ fragments. There are two pronounced structures: the delay-dependent KER and the constant KER as a function of time delay. The delay-dependent structure ranging between 4 eV and 7 eV decreases from high Coulombic energy at small delays towards low KER at large delays, containing the information of ICD lifetime. (b) The projection of KER spectrum into time delay axis. The red line is the exponential fitting for the experimental results. (c) Time-integrated KER between 400 and 600fs. Adapted from Ref. [42].

The KER spectrum of coincident $Ne^{2+} + Ne^+$ fragments as a function of pump-probe time delay is presented in Fig. 8(a). There are two pronounced structures in the spectrum. The time-independent structure above 7 eV originates from the multiphoton ionization within one EUV pulse from the ground state directly to a repulsive $Ne^{2+} + Ne^+$ state, which is not concerning too much. The time-dependent structure below 7 eV presents a strong signature of the Coulomb explosion, which KERs decrease with increasing pump-probe time delay. The ion yield of Coulomb explosion channel at different time delay contains the information of the lifetime of ICD. All events between 3 eV and 11 eV are projected onto delay axis (Fig. 8(b)), and evolution of ion yield as time delay is fitted by an exponential function to obtain the ICD lifetime of 150 ± 50 fs. The offset is due to the background stemming from the direct multi-photon absorption. Comparing with different theoretical methods, the ICD lifetime agrees best with the theory including not only the electronic dynamics but also the nuclear motion prior to the ICD process, which indicates that the accompanied nuclear motion is equally essential as well as electron motion to determine the ICD lifetime for small systems like Ne_2. Employing with the EUV-pump — EUV-probe approach, the direct measurement of ICD lifetime in real time is performed for the first time in FEL facility, opening up the possibility to map out the charge transfer and energy flow through transient processes in molecular reactions.

3.6. *Electron rearrangement dynamics in dissociating I_2^{n+}*

Electron rearrangement, migration and relaxation is relevant to "making and breaking" of bonds in chemical reactions. The understanding of intramolecular electron rearrangement and relaxation in strong short-wavelength radiation is particularly important for the most far-reaching application of FEL — coherent diffraction imaging. The coherent diffraction imaging relies on the "diffract before destruction" concept, requiring the imaged molecule with the same structure as original one upon irradiation [44]. However, upon illumination of strong short-wavelength light, the molecule would undergo rapid destruction due to the electron rearrangement and subsequent structural change. Thus, the investigation of intramolecular charge rearrangement and accompanied nuclear evolution in intense external field establishes the cornerstone of the single-molecule diffraction coherent imaging. Here, employing the EUV-pump — EUV-probe approach, the time scale of the photo-induced charge redistribution in dissociative molecule is obtained in dissociating I_2^{n+} [45]. And the critical internuclear distance R_{cr} up to which the electron rearrangement across the entire molecular ion can take place is experimentally determined.

In experiment, a pump pulse of 87 eV photon energy creates $4d$-type vacancy in I_2 and initiates the dissociation. Afterward, the I_2^{*+} undergoes the Auger process leading to I_2^{2+} or cascade Auger process leading to I_2^{3+}. With different time delay, corresponding to different internuclear distance, the dissociative states are probed by the second EUV pulse. The second photon creates another $4d$-type vacancy with a following Auger decay, undergoing dissociation into $I^{P+} + I^{Q+}$ fragments. The delay-dependent KER spectra of asymmetric charged fragments ($|P - Q| > 2$) are particularly of interest because the charge distribution mostly tends to be balanced between the two dissociative fragments. And asymmetric channels exhibit depletion in small time delay in contract to the symmetric charged fragments, which show a maximum at small delay time. The exemplary coincident $I^+ + I^{4+}$ channel and $I^+ + I^{5+}$ channel are shown in Fig. 9, which are produced by further ionization of precursor states I_2^{3+} and I_2^{2+}. The ion yield depletion at small delay hints the transition from molecular structure to isolated atoms during the dissociative process. The time-dependent ionic yield is explained that the probe photon further ionizes precursor ionic states, producing more positive charge in one of ions. At small internuclear distance, the charge could transfer from more positive charged one to the other one, resulting in the charge balance over two dissociating fragments. However, the

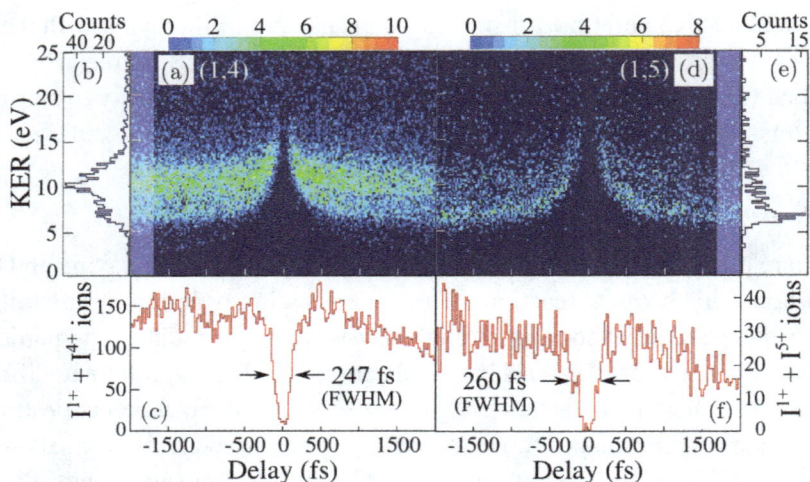

Fig. 9. Delay-dependent KER spectra of coincident $I^+ + I^{4+}$ fragments (a) and $I^+ + I^{5+}$ fragments (d). (b), (e) The delay-integrated KERs at large delay (shaded area). (c), (f) The projection of all events onto delay axes. The delay-dependent KERs manifest a yield minimum at small delay. Adapted from Ref. [45].

electron rearrangement is only possible when the orbital overlap of two ions is large enough at small time delay. Therefore the asymmetric charged fragments are only produced at large time delay where the orbital overlap between the two ions is small, i.e. long time delays. From the width of ion yield depletion, the critical value R_{cr} up to which the electron rearrangement may occur is experimentally determined. The time-resolved experiment addresses the question concerning the timescale of ultrafast electron transfers upon intense short-wavelength FEL, which establishs the basement of destruction mechanism in single-molecule coherence diffraction imaging.

4. Optical-pump — X-ray-probe (OPXP) Experiments

Besides performing the EUV-pump–EUV-probe experiment by wavefront-seperation method, the ultrashort FEL pulse can alternatively combine with UV-, NIR-, terahertz-, from either optical laser or FEL facilities for pump or probe pulse [46–50]. The typical setup of Optical-pump–X-ray-probe (OPXP) is described in Ref. [51]. In the OPXP scheme, a femtosecond IR or UV pulse populates the excited states by single or multi-photon processes, simultaneously initiating the reaction. Afterwards, a time-delay X-ray probe pulse ionizes the inner-shell electron of molecules, and breaks the molecule into several fragments. In previously mentioned EUV-pump–EUV-probe scheme, since the delay-line is provided by means of the split-mirror providing an inherent temporal stability, the temporal resolution is determined by the longitude coherence length of FEL pulse. Comparatively, the time resolution of OPXP scheme is limited by time jitter between an amplified Ti:Sapphire laser and FEL machine. It is much more difficult to synchronize physically different lasers that are not from identical radiation sources. The first-step synchronization includes that a conventional feedback is utilized to stabilize the laser arrival time with the radio frequency driving the accelerator, which gives the time jitter down to 1 picosecond. The further synchronization relies on the measurement of the arrival times of electron bunch relative to the external femtosecond pulse, leading to sub-100 fs time resolution [52–54]. In addition, by analyzing experiment data the actual time jitter can be reconstructed and more precisely determined, allowing to correct the time-dependent data. The OPXP has various applications not only in time-resolved experiments, also in impulsively molecular alignment which allows to measuring the photoelectron angle dependence in the molecular frame [21, 55]. Here, we select recent time-resolved experiments on polyatomic molecules with OPXP scheme performing on the LCLS at SLAC.

4.1. *Probing a prototypical photoinduced ring opening*

Upon illumination by X-ray pulse, the molecular response with X-ray predominantly involves the core electron localized at atomic sites since the absorption cross section of core-electron is much larger than valence-electrons. The interaction with X-ray does not perturb the valence electrons and affect the molecular geometry, opening a way to isolate molecular dynamics at specific site. In the interaction process, the X-ray pulse ionizes core electrons in selected elements of molecules and initiates the subsequent Auger process by ejecting one or more electron. The charge and energy deposited in the element rapidly distribute over the entire molecule, leading to fragmentation. The kinetic energy of charged fragments are considered to depend on the excitation state of the parent molecule. In experiment, the kinetic energy of fragments are measured to inspect the evolution of instantaneous geometry upon photoexcitation. The isomerization from 1,3-cyclohexadiene (CHD) to 1,3,5-hexatrience (HT) is a prototypical example to trace a photo-induced molecular reaction due to previous extensive studies. As shown in the inset of Fig. 10, the UV pulse launches the wavepacket on S_1 potential energy surface of CHD that rapidly crosses onto S_2 by nonradiative transition, and then accelerates to the S_2/S_0 conical intersection. Across the conical intersection, the wavepacket continue to fall to the HT potential energy well, leading to the ring opening. The ring opening process has received considerable attention in experiments and calculations in part because of its board applications such as the crucial role of vitamin D biosynthesis. It has been known by previous studies that the ring opening takes less than 200 fs through the S_2/S_0 conical intersection.

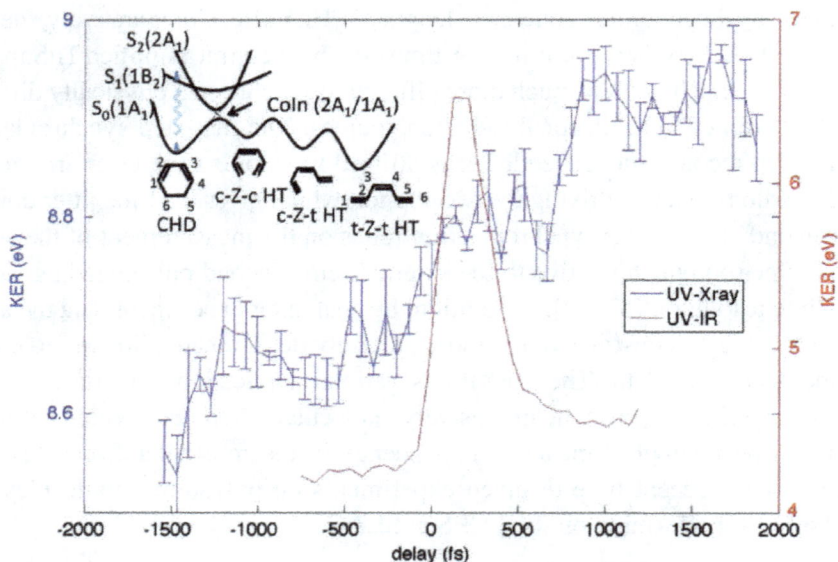

Fig. 10. The average ion-KER of per fragment following the UV-excitation. The blue and red line depict temporal evolutions in UV–X-ray and UV–IR experiments. The inset shows the photoexcited ring opening from 1,3-cyclohexadiene (CHD) to 1,3,5-hexatriene (HT). Adapted from Ref. [56].

In experiment, a 50 fs, 266 nm ultraviolet pulse initiates ring opening in CHD, and a time-delayed X-ray of 850 eV ionizes $1s$ electron of carbon, leading to dissociation. The fragments are detected by velocity-map imaging detector (VMI). The average ion-KER per fragment as a function of pump-probe time delay is presented in Fig. 10. As the blue line in Fig. 10, the average ion-KER increases from UV-excitation until $\sim$1 ps, and then keeps constant at long UV–X-ray time delay. The evolution of average ion-KER following the UV-excitation can be explained that the geometry changes along the photoexcited molecular reaction. When the X-ray probe pulse ionizes the inner-shell electron and initiates the Auger decay, the dicationic states are created by the Auger transition, resulting in Comloub explosion into several fragments. The calculation demonstrates that the distribution of the dicationic states is geometry- and excitation dependent. The ion-KER of fragmentation resulting from dicationic states represents transient molecular geometry during the ring opening process. The UV-pump–IR-probe shown as red line in Fig. 10 exhibits a pronounced different behavior with UV–X-ray experiment, which has ion-KER maximum at a time delay of $\sim$200 fs [57]. The time-delay shift of the maximum from zero is due to the interaction of strong-field IR pulse with valence electrons, which undergo an enhancement of the strong-field double-ionization at the S_2/S_0 conical intersection. The experiment manifests the capability of X-ray FEL to detect the evolution of molecular structure in photoexcited molecular reaction.

4.2. *Imaging charge transfer in iodomethane*

Since the discovery of X-ray radiation, the most fascinating application is to image the matter structure on the order of atomic length. The conventional X-ray crystallographic imaging on synchrotron radiation needs growing crystal large enough to obtain an adequate integrated signal. The FEL facility delivers intense pulse, allowing to obtain the diffraction signal from a single non-crystalline macro-molecule. The technique, so called coherent diffraction imaging, may revolutionize the biology and material science. One possible limitation of the state-of-the-art technique is radiation damage, thus requiring "diffraction before destruction". However, when a molecule is irradiated by the X-ray pulse, the electronic density and atomic geometry will rapidly change in few femtoseconds, which means that the reconstructed molecule is not the original interact one any more. The sample molecular system is a paradigmatic example to investigate the X-ray-induced radiation damage [58, 59]. In recent experiment at LCLS, combining of NIR-pump–X-ray-probe with ion-imaging spectrometer, the charge transfer dynamics around the whole molecule is investigated, which help us to understand the underlying mechanism of the radiation damage [60].

In experiment, the methyl iodide (CH_3I) is chosen for several reasons. Firstly, the methyl iodide contains the heave atom iodine, whose absorption cross section is much higher than the other constitutes. The high cross section localizes the photoabsorption in site- and shell-specified iodine atom. The poly-atomic molecule contains the carbon atom and several hydrogen atoms, allowing to investigate the influence of the molecular environment on the charge rearrangement and intra-molecule decay mechanism. The basic concept of the experiment is described in Fig. 11. The NIR dissociates CH_3I by multi-photon ionization. Subsequently, the CH_3I dissociates into the CH_3 and atomic I fragment, either neutral or singly charged. A time-delayed X-ray pulse with photon energy of 1500 eV ionizes the inner-shell electron in iodine fragment, and initiates the Auger process. At small time-delay in Fig. 11, where the two fragments are close to each other, the molecule can be described by the molecular orbital picture. The molecular Auger decay will involve the delocalized valence electron, resulting in charge transfer from the CH_3 ion to iodine. At large internuclear distance, the two fragments will be treated as two individual sites where the Auger process takes place in the individual iodine ion. The critical distance at which charge transfer can be negligible may distinguish where the valence electron can be treated as a delocalized electron or a localized electron in an individual fragment.

By measuring the charge states and kinetic energy release (KER) as a function of NIR and X-ray delay-time, the charge redistribution is probed at different internuclear separations. The KER spectrum as a function of time-delay of I^{6+}

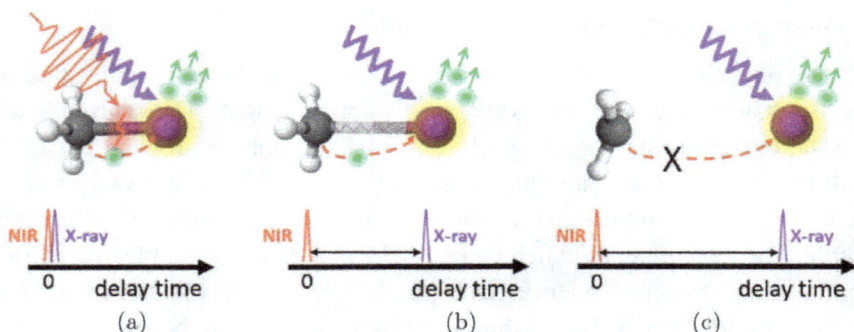

Fig. 11. The Schematic of NIR-pump — X-ray-probe experiment. The NIR pulse initiates the dissociation of CH_3I into CH_3 and I fragment. At various internuclear distance, corresponding different delay time, X-ray pulse ionizes the inner-shell electron of iodine atom and initiates the subsequent Auger process. (b) At small internuclear distance, the delocalized valence electron will be involved in the molecular Auger process, resulting in charge transfer from CH_3 to I fragment. (c) At large internuclear distance, the atomic Auger process occurs within the individual iodine where the charge transfer cannot take place any more. Adapted from Ref. [60].

is shown in Fig. 12(a). The three major features (marked 1, 2 and 3 in Fig. 12) can be identified. The broad band 1 whose peak locates at around 7 eV originates from the molecules that only interact with X-ray photons without interaction with NIR photon, and therefore shows delay-independent feature. The structure 2 shows a Coulomb-explosion feature that the KERs decrease with increasing time-delay. The delay-dependent feature attributes to $CH_3^+ + I^+$ dissociation channels. The simulated dash line in Fig. 12 can be explained that the molecule is dissociated along $CH_3^+ + I^+$ potential curve by multi-photon ionization of NIR, and subsequently ionized by X-ray pulse into $I^{6+} + C^+$ (lower line) and $I^{6+} + C^{2+}$ (higher line) fragments. The time-dependent structure 3 is worthy of special attention. The structure 3 comes from the further X-ray ionization of the $CH_3 + I^+$ intermediate state. The structure 3 does not show the Coulomb-explosion feature because the localized X-ray absorption in iodine atom leads to $CH_3 + I^{6+}$ with very low kinetic energy. But at small time-delay, corresponding small internuclear distance where the methyl and iodine fragment are close, X-ray inner-shell ionization of iodine initiates the molecular Auger process during dissociation. In molecular Auger process, the charge on the iodine fragment transfers to the initially neutral methyl group, and the charged methyl and iodine ion repel each other. Thus, the charged methyl group and highly charged iodine ion dissociate along a Coulomb-explosion-like potential curve with relative high KERs, leading to the depletion of the ion yield of I^{6+} at small time-delay. As the internuclear distance increasing the ion yield of low energy iodine increases at beginning 100 fs, and reaches constant after 300 fs. This indicates that at large pump-probe delay where the methyl group and the iodine fragment are far away, the inner shell ionization

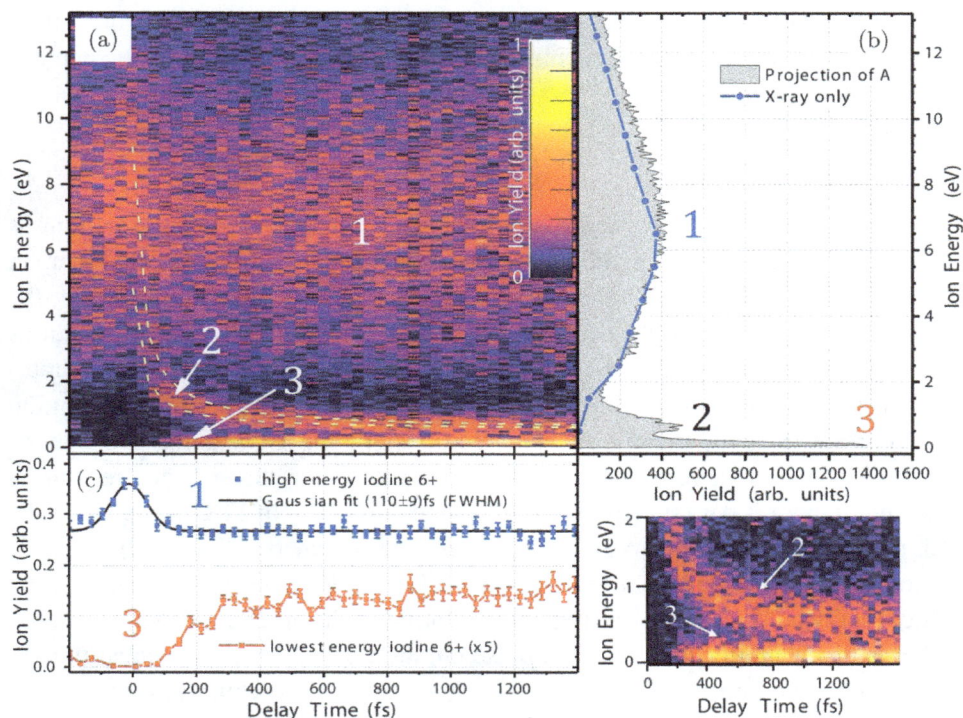

Fig. 12. The kinetic energy release spectrum of I^{6+} as a function of NIR and X-ray time delay. (a) Measured KER of I^{6+} as a function of NIR and X-ray time-delay. Marked 1, 2 and 3 indicate different dissociation channels. (b) Delay-integrated KER spectrum. Peak 1 corresponds to the reaction channel that CH_3I only interacts with X-ray pulse without interaction with NIR pulse. The peak 2 and 3 is due to the interaction with both NIR and X-ray pulse. (c) The delay-dependent ion yield of channel 1 and channel 3. The maximum at channel 1 is due to the overlapping of NIR and X-ray pulses. The yield evolution of channel 3 from zero to a constant can be explained that the CH_3 and I^+ fragments evolve from an entire molecule to two separated atoms. Adapted from Ref. [60].

of iodine only can trigger the atomic Auger process in individual iodine. Hence, the neutral methyl and I^{6+} fragment dissociates to low energy regime at large time delay. Depending on tracing the time-dependent iodine ion yield, two regimes in the dissociation process can be considered. At small internuclear distance where the CH_3 and I^+ just start to dissociate, the system can be described as a molecular orbital that the charge rearrangement is feasible. In the case of large internuclear distance, the CH_3 and I^+ can be treated as two individual atoms, where the Auger process localize in one atom and the charge transfer between two fragment group cannot occur. The experiment explains that how the highly charged constituent produced by X-ray relaxes in molecular environment, and defines the time scale of electron rearrangement from a site-specific ionized constitute around the whole molecule.

The result indicates that the electron rearrangement upon strong FEL illumination occurs in ultrashort time scale, and hints the strategy to prevent the radiation damage in coherent diffraction imaging.

5. Summary and Outlook

At the beginning of the chapter, the current status of the FEL's constructions is summarized. We emphasize the molecular dynamics in simple molecules upon illumination of the intense short-wavelength radiation, which establishes the cornerstone of the more sophisticated applications like the structural determinations of complex biomolecules and the electronic and conformational evolution in photo-excited chemical reactions. In the first section, we conclude the temporal resolution of time-resolved experiments on FEL facilities. In the EUV-pump–EUV-probe experiment, due to the inherent stability of the split-mirror, the temporal resolution of the equipment is determined by the longitude coherence of FEL pulse. Since the FEL pulse results from the SASE (Self-Amplified Spontaneous Emission), there are random sharp peaks under the envelope of individual FEL pulse. Interestingly, the random spiky pulse will provide the time resolution on the order of few femtosecond, which is much shorter than the FEL pulse duration of $\sim$30 fs. The optical-pump–X-ray-probe experiment does not easily achieve such high temporal resolution as split-mirror geometry since the jitter between the different radiation sources limits the temporal resolution on the order of 1 picosecond.

In the second section, we introduce the EUV-pump–EUV-probe method equipped with the split-mirror at FLASH in Hamburg, and applications on molecular dynamics. Employing the EUV-pump–EUV-probe, the demonstrative experiment on nuclear wavepacket of D_2 molecule was implemented, and the nuclear motion in photo-induced reactions can be monitored in real time. In addition, by recording the KER and fragment angular distribution as a function of time delay, the electronic dynamics in molecular coordinate can be extracted, revealing the charge rearrangement and energy flow in photo-chemical reaction. The research will expand our understanding of correlated dynamics in molecular reactions in the strong EUV field.

Compared with the EUV field, which easily interacts with the valence to trigger the molecular reactions, the X-ray laser tend to interact inner-shell electrons and form core holes in molecules. The optical-pump–X-ray-probe approach, which is operated on LCLS at Stanford, firstly launches a molecule to a excited state by an optical pulse, and at different time delay X-ray pulse ionizes one constitute of the molecule, providing the site-specific information of structural dynamics in chemical reaction. Here, we demonstrate the capability to probe the structural change and electron redistribution with the optical-pump–X-ray-probe experiment. Moreover, the investigation of electronic and nuclear motion in strong X-ray field helps us to understand the mechanism of radiation damage in X-ray coherence diffraction imaging, and to validate interpretation of imaging data.

In the future, the development of FEL facilities will provide high-quality pulse with shorter wavelength and pulse-duration. The new generation FELs (including FERMI in Italy and XFEL in Hamburg) will be operated in seeding-mode, delivering the Fourier-transformation-limited pulse. The seeded FEL will deliver the pulse with shot-to-shot identical temporal profile with high spectroscopic stability and resolution, which is expected to increase the photon numbers per pulse and considerably improve the temporal coherence. The optical-pump–X-ray-probe experiment will also benefit from the seeding technique since the same ultrashort laser are ultilized for exciting samples as well as seeding the FEL, thus suppressing the jitter between the 2-color pulses. In addition, the increasing repetition rate is expected to be operational in future FEL, resulting in the high statistical rate that is especially crucial in coincident experiments to give us multiple views of system under study. In detection side, the improved experimental endstation will integrate more functions including ion, electron and photon detection simultaneously. The new-developed endstation will realize various types of experiments including the time-resolved measurement, ion/electron coincident detection, fluorescence spectroscopy and diffraction imaging to support more applications in a broad of disciplines. The development of sample preparation provides access to a variety of samples for FEL. Hence, not only the gas-phase sample, but the nanoparticles and biomolecule can be adapted with the FEL facilities.

The revolution of the FEL techniques will allow us to discover and investigate new phenomena, which cannot be observed in previous light radiations, for instance the nonsequential ionization of core electrons and subsequent relaxation. Moreover, since high efficient detection the different reactive pathways can be readily separated and disentangled with multi-dimensional correlation. Thus, the reliable identification of intermediate states and unambiguous separation of different reaction pathways will be realized by means of high repetition rate and coincident measurements, allowing us to concentrate the processes of specific interest. With the improvement of detection technique, the experiments can be extend from the simple reaction in small molecule to the complicated reaction in large and complex molecular system, which has broad applications and significance ranging from chemistry to material science and biomolecule. In conclusion, the boost of FEL technique will not only push the limits of the understanding on the light-matter interaction, also motivate some promising applications like coherent diffraction imaging and molecule-reaction movie, finally manifest itself in science and everyday life.

Acknowledgments

We acknowledge the supports from the National Basic Research Program of China (2013CB922200); National Natural Science Foundation of China (11420101003, 11274232, 61308068), Shanghai Committee of Science and Technology (13ZR1463000), Shanghai Pujiang Program (13PJ1407500).

References

1. W. Ackermann, G. Asova, V. Ayvazyan, A. Azima, N. Baboi, J. Baehr, V. Balandin, B. Beutner, A. Brandt, A. Bolzmann, R. Brinkmann, O. I. Brovko, M. Castellano, P. Castro, L. Catani, E. Chiadroni, S. Choroba, A. Cianchi, J. T. Costello, D. Cubaynes, J. Dardis, W. Decking, H. Delsim-Hashemi, A. Delserieys, G. Di Pirro, M. Dohlus, S. Duesterer, A. Eckhardt, H. T. Edwards, B. Faatz, J. Feldhaus, K. Floettmann, J. Frisch, L. Froehlich, T. Garvey, U. Gensch, C. Gerth, M. Goerler, N. Golubeva, H.-J. Grabosch, M. Grecki, O. Grimm, K. Hacker, U. Hahn, J. H. Han, K. Honkavaara, T. Hott, M. Huening, Y. Ivanisenko, E. Jaeschke, W. Jalmuzna, T. Jezynski, R. Kammering, V. Katalev, K. Kavanagh, E. T. Kennedy, S. Khodyachykh, K. Klose, V. Kocharyan, M. Koerfer, M. Kollewe, W. Koprek, S. Korepanov, D. Kostin, M. Krassilnikov, G. Kube, M. Kuhlmann, C. L. S. Lewis, L. Lilje, T. Limberg, D. Lipka, F. Loehl, H. Luna, M. Luong, M. Martins, M. Meyer, P. Michelato, V. Miltchev, W. D. Moeller, L. Monaco, W. F. O. Mueller, A. Napieralski, O. Napoly, P. Nicolosi, D. Noelle, T. Nunez, A. Oppelt, C. Pagani, R. Paparella, N. Pchalek, J. Pedregosa-Gutierrez, B. Petersen, B. Petrosyan, G. Petrosyan, L. Petrosyan, J. Pflueger, E. Ploenjes, L. Poletto, K. Pozniak, E. Prat, D. Proch, P. Pucyk, P. Radcliffe, H. Redlin, K. Rehlich, M. Richter, M. Roehrs, J. Roensch, R. Romaniuk, M. Ross, J. Rossbach, V. Rybnikov, M. Sachwitz, E. L. Saldin, W. Sandner, H. Schlarb, B. Schmidt, M. Schmitz, P. Schmueser, J. R. Schneider, E. A. Schneidmiller, S. Schnepp, S. Schreiber, M. Seidel, D. Sertore, A. V Shabunov, C. Simon, S. Simrock, E. Sombrowski, A. A. Sorokin, P. Spanknebel, R. Spesyvtsev, L. Staykov, B. Steffen, F. Stephan, F. Stulle, H. Thom, K. Tiedtke, M. Tischer, S. Toleikis, R. Treusch, D. Trines, I. Tsakov, E. Vogel, T. Weiland, H. Weise, M. Wellhoeffer, M. Wendt, I. Will, A. Winter, K. Wittenburg, W. Wurth, P. Yeates, M. V Yurkov, I. Zagorodnov, and K. Zapfe, "Operation of a free-electron laser from the extreme ultraviolet to the water window," *Nat. Photonics*, **1**, 6, 336–342 (2007).
2. http://photon-science.desy.de/facilities/flash/flash_parameters/index_eng.html
3. P. Emma, R. Akre, J. Arthur, R. Bionta, C. Bostedt, J. Bozek, a. Brachmann, P. Bucksbaum, R. Coffee, F.-J. Decker, Y. Ding, D. Dowell, S. Edstrom, a. Fisher, J. Frisch, S. Gilevich, J. Hastings, G. Hays, P. Hering, Z. Huang, R. Iverson, H. Loos, M. Messerschmidt, a. Miahnahri, S. Moeller, H.-D. Nuhn, G. Pile, D. Ratner, J. Rzepiela, D. Schultz, T. Smith, P. Stefan, H. Tompkins, J. Turner, J. Welch, W. White, J. Wu, G. Yocky, and J. Galayda, "First lasing and operation of an ångstrom-wavelength free-electron laser," *Nat. Photonics*, **4**, 9, 641–647 (2010).
4. http://www-ssrl.slac.stanford.edu/lcls/users/proposals.html
5. T. Shintake, H. Tanaka, T. Hara, T. Tanaka, K. Togawa, M. Yabashi, Y. Otake, Y. Asano, T. Bizen, T. Fukui, S. Goto, A. Higashiya, T. Hirono, N. Hosoda, T. Inagaki, S. Inoue, M. Ishii, Y. Kim, H. Kimura, M. Kitamura, T. Kobayashi, H. Maesaka, T. Masuda, S. Matsui, T. Matsushita, X. Marechal, M. Nagasono, H. Ohashi, T. Ohata, T. Ohshima, K. Onoe, K. Shirasawa, T. Takagi, S. Takahashi, M. Takeuchi, K. Tamasaku, R. Tanaka, Y. Tanaka, T. Tanikawa, T. Togashi, S. Wu, A. Yamashita, K. Yanagida, C. Zhang, H. Kitamura, and T. Ishikawa, "A compact free-electron laser for generating coherent radiation in the extreme ultraviolet region," *Nat. Photonics*, **2**, 9, 555–559 (2008).

6. T. Ishikawa, H. Aoyagi, T. Asaka, Y. Asano, N. Azumi, T. Bizen, H. Ego, K. Fukami, T. Fukui, Y. Furukawa, S. Goto, H. Hanaki, T. Hara, T. Hasegawa, T. Hatsui, A. Higashiya, T. Hirono, N. Hosoda, M. Ishii, T. Inagaki, Y. Inubushi, T. Itoga, Y. Joti, M. Kago, T. Kameshima, H. Kimura, Y. Kirihara, A. Kiyomichi, T. Kobayashi, C. Kondo, T. Kudo, H. Maesaka, X. M. Marechal, T. Masuda, S. Matsubara, T. Matsumoto, T. Matsushita, S. Matsui, M. Nagasono, N. Nariyama, H. Ohashi, T. Ohata, T. Ohshima, S. Ono, Y. Otake, C. Saji, T. Sakurai, T. Sato, K. Sawada, T. Seike, K. Shirasawa, T. Sugimoto, S. Suzuki, S. Takahashi, H. Takebe, K. Takeshita, K. Tamasaku, H. Tanaka, R. Tanaka, T. Tanaka, T. Togashi, K. Togawa, A. Tokuhisa, H. Tomizawa, K. Tono, S. Wu, M. Yabashi, M. Yamaga, A. Yamashita, K. Yanagida, C. Zhang, T. Shintake, H. Kitamura, and N. Kumagai, "A compact X-ray free-electron laser emitting in the sub-angstrom region," *Nat Photonics* **6**, 8, 540–544 (2012).

7. M. Yabashi, H. Tanaka, T. Tanaka, H. Tomizawa, T. Togashi, M. Nagasono, T. Ishikawa, J. R. Harries, Y. Hikosaka, a Hishikawa, K. Nagaya, N. Saito, E. Shigemasa, K. Yamanouchi, and K. Ueda, "Compact XFEL and AMO sciences: SACLA and SCSS," *J. Phys. B At. Mol. Opt. Phys.* **46**, 16, 164001 (2013).

8. E. Allaria, R. Appio, L. Badano, W. A. Barletta, S. Bassanese, S. G. Biedron, A. Borga, E. Busetto, D. Castronovo, P. Cinquegrana, S. Cleva, D. Cocco, M. Cornacchia, P. Craievich, I. Cudin, G. D'Auria, M. Dal Forno, M. B. Danailov, R. De Monte, G. De Ninno, P. Delgiusto, A. Demidovich, S. Di Mitri, B. Diviacco, A. Fabris, R. Fabris, W. Fawley, M. Ferianis, E. Ferrari, S. Ferry, L. Froehlich, P. Furlan, G. Gaio, F. Gelmetti, L. Giannessi, M. Giannini, R. Gobessi, R. Ivanov, E. Karantzoulis, M. Lonza, A. Lutman, B. Mahieu, M. Milloch, S. V Milton, M. Musardo, I. Nikolov, S. Noe, F. Parmigiani, G. Penco, M. Petronio, L. Pivetta, M. Predonzani, F. Rossi, L. Rumiz, A. Salom, C. Scafuri, C. Serpico, P. Sigalotti, S. Spampinati, C. Spezzani, M. Svandrlik, C. Svetina, S. Tazzari, M. Trovo, R. Umer, A. Vascotto, M. Veronese, R. Visintini, M. Zaccaria, D. Zangrando, and M. Zangrando, "Highly coherent and stable pulses from the FERMI seeded free-electron laser in the extreme ultraviolet," *Nat. Photonics*, **6**, 10, 699–704 (2012).

9. H. N. Chapman, P. Fromme, A. Barty, T. a. White, R. a. Kirian, A. Aquila, M. S. Hunter, J. Schulz, D. P. DePonte, U. Weierstall, R. B. Doak, F. R. N. C. Maia, A. V. Martin, I. Schlichting, L. Lomb, N. Coppola, R. L. Shoeman, S. W. Epp, R. Hartmann, D. Rolles, A. Rudenko, L. Foucar, N. Kimmel, G. Weidenspointner, P. Holl, M. Liang, M. Barthelmess, C. Caleman, S. Boutet, M. J. Bogan, J. Krzywinski, C. Bostedt, S. Bajt, L. Gumprecht, B. Rudek, B. Erk, C. Schmidt, A. Hömke, C. Reich, D. Pietschner, L. Strüder, G. Hauser, H. Gorke, J. Ullrich, S. Herrmann, G. Schaller, F. Schopper, H. Soltau, K.-U. Kühnel, M. Messerschmidt, J. D. Bozek, S. P. Hau-Riege, M. Frank, C. Y. Hampton, R. G. Sierra, D. Starodub, G. J. Williams, J. Hajdu, N. Timneanu, M. M. Seibert, J. Andreasson, A. Rocker, O. Jönsson, M. Svenda, S. Stern, K. Nass, R. Andritschke, C.-D. Schröter, F. Krasniqi, M. Bott, K. E. Schmidt, X. Wang, I. Grotjohann, J. M. Holton, T. R. M. Barends, R. Neutze, S. Marchesini, R. Fromme, S. Schorb, D. Rupp, M. Adolph, T. Gorkhover, I. Andersson, H. Hirsemann, G. Potdevin, H. Graafsma, B. Nilsson, and J. C. H. Spence, "Femtosecond X-ray protein nanocrystallography," *Nature*, **470**, 7332, 73–77 (2011).

10. M. M. Seibert, T. Ekeberg, F. R. N. C. Maia, M. Svenda, J. Andreasson, O. Jonsson, D. Odic, B. Iwan, A. Rocker, D. Westphal, M. Hantke, D. P. DePonte, A. Barty, J. Schulz, L. Gumprecht, N. Coppola, A. Aquila, M. Liang, T. A. White, A. Martin, C. Caleman, S. Stern, C. Abergel, V. Seltzer, J.-M. Claverie, C. Bostedt, J. D. Bozek, S. Boutet, A. A. Miahnahri, M. Messerschmidt, J. Krzywinski, G. Williams, K. O. Hodgson, M. J. Bogan, C. Y. Hampton, R. G. Sierra, D. Starodub, I. Andersson, S. Bajt, M. Barthelmess, J. C. H. Spence, P. Fromme, U. Weierstall, R. Kirian, M. Hunter, R. B. Doak, S. Marchesini, S. P. Hau-Riege, M. Frank, R. L. Shoeman, L. Lomb, S. W. Epp, R. Hartmann, D. Rolles, A. Rudenko, C. Schmidt, L. Foucar, N. Kimmel, P. Holl, B. Rudek, B. Erk, A. Homke, C. Reich, D. Pietschner, G. Weidenspointner, L. Struder, G. Hauser, H. Gorke, J. Ullrich, I. Schlichting, S. Herrmann, G. Schaller, F. Schopper, H. Soltau, K.-U. Kuhnel, R. Andritschke, C.-D. Schroter, F. Krasniqi, M. Bott, S. Schorb, D. Rupp, M. Adolph,

T. Gorkhover, H. Hirsemann, G. Potdevin, H. Graafsma, B. Nilsson, H. N. Chapman, and J. Hajdu, "Single mimivirus particles intercepted and imaged with an X-ray laser," *Nature*, **470**, 7332, 78–81 (2011).

11. Y. H. Jiang, a Rudenko, M. Kurka, K. U. Kühnel, L. Foucar, T. Ergler, S. Lüdemann, K. Zrost, T. Ferger, D. Fischer, a Dorn, J. Titze, T. Jahnke, M. Schöffler, S. Schössler, T. Havermeier, M. Smolarski, K. Cole, R. Dörner, T. J. M. Zouros, S. Düsterer, R. Treusch, M. Gensch, C. D. Schröter, R. Moshammer, and J. Ullrich, "EUV-photon-induced multiple ionization and fragmentation dynamics: from atoms to molecules," *J. Phys. B At. Mol. Opt. Phys.* **42**, 13, 134012 (2009).

12. A. Rudenko, Y. H. Jiang, M. Kurka, K. U. Kühnel, L. Foucar, O. Herrwerth, M. Lezius, M. F. Kling, C. D. Schröter, R. Moshammer, and J. Ullrich, "Exploring few-photon, few-electron reactions at FLASH: from ion yield and momentum measurements to time-resolved and kinematically complete experiments," *J. Phys. B At. Mol. Opt. Phys.* **43**, 19, 194004 (2010).

13. N. Berrah, J. Bozek, and J. Costello, "Non-linear processes in the interaction of atoms and molecules with intense EUV and X-ray fields from SASE free electron lasers (FELs)," *J. Mod. Opt.* **57**, 12, 37–41 (2010).

14. M. Yabashi, H. Tanaka, T. Tanaka, H. Tomizawa, T. Togashi, M. Nagasono, T. Ishikawa, J. R. Harries, Y. Hikosaka, a Hishikawa, K. Nagaya, N. Saito, E. Shigemasa, K. Yamanouchi, and K. Ueda, "Compact XFEL and AMO sciences: SACLA and SCSS," *J. Phys. B At. Mol. Opt. Phys.* **46**, 16, 164001 (2013).

15. J. Ullrich, A. Rudenko, and R. Moshammer, "Free-electron lasers: new avenues in molecular physics and photochemistry," *Annu. Rev. Phys. Chem.* **63**, 635–60 (2012).

16. J. Feldhaus, M. Krikunova, M. Meyer, T. Möller, R. Moshammer, a Rudenko, T. Tschentscher, and J. Ullrich, "AMO science at the FLASH and European XFEL free-electron laser facilities," *J. Phys. B At. Mol. Opt. Phys.* **46**, 16, 164002 (2013).

17. L. Fang, T. Osipov, B. F. Murphy, a Rudenko, D. Rolles, V. S. Petrovic, C. Bostedt, J. D. Bozek, P. H. Bucksbaum, and N. Berrah, "Probing ultrafast electronic and molecular dynamics with free-electron lasers," *J. Phys. B At. Mol. Opt. Phys.* **47**, 12, 124006 (2014).

18. Y. Jiang, a. Rudenko, M. Kurka, K. Kühnel, T. Ergler, L. Foucar, M. Schöffler, S. Schössler, T. Havermeier, M. Smolarski, K. Cole, R. Dörner, S. Düsterer, R. Treusch, M. Gensch, C. Schröter, R. Moshammer, and J. Ullrich, "Few-Photon Multiple Ionization of N_2 by Extreme Ultraviolet Free-Electron Laser Radiation," *Phys. Rev. Lett.* **102**, 12, 123002 (2009).

19. A. A. Sorokin, S. V Bobashev, K. Tiedtke, and M. Richter, "Multi-photon ionization of molecular nitrogen by femtosecond soft x-ray FEL pulses," *J. Phys. B At. Mol. Opt. Phys.*, **39**, 14, L299–L304 (2006).

20. T. Sato, T. Okino, K. Yamanouchi, A. Yagishita, F. Kannari, K. Yamakawa, K. Midorikawa, H. Nakano, M. Yabashi, M. Nagasono, and T. Ishikawa, "Dissociative two-photon ionization of N_2 in extreme ultraviolet by intense self-amplified spontaneous emission free electron laser light," *Appl. Phys. Lett.* **92**, 15, 154103 (2008).

21. J. P. Cryan, J. M. Glownia, J. Andreasson, a. Belkacem, N. Berrah, C. I. Blaga, C. Bostedt, J. Bozek, C. Buth, L. F. DiMauro, L. Fang, O. Gessner, M. Guehr, J. Hajdu, M. P. Hertlein, M. Hoener, O. Kornilov, J. P. Marangos, a. M. March, B. K. McFarland, H. Merdji, V. S. Petrović, C. Raman, D. Ray, D. Reis, F. Tarantelli, M. Trigo, J. L. White, W. White, L. Young, P. H. Bucksbaum, and R. N. Coffee, "Auger Electron Angular Distribution of Double Core-Hole States in the Molecular Reference Frame," *Phys. Rev. Lett.* **105**, 8, 083004 (2010).

22. L. Fang, M. Hoener, O. Gessner, F. Tarantelli, S. T. Pratt, O. Kornilov, C. Buth, M. Gühr, E. P. Kanter, C. Bostedt, J. D. Bozek, P. H. Bucksbaum, M. Chen, R. Coffee, J. Cryan, M. Glownia, E. Kukk, S. R. Leone, and N. Berrah, "Double Core-Hole Production in N_2: Beating the Auger Clock," *Phys. Rev. Lett.* **105**, 8, 083005 (2010).

23. T. Pfeifer, Y. Jiang, S. Düsterer, R. Moshammer, and J. Ullrich, "Partial-coherence method to model experimental free-electron laser pulse statistics," *Opt. Lett.* **35**, 20, 3441–3 (2010).

24. Y. H. Jiang, T. Pfeifer, A. Rudenko, O. Herrwerth, L. Foucar, M. Kurka, K. U. Kühnel, M. Lezius, M. F. Kling, X. Liu, K. Ueda, S. Düsterer, R. Treusch, C. D. Schröter, R. Moshammer, and J. Ullrich, "Temporal coherence effects in multiple ionization of N_2 via XUV pump-probe autocorrelation," *Phys. Rev. A*, **82**, 4, 041403 (2010).

25. K. Meyer, C. Ott, P. Raith, A. Kaldun, Y. Jiang, A. Senftleben, M. Kurka, R. Moshammer, J. Ullrich, and T. Pfeifer, "Noisy Optical Pulses Enhance the Temporal Resolution of Pump-Probe Spectroscopy," *Phys. Rev. Lett.* **108**, 9, 098302 (2012).

26. Y. H. Jiang, A. Rudenko, J. F. Pérez-Torres, O. Herrwerth, L. Foucar, M. Kurka, K. U. Kühnel, M. Toppin, E. Plésiat, F. Morales, F. Martín, M. Lezius, M. F. Kling, T. Jahnke, R. Dörner, J. L. Sanz-Vicario, J. van Tilborg, A. Belkacem, M. Schulz, K. Ueda, T. J. M. Zouros, S. Düsterer, R. Treusch, C. D. Schröter, R. Moshammer, and J. Ullrich, "Investigating two-photon double ionization of D_2 by XUV-pump–XUV-probe experiments," *Phys. Rev. A*, **81**, 5, 051402 (2010).

27. F. Sorgenfrei, W. F. Schlotter, T. Beeck, M. Nagasono, S. Gieschen, H. Meyer, a Föhlisch, M. Beye, and W. Wurth, "The extreme ultraviolet split and femtosecond delay unit at the plane grating monochromator beamline PG2 at FLASH," *Rev. Sci. Instrum.* **81**, 4, 043107 (2010).

28. J. Ullrich, R. Moshammer, A. Dorn, R. Dörner, L. P. H. Schmidt, and H. Schmidt-Böcking, "Recoil-ion and electron momentum spectroscopy: reaction-microscopes," *Reports Prog. Phys.* **66**, 9, 1463 (2003).

29. Y. Jiang, T. Pfeifer, a. Rudenko, O. Herrwerth, L. Foucar, M. Kurka, K. Kühnel, M. Lezius, M. Kling, X. Liu, K. Ueda, S. Düsterer, R. Treusch, C. Schröter, R. Moshammer, and J. Ullrich, "Temporal coherence effects in multiple ionization of N_2 via XUV pump-probe autocorrelation," *Phys. Rev. A*, **82**, 4, 041403 (2010).

30. M. Magrakvelidze, O. Herrwerth, Y. Jiang, a. Rudenko, M. Kurka, L. Foucar, K. Kühnel, M. Kübel, N. Johnson, C. Schröter, S. Düsterer, R. Treusch, M. Lezius, I. Ben-Itzhak, R. Moshammer, J. Ullrich, M. Kling, and U. Thumm, "Tracing nuclear-wave-packet dynamics in singly and doubly charged states of N_2 and O_2 with XUV-pump–XUV-probe experiments," *Phys. Rev. A*, **86**, 1, 013415 (2012).

31. J. Pérez-Torres, F. Morales, J. Sanz-Vicario, and F. Martín, "Asymmetric electron angular distributions in resonant dissociative photoionization of H_2 with ultrashort xuv pulses," *Phys. Rev. A*, **80**, 1, 011402 (2009).

32. A Yamada, H. Fukuzawa, K. Motomura, X.-J. Liu, L. Foucar, M. Kurka, M. Okunishi, K. Ueda, N. Saito, H. Iwayama, K. Nagaya, A. Sugishima, H. Murakami, M. Yao, A. Rudenko, K. U. Kühnel, J. Ullrich, R. Feifel, A. Czasch, R. Dörner, M. Nagasono, A. Higashiya, M. Yabashi, T. Ishikawa, H. Ohashi, H. Kimura, and T. Togashi, "Ion-ion coincidence studies on multiple ionizations of N_2 and O_2 molecules irradiated by extreme ultraviolet free-electron laser pulses," *J. Chem. Phys.* **132**, 20, 204305 (2010).

33. D. Polli, P. Altoe, O. Weingart, K. M. Spillane, C. Manzoni, D. Brida, G. Tomasello, G. Orlandi, P. Kukura, R. A. Mathies, M. Garavelli, and G. Cerullo, "Conical intersection dynamics of the primary photoisomerization event in vision," *Nature*, **467**, 7314, 440–443 (2010).

34. M. Lammers, H. Neumann, J. W. Chin, and L. C. James, "Acetylation regulates Cyclophilin A catalysis, immunosuppression and HIV isomerization," *Nat. Chem. Biol.* **6**, 5, 331–337 (2010).

35. T. Kobayashi, T. Saito, and H. Ohtani, "Real-time spectroscopy of transition states in bacteri-orhodopsin during retinal isomerization," *Nature*, **414**, 6863, 531–534 (2001).

36. Y. H. Jiang, a Senftleben, M. Kurka, a Rudenko, L. Foucar, O. Herrwerth, M. F. Kling, M. Lezius, J. V Tilborg, a Belkacem, K. Ueda, D. Rolles, R. Treusch, Y. Z. Zhang, Y. F. Liu, C. D. Schröter, J. Ullrich, and R. Moshammer, "Ultrafast dynamics in acetylene clocked in a femtosecond XUV stopwatch," *J. Phys. B At. Mol. Opt. Phys.* **46**, 16, 164027 (2013).

37. Y. H. Jiang, a. Rudenko, O. Herrwerth, L. Foucar, M. Kurka, K. U. Kühnel, M. Lezius, M. F. Kling, J. Van Tilborg, a. Belkacem, K. Ueda, S. Düsterer, R. Treusch, C. D. Schröter, R. Moshammer,

and J. Ullrich, "Ultrafast extreme ultraviolet induced isomerization of acetylene cations," *Phys. Rev. Lett.* **105**, 26, 263002 (2010).

38. M. E.-A. Madjet, O. Vendrell, and R. Santra, "Ultrafast Dynamics of Photoionized Acetylene," *Phys. Rev. Lett.* **107**, 26, 263002 (2011).

39. L. S. Cederbaum, J. Zobeley, and F. Tarantelli, "Giant Intermolecular Decay and Fragmentation of Clusters," *Phys. Rev. Lett.* **79**, 24, 4778–4781 (1997).

40. S. Marburger, O. Kugeler, U. Hergenhahn, and T. Möller, "Experimental Evidence for Interatomic Coulombic Decay in Ne_2 Clusters," *Phys. Rev. Lett.* **90**, 20, 203401 (2003).

41. T. Jahnke, a. Czasch, M. Schöffler, S. Schössler, a. Knapp, M. Käsz, J. Titze, C. Wimmer, K. Kreidi, R. Grisenti, a. Staudte, O. Jagutzki, U. Hergenhahn, H. Schmidt-Böcking, and R. Dörner, "Experimental Observation of Interatomic Coulombic Decay in Neon Dimers," *Phys. Rev. Lett.* **93**, 16, 163401 (2004).

42. K. Schnorr, A. Senftleben, M. Kurka, A. Rudenko, L. Foucar, G. Schmid, A. Broska, T. Pfeifer, K. Meyer, D. Anielski, R. Boll, D. Rolles, M. Kuebel, M. F. Kling, Y. H. Jiang, S. Mondal, T. Tachibana, K. Ueda, T. Marchenko, M. Simon, G. Brenner, R. Treusch, S. Scheit, V. Averbukh, J. Ullrich, C. D. Schroeter, and R. Moshammer, "Time-Resolved Measurement of Interatomic Coulombic Decay in Ne_2," *Phys. Rev. Lett.* **111**, 9, 093402 (2013).

43. K. Schnorr, PhD dissertation, Heidelberg University, 2014.

44. R. Neutze, R. Wouts, D. van der Spoel, E. Weckert, and J. Hajdu, "Potential for biomolecular imaging with femtosecond X-ray pulses," *Nature*, **406**, 6797, 752–757 (2000).

45. K. Schnorr, A. Senftleben, M. Kurka, A. Rudenko, G. Schmid, T. Pfeifer, K. Meyer, M. Kübel, M. F. Kling, Y. H. Jiang, R. Treusch, S. Düsterer, B. Siemer, M. Wöstmann, H. Zacharias, R. Mitzner, T. J. M. Zouros, J. Ullrich, C. D. Schröter, and R. Moshammer, "Electron Rearrangement Dynamics in Dissociating I_2^{n+} Molecules Accessed by Extreme Ultraviolet Pump-Probe Experiments," *Phys. Rev. Lett.* **113**, 7, 073001 (2014).

46. U. Fruehling, M. Wieland, M. Gensch, T. Gebert, B. Schuette, M. Krikunova, R. Kalms, F. Budzyn, O. Grimm, J. Rossbach, E. Ploenjes, and M. Drescher, "Single-shot terahertz-field-driven X-ray streak camera," *Nat. Photonics*, **3**, 9, 523–528 (2009).

47. P. Johnsson, A. Rouzée, W. Siu, Y. Huismans, F. Lépine, T. Marchenko, S. Düsterer, F. Tavella, N. Stojanovic, H. Redlin, A. Azima, and M. J. J. Vrakking, "Characterization of a two-color pump–probe setup at FLASH using a velocity map imaging spectrometer," *Opt. Lett.* **35**, 24, 4163–4165 (2010).

48. S. Y. Liu, Y. Ogi, T. Fuji, K. Nishizawa, T. Horio, T. Mizuno, H. Kohguchi, M. Nagasono, T. Togashi, K. Tono, M. Yabashi, Y. Senba, H. Ohashi, H. Kimura, T. Ishikawa, and T. Suzuki, "Time-resolved photoelectron imaging using a femtosecond UV laser and a VUV free-electron laser," *Phys. Rev. A*, **81**, 3, 031403 (2010).

49. M. Krikunova, T. Maltezopoulos, P. Wessels, M. Schlie, A. Azima, M. Wieland, and M. Drescher, "Ultrafast photofragmentation dynamics of molecular iodine driven with timed XUV and near-infrared light pulses," *J. Chem. Phys.* **134**, 2, 024313 (2011).

50. M. Krikunova, T. Maltezopoulos, P. Wessels, M. Schlie, A. Azima, T. Gaumnitz, T. Gebert, M. Wieland, and M. Drescher, "Strong-field ionization of molecular iodine traced with XUV pulses from a free-electron laser," *Phys. Rev. A*, **86**, 4, 043430 (2012).

51. J. M. Glownia, J. Cryan, J. Andreasson, A. Belkacem, N. Berrah, C. I. Blaga, C. Bostedt, J. Bozek, L. F. DiMauro, L. Fang, J. Frisch, O. Gessner, M. Gühr, J. Hajdu, M. P. Hertlein, M. Hoener, G. Huang, O. Kornilov, J. P. Marangos, a M. March, B. K. McFarland, H. Merdji, V. S. Petrovic, C. Raman, D. Ray, D. a Reis, M. Trigo, J. L. White, W. White, R. Wilcox, L. Young, R. N. Coffee, and P. H. Bucksbaum, "Time-resolved pump-probe experiments at the LCLS," *Opt. Express*, **18**, 17, 17620–30 (2010).

52. A. Cavalieri, D. Fritz, S. Lee, P. Bucksbaum, D. Reis, J. Rudati, D. Mills, P. Fuoss, G. Stephenson, C. Kao, D. Siddons, D. Lowney, a. MacPhee, D. Weinstein, R. Falcone, R. Pahl, J. Als-Nielsen,

C. Blome, S. Düsterer, R. Ischebeck, H. Schlarb, H. Schulte-Schrepping, T. Tschentscher, J. Schneider, O. Hignette, F. Sette, K. Sokolowski-Tinten, H. Chapman, R. Lee, T. Hansen, O. Synnergren, J. Larsson, S. Techert, J. Sheppard, J. Wark, M. Bergh, C. Caleman, G. Huldt, D. van der Spoel, N. Timneanu, J. Hajdu, R. Akre, E. Bong, P. Emma, P. Krejcik, J. Arthur, S. Brennan, K. Gaffney, a. Lindenberg, K. Luening, and J. Hastings, "Clocking Femtosecond X Rays," *Phys. Rev. Lett.* **94**, 11, 114801 (2005).

53. B. Steffen, V. Arsov, G. Berden, W. A. Gillespie, S. P. Jamison, A. M. MacLeod, A. F. G. van der Meer, P. J. Phillips, H. Schlarb, B. Schmidt, and P. Schmüser, "Electro-optic time profile monitors for femtosecond electron bunches at the soft x-ray free-electron laser FLASH," *Phys. Rev. ST Accel. Beams*, **12**, 3, 32802 (2009).

54. A. Azima, S. Düsterer, P. Radcliffe, H. Redlin, N. Stojanovic, W. Li, H. Schlarb, J. Feldhaus, D. Cubaynes, M. Meyer, J. Dardis, P. Hayden, P. Hough, V. Richardson, E. T. Kennedy, and J. T. Costello, "Time-resolved pump-probe experiments beyond the jitter limitations at FLASH," *Appl. Phys. Lett.* **94**, 14, 144102 (2009).

55. J. P. Cryan, J. M. Glownia, J. Andreasson, a Belkacem, N. Berrah, C. I. Blaga, C. Bostedt, J. Bozek, N. a Cherepkov, L. F. DiMauro, L. Fang, O. Gessner, M. Gühr, J. Hajdu, M. P. Hertlein, M. Hoener, O. Kornilov, J. P. Marangos, a M. March, B. K. McFarland, H. Merdji, M. Messerschmidt, V. S. Petrović, C. Raman, D. Ray, D. a Reis, S. K. Semenov, M. Trigo, J. L. White, W. White, L. Young, P. H. Bucksbaum, and R. N. Coffee, "Molecular frame Auger electron energy spectrum from N_2," *J. Phys. B At. Mol. Opt. Phys.* **45**, 5, 055601 (2012).

56. V. S. Petrović, M. Siano, J. L. White, N. Berrah, C. Bostedt, J. D. Bozek, D. Broege, M. Chalfin, R. N. Coffee, J. Cryan, L. Fang, J. P. Farrell, L. J. Frasinski, J. M. Glownia, M. Gühr, M. Hoener, D. M. P. Holland, J. Kim, J. P. Marangos, T. Martinez, B. K. McFarland, R. S. Minns, S. Miyabe, S. Schorb, R. J. Sension, L. S. Spector, R. Squibb, H. Tao, J. G. Underwood, and P. H. Bucksbaum, "Transient X-Ray Fragmentation: Probing a Prototypical Photoinduced Ring Opening," *Phys. Rev. Lett.* **108**, 25, 253006 (2012).

57. V. S. Petrovic, S. Schorb, J. Kim, J. White, J. P. Cryan, J. M. Glownia, D. Broege, S. Miyabe, H. Tao, T. Martinez, and P. H. Bucksbaum, "Enhancement of strong-field multiple ionization in the vicinity of the conical intersection in 1 , 3-cyclohexadiene ring opening Enhancement of strong-field multiple ionization in the vicinity of the conical intersection in 1, 3-cyclohexadiene ring opening," *J. Chem. Phys.* **139**, 18, 184309 (2013).

58. B. Erk, D. Rolles, L. Foucar, B. Rudek, S. W. Epp, M. Cryle, C. Bostedt, S. Schorb, J. Bozek, A. Rouzee, A. Hundertmark, T. Marchenko, M. Simon, F. Filsinger, L. Christensen, S. De, S. Trippel, J. Kuepper, H. Stapelfeldt, S. Wada, K. Ueda, M. Swiggers, M. Messerschmidt, C. D. Schroeter, R. Moshammer, I. Schlichting, J. Ullrich, and A. Rudenko, "Ultrafast Charge Rearrangement and Nuclear Dynamics upon Inner-Shell Multiple Ionization of Small Polyatomic Molecules," *Phys. Rev. Lett.* **110**, 5, 053003 (2013).

59. B. Erk, D. Rolles, L. Foucar, B. Rudek, S. W. Epp, M. Cryle, C. Bostedt, S. Schorb, J. Bozek, A. Rouzee, A. Hundertmark, T. Marchenko, M. Simon, F. Filsinger, L. Christensen, S. De, S. Trippel, J. Kuepper, H. Stapelfeldt, S. Wada, K. Ueda, M. Swiggers, M. Messerschmidt, C. D. Schroeter, R. Moshammer, I. Schlichting, J. Ullrich, and A. Rudenko, "Inner-shell multiple ionization of polyatomic molecules with an intense x-ray free-electron laser studied by coincident ion momentum imaging," *J. Phys. B At. Mol. Opt. Phys.* **46**, 16, 164031 (2013).

60. B. Erk, R. Boll, S. Trippel, D. Anielski, L. Foucar, B. Rudek, S. W. Epp, R. Coffee, S. Carron, S. Schorb, K. R. Ferguson, M. Swiggers, J. D. Bozek, M. Simon, T. Marchenko, J. Kupper, I. Schlichting, J. Ullrich, C. Bostedt, D. Rolles, and a. Rudenko, "Imaging charge transfer in iodomethane upon x-ray photoabsorption," *Science*, **345**, 6194, 288–291 (2014).

www.ingramcontent.com/pod-product-compliance
Lightning Source LLC
Chambersburg PA
CBHW081514190326
41458CB00015B/5363